Springer Tracts in Civil Engineering

Series Editors

Giovanni Solari, Wind Engineering and Structural Dynamics Research Group, University of Genoa, Genova, Italy

Sheng-Hong Chen, School of Water Resources and Hydropower Engineering, Wuhan University, Wuhan, China

Marco di Prisco, Politecnico di Milano, Milano, Italy

Ioannis Vayas, Institute of Steel Structures, National Technical University of Athens, Athens, Greece

CW01500636

Springer Tracts in Civil Engineering (STCE) publishes the latest developments in Civil Engineering - quickly, informally and in top quality. The series scope includes monographs, professional books, graduate textbooks and edited volumes, as well as outstanding PhD theses. Its goal is to cover all the main branches of civil engineering, both theoretical and applied, including:

- Construction and Structural Mechanics
- Building Materials
- Concrete, Steel and Timber Structures
- Geotechnical Engineering
- Earthquake Engineering
- Coastal Engineering; Ocean and Offshore Engineering
- Hydraulics, Hydrology and Water Resources Engineering
- Environmental Engineering and Sustainability
- Structural Health and Monitoring
- Surveying and Geographical Information Systems
- Heating, Ventilation and Air Conditioning (HVAC)
- Transportation and Traffic
- Risk Analysis
- Safety and Security

Indexed by Scopus

To submit a proposal or request further information, please contact: Pierpaolo Riva at Pierpaolo.Riva@springer.com, or Li Shen at Li.Shen@springer.com

More information about this series at http://www.springer.com/series/15088

Zheng Lu · Sami F. Masri · Xilin Lu

Particle Damping Technology Based Structural Control

机械工业出版社
CHINA MACHINE PRESS

Zheng Lu
Department of Disaster Mitigation
for Structures
Tongji University
Shanghai, China

Sami F. Masri
Sonny Astani Department of Civil
and Environmental Engineering
University of Southern California
Los Angeles, CA, USA

Xilin Lu
Department of Disaster Mitigation
for Structures
Tongji University
Shanghai, China

ISSN 2366-259X   ISSN 2366-2603   (electronic)
Springer Tracts in Civil Engineering
ISBN 978-981-15-3501-7   ISBN 978-981-15-3499-7   (eBook)
https://doi.org/10.1007/978-981-15-3499-7

Jointly published with China Machine Press, Beijing, China
The print edition is not for sale in China. Customers from China please order the print book from: China Machine Press.
ISBN of the Co-Publisher's edition: 978-7-111-59964-7

This Springer imprint is published by the registered company Springer Nature Singapore Pte Ltd.
The registered company address is: 152 Beach Road, #21-01/04 Gateway East, Singapore 189721, Singapore

Preface

In recent years, natural disasters, such as earthquakes and typhoons, have occurred frequently, causing serious damage to the structure. In order to enhance the resistance of structures against natural catastrophes, especially to attenuate their structural response under earthquake and wind loads, vibration control technology is applied on structures as an effective method. Particle damping technology is a kind of vibration reduction technology which consumes the vibration energy of the system by friction and impact between tiny particles filled in a finite enclosed space of the vibration cavity. It has been widely used in mechanical engineering, aerospace engineering and other fields, and its application on civil engineering has been expanding.

The Chinese Book is funded by National Science and Technology Academic Monograph Publishing Fund (2017). The English version is the update version. This book systematically introduces particle damping technology, which can be utilized to effectively mitigate the seismic-induced and wind-induced vibration of various structures. The research achievements pertaining to this technology are comprehensively presented in this book. The 9 chapters in the book can be divided into 3 main aspects, first of all, in Chaps. 1 and 2, the book gives a detailed introduction and comprehensive description of structural vibration control technology and particle damping technology utilized in civil engineering, and provides a newly efficacious approach for the mitigation of earthquake-induced and wind-induced vibrations generated in high-rise buildings. The second part is illustrated in Chaps. 3–7, systematically theoretical analysis and practical simulation method for particle impact dampers are proposed, as well as demonstrated the extensive shaking table test and wind tunnel test on particle impact dampers applied to civil structures, which are conducive to understanding the damping mechanisms and optimal working conditions of particle impact dampers. In Chaps. 8 and 9, profound discussions are carried out on the future perspectives of particle damping technology, especially emphasized on studying and developing semi-active particle damping technology, and established the guidelines for optimization design of particle impact dampers.

Some distinct features of this book can be summarized as follows:

- The elaborated simulation model based on discrete element method and the simplified simulation model based on equivalent principles are demonstrated, the models are verified by corresponding experimental results.
- The steady-state responses of particle impact dampers under various scenarios are illustrated.
- The vibration attenuation effects of particle dampers subjected to different dynamic loads have been intensively studied, by which the damping mechanism and the "global" measures that are used to evaluate the damping performance are proposed.
- The shaking table test and wind tunnel test have been conducted to verify the proposed simulation methods, and the satisfactory damping performance of particle dampers has been corroborated.
- The optimization design guidelines of particle impact dampers are provided.
- The semi-active particle damping technology in which the primary system is subjected to non-stationary excitations is preliminary explored.

Shanghai, China   Zheng Lu
Los Angeles, USA   Sami F. Masri
Shanghai, China   Xilin Lu

Contents

Chapter 1
Introduction to Structural Vibration Control Technology

Structural vibration control is to control the vibration of the structure under earthquake and wind by changing the stiffness, mass, damping and shape of the structure and providing a certain amount of passive or active reaction forces. The concept of structural vibration control technology was put forward more than a hundred years ago by Japanese engineering professor, John Milne [1]. John built a timber structure and placed it on the ball bearing to prove that the structure could be isolated from the seismic shaking. During the World War II, the concepts of structural vibration control, such as structural vibration isolation, structural energy dissipation and structural vibration damping, were greatly developed and effectively applied to aircraft structures.

In 1960s, the concept of structural vibration control began to enter the field of civil engineering and was developed towards various directions. The structural vibration control in civil engineering is to install devices (such as isolation pads or isolation blocks), components (such as energy dissipation braces, energy dissipation shear walls, energy dissipation nodes, damper, etc.) or some kinds of substructures (such as tuned mass, etc.), or to apply external forces (such as external energy input) to change or adjust the dynamic characteristics of the structure. The aim is to properly control the vibration responses of the building, ensure the safety of structures, give people feeling of comfort and safety, and guarantee the function of equipment.

According to external energy supply, structural control can be divided into passive control, active control, hybrid control and semi-active control [1]. For passive control, the control force is passively produced without external energy when the device vibrates and deforms along with the structure. The active control is controlled by external energy. The control force is actively implemented by using external energy according to certain rules. The hybrid control is to apply active and passive controls simultaneously on the structure so as to give full play to the advantages of various control devices. It has some obvious characteristics including good control effect, low cost, low energy consumption and easy engineering application. The need for energy of semi-active control system is much smaller than that of typical active control system. The control force is passively generated by the movement of the

© China Machine Press and Springer Nature Singapore Pte Ltd. 2020
Z. Lu et al., *Particle Damping Technology Based Structural Control*,
Springer Tracts in Civil Engineering,
https://doi.org/10.1007/978-981-15-3499-7_1

control device itself. However, in the control process, the control device can actively adjust its own parameters by using external energy. In general, the system does not add mechanical energy, thus it can ensure the stability of the system. Typically, semi-active control is considered as a kind of controllable passive control device.

1.1 Passive Control

Passive control is widely used in engineering because of its simple concept and clear mechanism. Common passive controls include base isolation, seismic energy dissipation and passive tuned vibration absorption.

1.1.1 Base Isolation

The basic principle of base isolation is to extend the structural period and set the structure proper damping to reduce acceleration responses. At the same time, the top displacement of the structure is mainly provided by the isolation system between the bottom of the structure and the foundation. The structure itself does not produce large relative displacements [2]. The current base isolation technology can be divided into two types: elastic (viscous) isolation and base isolation with sliding devices. The elastic (viscous) isolation is to add a flexible layer with lower lateral stiffness between the bottom of the structure and the top surface of the foundation, so the period of system can be extended and the deformation is concentrated on the bottom layer. The upper structure is basically rigidly motioning. The flexible bottom layer plays the role of low-pass filtering to ensure that the fundamental frequency of the structure is much lower than the frequency when the base is fixed and lower than the excellent frequency range of input seismic motion. The most widely used rubber bearing base-isolation system is a typical kind of elastic (viscous) isolation. The base isolation with sliding devices means that the friction material with small friction coefficient is disposed between the structure and the foundation. When the inertia force of the structure during the earthquake is greater than the friction force of the system, there will exit slides between structure and foundation. On the one hand, the effect of horizontal earthquake transmitted to the structure is limited. On the other hand, the seismic energy is dissipated. Base isolation is mainly used for low- rise structures and bridges with high frequencies.

1.1.2 Seismic Energy Dissipation

Seismic energy dissipation is to design some energy-dissipating structural components (such as braces, shear walls, connectors, etc.) or to install some energy dissipation devices in certain parts of the structure (such as interlayer space, joints, bonding
joints, etc.). In the event of wind or minor earthquakes, those energy-dissipating
components and devices have sufficient initial stiffness and stay in the elastic state,
so the structure still has sufficient lateral stiffness to meet the use requirements.
When moderate or major earthquakes occur, as the lateral deformation of the structure increases, energy-dissipating components or devices take the lead to enter the
inelastic state, produce large damping forces and dissipate seismic energy greatly.
Thereby apparent inelastic state of major structure can be avoided, which makes the
seismic response of the structure decay rapidly and protects the major structure and
components during earthquakes. Currently, there are four types of commonly used
dampers: viscoelastic damper, viscous damper, friction damper, and metal damper.
The first two types are also named as speed-dependent energy dissipation devices, and
the latter two types are called as displacement-dependent energy dissipation devices.
Many scholars have made detailed reviews of these energy absorptions [3–5].

(1) *Viscoelastic damper*

The viscoelastic damper generally consists of viscoelastic material and constraining
steel plate. The viscoelastic material and the constraining steel plate are combined
together in layers to dissipate energy by shear hysteresis deformation of the viscoelastic material. The viscoelastic material is a kind of macromolecular polymer
which has both elastic and viscous properties. The former can provide rigidity and
the latter can provide damping, so the viscoelastic material has ability of energy
dissipation. The viscoelastic damper has sufficient reliability with low cost and convenient installation, so it is suitable for structural vibration control caused by various
dynamic loads.

The application of viscoelastic dampers in vibration control can be traced back
to the fatigue vibration control of aircraft structures in 1950s. The application in
structural engineering began at the 110-story New York World Trade Center in the
United States. The structure was built in 1969 and approximately 11,000 viscoelastic
dampers were installed in each layer to reduce the response caused by wind. In
addition, viscoelastic dampers were also used in the Columbia Seafirst and Two
Union Square buildings in Seattle.

The properties of viscoelastic materials are closely related to vibration frequency,
strain and temperature. In general, the relationship between shear stress and shear
strain can be explained by the equation below [6]:

$$\tau(t) = G'(\omega)\gamma(t) + \frac{G''(\omega)}{\omega}\dot{\gamma}(t) \tag{1.1}$$

$G'(\omega)$ is the storage elastic modulus of the viscoelastic material and $G''(\omega)$ is
the loss elastic modulus of the viscoelastic material. The analytical expressions of

$G'(\omega)$ and $G''(\omega)$ are shown in Ref. [7–9]. According to constitutive relation shown in Eq. (1.1), the force-displacement relationship of the viscoelastic damper is present below:

$$F(t) = k_d(\omega)X + c_d(\omega)\dot{X} \tag{1.2}$$

where $k_d(\omega)$ can be calculated by $AG'(\omega)/\delta$, and $c_d(\omega)$ can be calculated by $AG''(\omega)/\omega\delta$. A represents the shear area and δ represents the thickness.

The linear structure is still linear after the installation of viscoelastic dampers. The function of the damper is to increase damping and lateral stiffness of the structure, which brings great convenience to the analysis [10]. According to the modal strain energy method [11], Chang et al. [12] proposed a method to calculate the vibration damping ratio and frequency of the controlled structure after installing the viscoelastic damper, which can facilitate the structural analysis.

The viscoelastic material is thermo-sensitive. Chang conducted theoretical and experimental researches on the relationship between the mechanical properties of viscoelastic dampers and temperature [13, 14]. The result shows that if the temperature variation has little effect on the natural vibration frequency of the viscoelastic damping system, little effect will be shown on the vibration damping capacity of the viscoelastic damper when the stiffness of the damper is large enough.

There exists correlation between the elastic modulus storage and the elastic modulus loss of the viscoelastic material and the excitation frequency, which brings difficulties to the nonlinear analysis of energy dissipation system. In order to solve these problems, Markris proposed a complex parameter model of viscoelastic material. The parameters in the model are all complex numbers, but they are independent of the excitation frequency. The complex parameter model brings great convenience to frequency domain analysis of the viscoelastic damping system [15].

In order to verify theoretical research results of viscoelastic dampers and their feasibility in engineering, scholars around the world conducted a great deal of experimental research.

In 1993, Blondet et al. conducted an experiment on two full-scale viscoelastic energy dampers and six models. During the experiment, six models were destroyed when the strain reached up to 300%. The damage mainly occurred at the bond between the viscoelastic material and the steel plate [16]. In China, Beijing University of Technology, Harbin Institute of Technology, Guangzhou University and Southeast University also carried out systematic performance tests on full-scale viscoelastic dampers or models of viscoelastic dampers made of different viscoelastic materials [17–20].

Chang et al. conducted a dynamic test of two 2:5 steel frame models in 1994. One of the models was an uncontrolled structure and the other was a controlled structure with a viscoelastic damper [12]. Foutch et al. conducted a shaking table test on two reinforced concrete models with viscoelastic energy dampers installed in the US Military Construction Engineering Laboratory in 1993 [21]. The results show that viscoelastic energy dampers have good damping effect on steel structures and reinforced concrete structures under any earthquake action. At the same time, dampers can effectively reduce the structural damage because they dissipate energy in the cracking stage of reinforced concrete structures. Beijing University of Technology, Harbin Institute of Technology, and Xi'an University of Architecture and Technology respectively conducted shaking table tests on the steel structure model and reinforced concrete structure model with viscoelastic dampers. These tests also achieved good control effects [17, 22, 23].

It should be pointed out that when the temperature is constant, the viscoelastic material reacts linearly in a large range of strain. However, in the case of large strain, the temperature of viscoelastic material will increase due to energy dissipation, which leads to the change of viscoelastic material's mechanical property. Thus the overall reaction is non-linear. For this reason, if the viscoelastic damper is likely to have large strain, the traditional frequency domain method cannot be used to analyze the dynamic response of the energy consuming system.

(2) *Viscous damper*

Viscous dampers were originally used in the vibration reduction of military launches and some industrial machinery such as missile launchers and artillery [24, 25]. Then it was gradually applied to the energy dissipation and vibration reduction in civil engineering [26]. Viscous dampers are mainly divided into two categories types: viscoelastic fluid damper [27] and viscous damping wall [28, 29].

The viscoelastic fluid damper was first used in 1862 when the British army used this energy-dissipating device on the launcher of the cannon to reduce the displacement of the launch pad caused by the projectile [30]. At the end of the World War I, viscoelastic fluid damper is used on launchers to allow the use of larger shells and greater launching forces because it could reduce the rebound force. The use of this type of damper in automobiles in 1920s and 1930s to reduce vibration promoted the innovation of viscoelastic fluid dampers for ensuring the long service life. During the Cold War, the United States and the Soviet Union further improved the performance of dampers because of military needs. Around 1990, after the end of the Cold War, the viscoelastic fluid damper began to turn to civilian use, and it began to be rapidly and extensively researched and applied in the field of civil engineering [25]. The viscoelastic fluid damper is mainly composed of a pot filled with highly viscous fluid and a piston that moves within the fluid. Under the external excitation, the relative movement between the piston and the pot causes the highly viscous fluid in the pot to flow from one side to the other side through a small hole in the piston or the edge of the piston, thereby generating viscous damping. After the viscous damper is installed on the structure, it can provide enough damping to reduce the seismic response of the structure and the wind-induced vibration response. It can also be used as auxiliary equipment of base isolation system to enhance the seismic capacity of the structure.

The viscous damping wall is an energy absorbing damper for building structures, which was first proposed by Japanese scholars (Arima and Miyazaki) in 1986 [28, 29]. It is mainly composed of the inner steel plate suspended on the upper floor, two outer steel plates fixed on the lower floor, and highly viscous fluid between the inner and outer steel plates. The relative velocity is generated between the upper and lower floors during the earthquake. The upper inner steel plate moves in the viscous liquid between the lower outer steel plates, which generates damping force, absorbs seismic energy, and reduces seismic response. By changing the viscosity of the viscous fluid, the distance between the inner and outer steel plates and the area of the steel plate, the viscous resistance and energy absorption capacity of the viscous damping wall can be adjusted. In addition to the viscous damper wall, there are usually other external protective walls made of reinforced concrete or fireproof material to withstand the adverse effects of the external environment.

Domestic and foreign scholars have made extensive research on the performance of viscous dampers [31–36]. It has been found that if the fluid in viscous damper is Newtonian fluid, the damping force is proportional to the relative speed of motion. If the piston moves over a wide frequency range, the viscous damper will exhibit the characteristics of viscoelastic fluid. In this regard, Makris and Constantinou proposed a generalized Maxwell model in 1991 to describe the mechanical properties of viscous dampers. Considering the simplification of expressions, the most common expression in civil engineering is as below [25]:

$$F = CV^\alpha \tag{1.3}$$

where C is the viscous damping coefficient, V is the movement speed of the damper piston relative to the damper housing and α is the constant ranging from 0.1 to 2.

According to the foreign experience, when buildings use viscous dampers to resist earthquakes, the value of α is usually between 0.4 and 0.5. When dampers are used for resisting wind, α is usually between 0.5 and 1.0. If dampers are used for both earthquake and wind, the smaller value between 0.5 and 1.0 will be taken [25].

In 1993, Constantinou and Symans carried out a system performance test on a viscous damper which was installed in a 1:4 three-layer steel structure model to investigate the damping effect of the viscous damper [26]. Reinhom et al. tested the damping capacity of a viscous damper on a 1:3 reinforced concrete frame model in 1995 [37]. In 1988, Arima and Miyazaki et al. studied the mechanical properties of viscous damping walls and tested dynamic responses of viscous damping walls in a five-layer steel frame model and a four-layer full-scale structure [28]. Harbin Institute of Technology, Tongji University, and Southeast University also conducted experimental researches on the mechanical properties of viscous dampers [35, 36, 38]. Harbin Institute of Technology and Tongji University conducted shaking table tests on the structural model with viscous damper [38, 39]. Tsinghua University tested a small-scale structural model with viscous damping wall [40]. These studies show that the viscous damper has excellent energy dissipation and does not cause large changes in temperature. What's more, viscous damper only provides damping forces to the structure and does not increase structure's stiffness basically. If the structure is added with viscous damper reasonably, the displacement reaction and the internal force reaction can be reduced simultaneously.

(3) *Friction damper*

Friction damper is the mechanism that produces sliding and frictional forces by metal friction plates under a certain degree of pretightening forces. The mechanism causes the friction damper reciprocate sliding due to the vibration deformation, so the sliding friction force will dissipate energy to achieve the purpose of vibration reduction. The friction force is easy to control and can be easily controlled by adjusting the pretightening force, and its performance is not sensitive to ambient temperature and frictional heat.

The scholars from various countries have carried out in-depth research on the friction damper according to different requirements of different structures. By changing the structure of the friction damper, the friction surface material and the connection of the structure, many achievements have been made in theory, experiment and application. In 1982, Pall from Canada proposed a cross-core friction damper. The outline of the damper is parallelogram and it is connected to the structure by X-shaped diagonal braces. The performance is more stable than ordinary friction dampers, and the diagonal braces are not limited by the critical force. The test proves its great energy dissipation ability [41]. Grigorian et al. also proposed two simple friction dampers that were similar to viscoelastic energy dissipating devices [42]. In 1990, Aiken and Kelly et al. proposed a resettable Sumitomo energy-absorbing system [43]. Ou et al. from China studied and improved the friction damper, and proposed a kind of T-core plate friction damper and pseudo-viscous friction damper [44, 45]. Friction dampers mostly use frictional interface materials such as steel-steel, steel-copper and so on. The performance of the friction interface material has great influence on the performance of the energy consuming device.

The results of Scholl and Nims et al. show that as for friction dampers, the ratio of the initial sliding displacement to the inter-story yield displacement and the ratio of the stiffness of energy-dissipating braces to the structural interlayer stiffness are key factors to control the vibration [46, 47].

Under minor earthquakes, the friction damper does not slip and can only play a role as bracing, so the vibration control effect is not enough. To solve this problem, Tsiatas and Daly proposed to combine the friction damper and the viscous damper together to form a combined energy dissipating system. Under the action of wind and minor earthquake, only the viscous damper functions. In the case of major earthquakes, the friction damper can also participates in energy dissipation, which ensures better vibration control effect [48]. Lu Xilin et al. conducted the dynamic analysis of the structure with friction dampers and viscous dampers [49].

Friction dampers have also been used in different fields around the world. A number of buildings, such as the Concordia University Library in Canada and the Space Company's headquarters building, used Pall friction dampers to enhance seismic capacity [50]. The Sumitomo friction damper is widely used in Japan. A 31-story steel structure in Omiya, a 22-story steel structure in Tokyo, and a 6-story reinforced concrete structure all used this system [43]. In 1997, the friction damper was used to enhance a government building's seismic capacity in Northeast China. In 2001, a newly built Middle School canteen and chemical test building in Yunnan also used T-core friction dampers and quasi-viscous friction dampers to enhance earthquake resistance [51–53].

(4) *Metal damper*

The energy dissipation mechanism of metal damper is that the metal undergoes plastic-yielding hysteresis deformation to dissipate energy during the structural vibration, thereby achieving the purpose of vibration reduction. The metal damper is characterized by stable hysteresis characteristics, good low-cycle fatigue performance and low cost. What's more, it is independent to ambient temperature.

In early 1970s, Kelly et al. first put forward metal dampers. Later, scholars from various countries conducted both theoretical and experimental researches and developed energy-dissipating devices with various materials in various structures [54–56]. The soft steel has advantages of low yield point, mass fracture deformation and good low-cycle fatigue performance, and it is particularly suitable for making into metal yielding energy-dissipating devices due to its easy collection. At present, triangular and X-shaped steel plates are 2 typical metal dampers. These two metal dampers are deformed by equal curvature with uniformly-distributed bending stress and can yield simultaneously. In addition, lead and shape memory alloys also have good energy dissipation ability and can also be used to make metal dampers [57, 58].

In order to establish the hysteretic model of metal dampers, the hysteretic model based on the constitutive relationship of materials and experimental research are two important methods [59–61].

The metal damper is a kind of non-linear device that, when installed on the structure, will cause the controlled structure to exhibit significant nonlinear characteristics. Studies have shown that the ratio of the stiffness of the support to the stiffness of the damper, the ratio of the whole stiffness of the support & the energy damper to the interlayer stiffness, and the ratio of the yield displacement of the damper to the interlayer yield displacement are 3 major parameters of the vibration effect [46]. In order to analyze the vibration control effect, two methods can be used. One is to use the hysteretic model of dampers to make the dynamic time history analysis [62, 63]. The other is using the equivalent linearization method to get equivalent linear parameters of hysteretic curves of mental dampers, then calculating and analyzing the controlled structure [47, 64].

Metal dampers have been successfully used in civil engineering. A six-story government office building in New Zealand uses the steel pipe energy absorbing device in the prefabricated wall panel. A 29-story steel structure suspension building in Naples, Italy, installs a cone soft steel damper between the core tube and the suspended floor. Two structures in San Francisco and three structures in Mexico adopt X-shaped steel plate yielding energy dissipating devices for seismic retrofitting. The honeycomb energy dissipating device and the bell- shaped energy dissipating device which are developed by Kajima Company of Japan are respectively applied to a 15-storey steel office building and two adjacent buildings [65, 66].

1.1.3 Passive Tuned Vibration Absorption

Passive tuned vibration absorption control consists of the structure and the substructure attached to the major structure. The additional substructure has its own mass, stiffness and damping. By adjusting the mass and stiffness of the substructure, the natural frequency can be changed to be as close as possible to the fundamental frequency or excitation frequency of the major structure. Thus, when the main structure is forced to vibrate, the substructure generates an inertial force opposite to the vibration direction of the major structure to act on the structure. So the vibration reaction of the major structure is attenuated and controlled. The control is not provided by external energy. It is achieved by adjusting the frequency characteristics of the structure. The mass of the substructure can be solid mass, in which case the substructure is called as tuned mass damper (TMD). It can also be the mass of the liquid stored in a container whose tuned damping effect is generated by the oscillation of the liquid in the container. Dynamic pressure difference and viscous damping energy dissipation are two ways to ensure damping effects. In this case, the substructure is called as tuned liquid damper (TLD). TLD can be divided into two types. One is with liquid storage tank and the other is with U-shaped column. The liquid in the container can be replaced by solid particles. The energy dissipation of the solid particles is mainly achieved by the momentum exchange caused by the impact between particles & the major structure and the friction energy dissipation between the systems. If there is only one single particle, the substructure is called as Impact Damper. If there are multiple particles, the substructure is called as Particle Damper. The passive tuned vibration absorption has been successfully applied to seismic control, wind vibration control and wave induced vibration control of multi-rise structures, towering towers, long-span bridges, and offshore platforms.

1.2 Active Control

Active control is the control with external energy. The external energy can be actively input into the structural system. The input energy value is determined by certain control strategy. The response of the structure and the input energy value are tracked and predicted in real time to meet certain optimization criteria. That is, under the limited energy input conditions, the vibration of the structure is controlled to the utmost extent. Active control is the method of adjusting the structural vibration response according to actual needs. Theoretically, it is the most effective structural control method.

The concept of active control was first put forward by Zuk [67] and the systematic active control theory was formed in 1970s. In 1972, American Chinese scholar Yao [68] proposed that modern control theory could be applied in civil engineering, which was regarded as the beginning of research on vibration control of structures in civil engineering. However, around 1990, active control was actually studied and applied to the civil engineering structure from both theoretical and practical aspects.

The active control system is mainly composed of three parts: information acquisition (sensor), computer control system (controller) and active drive system (actuator). External actuators in the active control system can apply external forces to the structure in specified manners. The forces can be used to increase and dissipate energy of the structure. The signal transmitted to the actuator is the vibration response information of the structure and external disturbance and is measured by some physical sensors, such as optical sensors, mechanical sensors, electrical sensors, chemical sensors, and the like.

According to the working mode of the controller, active control can be divided into three types: open loop control, closed loop control and open-closed loop control. Open loop control, as shown in Fig. 1.1, means that the controller measures the external excitation of the structure through sensors and accordingly adjusts the control force applied by the actuator to the structure without reflecting the output structural response of the system. As for closed-loop control shown in Fig. 1.2, the controller measures the structural response through the sensor and adjusts the control force applied by the actuator to the structure without reflecting the input information of the external excitation of the structure. As for open-closed loop control, the control system simultaneously measures the input external excitation of the structure and the output structural response of the system through the sensor, and based on the

Fig. 1.1 Open loop control system

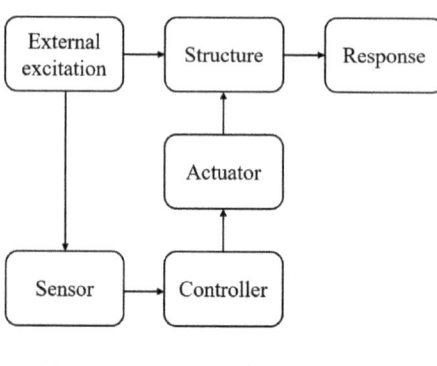

Fig. 1.2 Closed loop control system

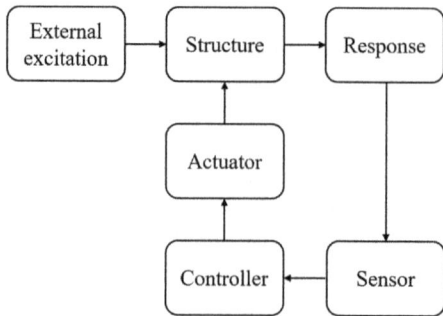

Fig. 1.3 Open-closed loop
control system

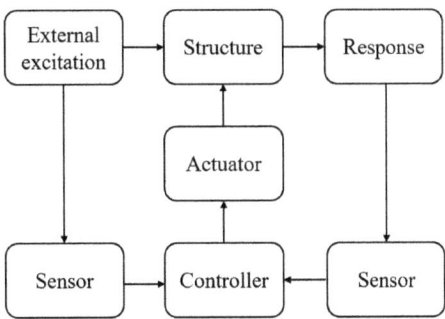

general information, the control system adjusts the force applied by the actuator to the structure, as shown in Fig. 1.3.

Since the closed-loop control system can track dynamic responses of the structure in real time, the active control of the structure generally adopts the closed loop control method, and sometimes open-closed control system is also adopted. The closed loop control system works by measuring dynamic responses of the structure by the sensor mounted on the structure. The controller calculates the required control force according to a certain kind of control rule, and the force is applied to the structure through the actuator. Thereby the whole system can achieve the purpose of reducing or suppressing structural dynamic responses.

Active control mainly includes active tendon system (ATS), active bracing system (ABS) and active mass damper system (AMD).

The basic principle of ATS is to set up crossed tendons between layers of the frame structure, install the hydraulic servo system on tendons and set some sensors around or in the structure. When the structure suffers from the earthquake or wind, the computer control center will start the servo system to apply forces to the tendons based on the signal collected by the sensor. In order to verify the effectiveness of ATS, a full-scale 6-story test structure was built in Tokyo, Japan [69]. ATS is located in the first floor and connected to the tendon by four drives to control the vibration in 2 directions of the structure. This test structure serves as a long- term observation building for ATS.

The active bracing system (ABS) places active bracing devices between floors of the structure. According to the inter-layer reaction information of the structure, the electro-hydraulic servo control mechanism can control the contraction movement of the braces and change the magnitude and direction of the bracing force to control structural vibration. Loh et al. [70] conducted a vibration table test of a full-scale three-layer steel structure model to verify the effectiveness of ABS. The model was 9 m high with an additional mass of 3468 kg per layer, and two sets of ABS control systems were arranged on the first floor. The test results show that the ABS control system has obvious effects, and the relative displacement of the controlled structure is reduced by more than 50%.

AMD is a kind of active control system formed by adding measurement, controller and drive mechanism based on the passive tuned mass damper (TMD). The structural response information is observed by the sensor. The computer center determines the value of the control force according to the control algorithm, and the servo system is activated to apply the force to the structure by means of additional mass.

In 1989, the Tokyo Kyobashi Seiwa Building in Japan adopted the active mass damper for the first time [71]. The Kyobashi Seiwa Building has 11 floors and a height of 33.1 m. It is a flat high-rise steel frame structure with an aspect ratio of 8.25. According to the traditional seismic and wind-resistant design, the horizontal stiffness is difficult to meet the requirements, so two AMD control devices are arranged on it. The central AMD has a mass of 4 tons which is used to control the lateral vibration of the structure, and the AMD at the end is 1 ton to reduce the torsional vibration of the structure.

Active control introduces the optimal control concept of modern control theory into the vibration control of building structures, which has outstanding effects. Practice has proved that the structural control system with active control can reduce the structural vibration response by 40–85%, compared with the traditional structural system without any control technology. Although active control has obvious superiority in vibration control and has made certain breakthroughs in theoretical research, experimental research and engineering application, there are still some defects to be improved. The first shortcoming is that active control systems are expensive and require huge external energy support. It is difficult to ensure energy supply under extreme conditions such as rare earthquakes. Secondly, the stability cannot be guaranteed. There are many factors that will affect the operation of the control system such as whether the control algorithm can be truly optimized, whether the data acquisition, analysis and control system can work reliably (whether the skew effect is improved), whether the external energy and power reserve are stable, and so on.

1.3 Hybrid Control

Hybrid control is the structural vibration control method that applies both active control and passive control to the same structure. Although active control has outstanding effects, it is expensive and technically demanding. Passive control is simple, reliable, low in cost and easy to implement, but the control range and effect are limited. By combining this two method together, the whole mechanism could be made more reasonable, economical and much safer. For example, when the structure is under the action of frequent earthquakes, the passive control system is mainly used to achieve the vibration dissipation. When the structure suffers from the rare earthquake, the active control system begins to work, and the structure simultaneously relies on the passive control and the active control to operate together. Thus the best vibration control effect can be achieved.

From the point of the effectiveness of passive and active control, there are two hybrid forms. The first one is the master-slave hybrid form. That is, one control system is dominant and the other is auxiliary. The second hybrid form is parallel, which means the active control and the passive control work independently and correct the structure together. At the present stage, the most studied form is the master-slave hybrid form in which the passive control is dominant and the active control is auxiliary. The two main hybrid methods are as follows.

The first method is to regard passive control as the protective measure of the structure under the action of the frequent earthquake and the active control is the protective measure against the rare earthquake. The active control system is the last line of defense of the structural damage.

The second method orders passive control as the main part of the whole control system. The active control provides limit control for the passive control system and provides the resilience required for passive control.

Now, several typical hybrid control devices are listed below. The first one is hybrid mass damper (HMD), which combines TMD and active controller together. The ability of HMD to reduce structural responses mainly depends on the motion of TMD. The force from the actuator is used to increase the effectiveness of HMD and change the robustness of the dynamic characteristics of the main structure.

The second typical device is hybrid base isolation. It is a kind of active control device added to the isolation layer to keep the relative displacement of the isolation layer within the allowable range and further reduce the seismic response of the superstructure. Combining laminated rubber bearings with HMD or variable damper is a kind of commonly used method.

The third one is to combine damper with active control. It is a kind of system controlled by both viscoelastic damper and ABS. On the one hand, due to the action of viscoelastic damper, the control force required by ABS is greatly reduced. On the other hand, due to the action of ABS, the damping ratio of viscoelastic damper is greatly improved and the shear force that viscoelastic damper suffers is reduced.

1.4 Semi-active Control

Semi-active control is a kind of vibration control technology that aims for the purpose by changing the stiffness or damping of the structure and adaptively adjusting the dynamic characteristics of the structure according to the structural reaction. It has the advantage that the control effect is close to the active control with little energy input. Because it is a kind of system with limited input and limited output, the control stability is sufficient. When the energy supply is interrupted, it can immediately change into the passive control system to play the controlling role. Thus this technology has broad application prospects. Because semi-active control is essentially a kind of parameter controlling which reduces the structural vibration by changing the stiffness or damping of the structure, semi-active control can be divided into active variable stiffness system (AVS), active variable damping system (AVD) and active variable stiffness/damping system (AVS/D).

1.4.1 Active Variable Stiffness System

The active variable stiffness system uses the variable stiffness device to switch the stiffness values of the controlled structure in real time according to the preset control rate in each sampling period, so the controlled structure can keep away from the resonance state as much as possible in each sampling period in order to achieve the purpose of vibration reduction.

In terms of theoretical research, the active variable stiffness system (AVS) was first proposed by Japanese scholar Kobori [72, 73]. The basic idea is to change the stiffness of the system according to the reaction of the structure through the computer-controlled rapid response locking device. Thus the natural vibration period of the structure can be different from the excellent period of ground motion to reduce the structural reaction. The active variable stiffness control system essentially uses the control algorithm to automatically adjust the switching state of the variable stiffness according to the displacement and velocity in the structural vibration process, so the structure is always in the non-resonant state. However, the time lag caused by the switch will affect its control effect, which is also the obstacle of the system's popularization and application in civil engineering. Therefore, it is necessary to develop a kind of semi-active control algorithm that is more efficient, more stable, more robust and insensitive to time lag.

As for experimental research, Kobori et al. [72] carried out a full-scale three-layer steel structure active variable stiffness control shaking table test. The result shows that the semi-active variable stiffness control technology is feasible to reduce the seismic response of the structure; Richer et al. [74] proposed energy dissipating restraint (EDR) devices, which could provide different stiffness during uploading and unloading; Nasu et al. [75] focused on a high-rise structure with active variable stiffness device. The conclusion is that AVS is simple in construction with low demand for external energy and can be used for buildings with different stiffness and in different height under different seismic intensities; Liu et al. [76] proposed the variable stiffness semi-active control system and its theory and successfully completed the first semi- active control vibration test in China. The good control effect was achieved which promoted the research and application of this technology in China; Li [76] focused on a large-scale aqueduct. The problem of continuous variable stiffness control using the isolated rubber bearing and variable stiffness control device in the foundation layer was discussed.

In terms of engineering applications, the world's first AVS semi-active control device is in Kashima Technical Research Institute in Tokyo, Japan. Hydraulic components are used to change the connection conditions of rigid braces and girders, and the inter-layer stiffness can be adjusted at any time to avoid the resonance state. After being subjected to the strong earthquake with peak acceleration of 0.25 g, the operation is normal and the damping effect is very obvious. The system consumes less energy, and the backup power supply is available for 3 min even when the electric supply is disturbed accidently [77].

1.4.2 Active Variable Damping System

The principle of the active variable damping system is proximately the same as that of the active variable stiffness system, except that the object controlled by the active variable damping control system is the damping of the structure. Active variable damping system is usually composed of hydraulic cylinders, pistons and electro-hydraulic servo valves. In practical application, the piston rod and the cylinder of the active variable damping system are respectively connected to two different members of the structure. When the structure vibrates under the action of earthquake or wind, the relative motion between the piston and the cylinder is generated by the structure. The value of the damping force that the system can provide mainly depends on the size of valve's opening.

Active Variable Damping System (AVD) was first put forward by Hrovat [78]. He studied the wind-induced vibration response of the control structure with variable damping systems, and observed that the effect of variable damping control is close to that of the active control through simulation analysis. The variable damper was tested by Kawashima et al. [79]. They studied the control effect of the bridge structure under seismic excitation with a hybrid control system consisting of a variable damper and a rubber support. The result shows that the control effect is obvious. Various variable damping devices have been developed, including semi-active fluid dampers, controllable friction dampers, semi-active tuned mass dampers, semi-active tuned liquid dampers, and electric/magneto-rheological dampers.

1.4.3 Active Variable Stiffness/Damping System

Chinese scholar Zhou Fulin [80] proposed to combine the active variable stiffness system (AVS) and the active variable damping system (AVD) together to form a new kind of semi-active control technology named as active variable stiffness/damping system (AVS/D). By increasing the variable damping terms., the energy dissipating effect of the active variable stiffness device in the energy release phase can be enhanced, which not only avoids the predominant period of ground motion, but also reduces the peak of the reaction and the external excitation in a wider frequency band. He et al. [81] designed a three-layer steel frame structure to test the effectiveness and practicability of the active variable stiffness/damping (AVS/D) control technique and the instantaneous optimal semi-active control algorithm. The experimental results show that, compared with the passive control method, the semi-active control technology can not only effectively control the displacement and acceleration response of the structure under earthquake action, but also overcome the negative control effects of variable stiffness control technology and variable damping control technology. The seismic excitation is very fierce. The implementation method is simple to realize, which provides a basis for further research and engineering applications in the future.

In addition, semi-active control also includes active tuned parameter mass damper (ATMD) and aerodynamic air damper (ADA). Based on TMD, the mechanism for adjusting stiffness and damping coefficient of TMD is added on ATMD to maximize the vibration absorption of the main structure and reduce the dynamic response. As for ADA, when the speed at the top of the structure is opposite to the wind, the air damper is opened to expand the windward side, and vice versa.

References

1. Housner, G.W., L.A. Bergman, T.K. Caughey, et al. 1997. Structural control: Past, present, and future. *Journal of Engineering Mechanics, ASCE* 123 (9): 897–971.
2. Tang, J.X., and Z.H. Liu. 1993. *Base isolation system of buildings*. Wu Han: Huazhong University of Science and Technology Press.
3. Hanson, R.D., et al. 1993. State-of-the-art and state-of-the-practice in seismic energy dissipation. In *Proceedings of ACT-17-1 on seismic isolation, energy dissipation and active control*, vol. 2, 449–471.
4. Ou, J.P., B. Wu, and T.T. Soong. 1996. Recent advances in research on and application of passive energy dissipation system. *Earthquake Engineering and Engineering Vibration* 16: 72–96.
5. Yan, F. 2004. *Experimental study on energy dissipation structure with viscous damping walls, in structural engineering*. Shang Hai: Tongji University.
6. Zhang, R.H., T.T. Soong, and P. Mahmoodi. 1989. Seismic response of steel frame structures with added viscoelastic dampers. *Earthquake Engineering and Structural Dynamics* 18: 389–396.
7. Tsai, C.S., and H.H. Lee. 1993. Application of viscoelastic dampers to high-rise buildings. *Jouranl of Structural Engineering, ASCE* 119: 1222–1233.
8. Kasai, K., J.A. Munshi, M.L. Lai, and B.F. Maison. 1993. Viscoelastic damper hysteretic model: theory, experimental and application. In *Proceedings of ACT-17-1 on seismic isolation, energy dissipation and active control*, vol. 2, 521–532.
9. Sheng, K.L., and T.T. Soong. 1995. Modeling of viscoelastic dampers for structural application. *Jouranl of Engineering Mechanics ASCE* 121: 694–701.
10. Zhang, R.H., and T.T. Soong. 1992. Seismic design of viscoelastic dampers for structural application. *Journal of Structural Engineering* 118: 1375–1392.
11. Soong, T.T., and M.L. Lai. 1991. Correlation of experimental results and predictions of viscoelastic damping of a model structure. *Proceeding of Damping* 91: 1–9.
12. Chang, K.C., M.L. Lai, T.T. Soong, D.S. Hao, Y.C. Yeh. 1993. *Seismic behavior and design guidelines for steel frame with added viscoelastic dampers*. New York: State University of New York at Buffalo.
13. Chang, K.C., M.L. Lai, and T.T. Soong. 1992. Effect of ambient temperature on viscoelastically damped structure. *Journal of Structure Engineering* 118 (7): 1955–1973.
14. Kasai, K., Y. Fu, and M.L. Lai. 1994. Finding of temperature-insensitive viscoelastic damper frames. *Proceedings of the First World Conference on Structural Control* 1: 3–12.
15. Markis, N. 1994. Complex-parameter Kelvin model for elastic foundation. *Earthquake Engineering and Structural Dynamics* 23: 251–264.
16. Blodnet, M. *Dynamic response of two viscoelastic dampers*. Project of No. ES-2046. Berkeley: California: Department of Civil Engineering, University of California.
17. Hao, D.S., H.T. Qin, and Y.Z. Ye. 1994. Experimental study on full-size steel frame with dampers. *Earthquake Resistant Engineering*.
18. Wu, B., and A. Guo. 1998. Study on properties of viscoelastic dampers. *Earthquake Engineering and Engineering Vibration* 18 (2).

19. Zhou, Y., Z. Xu, and X. Deng. 2000. Experimental study on properties of viscoelastic dampers. In *International symposium on structural control and health diagnosis*, Shenzhen, China.
20. Li, A., and W. Cheng. 2002. A review of seismic isolation, damping and vibration control techniques of structures. In *Progress in modern earthquake engineering, Nanjing, China*.
21. Foutch, D.A., S.L. Wood, and P.A. Brady. 1993. Seismic retrofit of nonductile reinforced concrete frames using viscoelastic dampers. In *Proceeding of ACT-17-1 on seismic isolation, energy dissipation and active control*, vol. 2.
22. Zou, X., and J. Ou. 1999. Experimental and analysis on seismic vibration—suppressed effects of viscoelastic energy dissipation. *Journal of Harbin University of Civil Engineering and Architecture* 32 (4).
23. Xu, Z., et al. 2001. The shaking table test of viscoelastic structure. *Journal of Building Structures* 22 (5).
24. Harris, C.M., and C.E. Crede. 1976. *Shock and vibration handbook*. N.Y.: McGraw-Hill.
25. Douglas, P.T. 2002. History, design, and applications of fluid dampers in structural engineering. In *Proceedings of structural engineers world congress, Japan*.
26. Constantinou, M.C., and M.D. Symans. 1993. Seismic response of structures with supplemental fluid dampers. *Structural Design of Tall Buildings* 2: 93–132.
27. Makris, N., and M.C. Constantinou. 1993. Analytical model of viscoelastic fluid dampers. *Journal of Structural Engineering, ASCE* 119: 3310–3325.
28. Arima, F., and M. Miyazaki. 1988. A study on buildings with large damping using viscous damping walls. In *Proceedings of ninth world conference on earthquake engineering, Japan*.
29. Miyazaki, M., and Y. Mitsusaka. 1992. Design of a building with 20% or greater damping. In *Proceedings of tenth world conference on earthquake engineering, Spain*.
30. Thomson, W. 1965. *Vibration theory and applications*. Englewood, Cliffs, New Jersey: Prentice-Hall.
31. Makris, N., and M.C. Constantinou. Fractional derivative model for viscous dampers. *Journal of Structural Engineering* 117 (9): 2708–2724.
32. Pekcan, G., J.B. Mander, and S.S. Chen. 1999. Fundamental considerations for the design of non-linear viscous dampers. *Earthquake Engineering and Structural Dynamics* 28: 1405–1425.
33. Zhao, Z., et al. 2000. The research of response of oleo damper excited by random excitation. *Earthquake Engineering and Structural Dynamics* 20 (1): 105–111.
34. Sadek, F., B. Mohraz, and M.A. Riley. 2000. Linear procedures for structures with velocity-dependent dampers. *Journal of Structural Engineering, ASCE* 126 (8).
35. Wen, D., et al. 2002. The experimental study on property of energy dissipation of viscous liquid damper. *World Information On Earthquake Engineering* 18 (4): 30–34.
36. Ye, Z., A. Li, and Y. Xu. 2002. Fluid viscous damper technology and its engineering application for structural vibration energy dissipation. *Journal of Southeast University* 32 (3).
37. Reinhorn, A.M., C. Li, and M.C. Constantinou. 1995. Experimental and analytical investigation of seismic retrofit of structures with supplement damping: part 1: Fluid viscous damping devices. In *NCEER Report 95-0001*. New York: State University of New York at Buffalo, Buffalo.
38. Ding, J. 2001. *Theory and experimental study on viscous fluid damping energy dissipation system for civil engineering*. Harbin: Harbin Institute of Technology.
39. He, Q. 2003. *Experimental study on mechanics performance and the effect of energy dissipation of a viscous damper*. Shanghai: Tongji University.
40. Tan, Z., and J. Qian. 1998. Study on earthquake response reduction of RC frames with viscous-damping walls. *Journal of Building Structures*.
41. Pall, A.S., and C. Marsh. 1982. Response of friction damped braced frames. *Journal of the Structural Division, ASCE* 108 (ST6): 2325–2336.
42. Grigorian, C.E., and E.P. Popov. 1993. Slotted bolted connections for energy dissipation. In *Proceeding of ACT-17-1 on seismic isolation, energy dissipation and active control*, vol. 2, 545–556.
43. Aiken, J.D., and J.M. Kelly. 1990. Earthquake simulator testing and analytical studies of two energy-absorbing system for multistory structure. In *Report No. UCB/EERC-90/03*. Berkeley, CA: Earthquake Engineering Research Center, University of California.

44. Wu, B., and J. Ou. 1999. Earthquake simulation testing and parameter study on high-rise building with pseudo-viscous frictional energy dissipators. *World Information on Earthquake Engineering* 15 (2).
45. Zou, X. 2000. *Experimental study and analysis of viscoelastic and frictional EDORs*. Harbin: Harbin Institute of Technology.
46. Scholl, R.E. 1993. Design criteria for yield and friction energy dissipation. In *Proceedings of ACT-17-1 on seismic isolation, energy dissipation and active control*, vol. 2.
47. Nims, D.K., P.J. Richter, and R.E. Bachman. 1993. The use of the energy dissipating retraint for seismic hazard mitigation. *Earthquake Spectra* 9: 467–486.
48. Tsiatas, G., and K. Daly. 1994. Controlling vibrations with combination viscour/friction mechanisms. In *Proceedings of first world conference on structural control*, vol. 1, WP4-3–WP4-11.
49. Zhou, Q. 2000. *Numerical simulation and experimental study on a one-bay steel frame with dampers*. Shanghai: Tongji University.
50. Pall, A.S., and R. Pall. 1993. Friction-dampers used for seismic control of new and existing buildings in Canada. In *Proceeding of ACT-17-1 on seismic isolation, energy dissipation and active control*, vol. 2, 675–686.
51. Wu, B., et al. 1998. Strengthening the earthquake resistance of a building using friction dampers. *Journal of Building Structures* 19 (5).
52. Ou, J., et al. 2001. Seismic analysis and design for dining-room structure of Zhenrong middle school with energy dissipators part I: Response spectra method. *Earthquake Engineering and Engineering Vibration* 21 (1).
53. Ou, J., et al. 2001. Seismic analysis and design for dining-room structure of Zhenrong middle school with energy dissipators part II: Seismic damage performance control design. *Earthquake Engineering and Engineering Vibration* 21 (1).
54. Kelly, J.M., R.L. Skinner, and A.J. Heine. 1972. Mechanics of energy absorption in special devices for use in earthquake-resistant structures. *Bulletin of New Zealand National Society for Earthquake Engineering* 5 (3): 63–88.
55. Aiken, J.D., D.K. Nims, A.S. Whittaker, and J.M. Kelly. 1993. Testing of passive energy dissipation systems. *Earthquake Spectra* 9 (3): 335–370.
56. Skinner, R.I., J.M. Kelly, A.J. Heine, and W.H. Robinson. 1980. Hysteresis dampers for the protection of structures from earthquake. *National Society for Earthquake Engineering* 13 (1): 22–26.
57. Monte, M.D., and H.A. Robison. 1998. Lead shear damper suitable for reducing the motion induced by wind and earthquake. In *Proceedings of the eleventh world conference on earthquake engineering, Acapulco, Mexico*.
58. Ni, L., et al. 2002. Investigation and experiment of SMA superelastic damping. *Earthquake Engineering and Engineering Vibration* 22 (6).
59. Dargush, G.F., and T.T. Soong. 1995. Behavior of metallic plate dampers in seismic passive energy dissipation system. *Earthquake Spectra* 11: 545–568.
60. Tsai, C.S., and K.C. Tsai. 1994. TPEA devices as seismic damper for high-rise buildings. *Journal of Engineering Mechanics, ASCE* 121: 1075–1081.
61. Ou, J., and B. Wu. 1995. Experimental comparison of the properties of friction and mild steel yielding energy dissipators and their effects on reducing vibration of structure under earthquakes. *Earthquake Engineering and Engineering Vibration* 15: 73–87.
62. Xia, C., and R.D. Hanson. 1992. Influence of ADAS element parameters on building seismic response. *Journal of Structural Engineering, ASCE* 118: 1903–1918.
63. Tsai, K.C., H.W. Chen, C.E. Hong, and Y.F. Su. 1993. Design of steel triangular plate energy absorbers for seismic-resistant construction. *Earthquake Spectra* 9: 505–528.
64. Ou, J., B. Wu, and X. Long. 1998. Seismic design methods of passive energy dissipation systems. *Earthquake Engineering and Engineering Vibration* 18 (2).
65. Perry, C.L., E.A. Fierro, H. Sedarat, and R.E. 1993. Scholl seismic upgrade in San Francisco using energy dissipation devices. *Earthquake Spectra* 9 (3): 559–579.
66. Ciampi, V. 1991. Use of energy dissipation devices, based on yielding of steel for earthquake protection of structures. In *Proceedings of international meeting on earthquake protection of buildings, Ancona, Italy*.

67. Zuk, W. 1968. Kinetic structures. *Civil Engineering* 39 (12): 62–64.
68. Yao, J.T.P. 1972. Concept of structural control. *Journal of the structural Division, ASCE* 98 (7): 1567–1574.
69. Soong, T.T., and W.F. Chen. 1990. *Active structural control: Theory and practice.* Longman Scientific & Technical, Wiley, 1282–1285.
70. Loh, C.H., P.Y. Lin, and N.H. Chung. 1999. Experimental verification of building control using active bracing system. *Earthquake Engineering and Structural Dynamics* 28 (10): 1099–1119.
71. Kobori, T. 1996. Future direction on research and development of seismic-response-controlled structures. *Microcomputers in Civil Engineering* 11 (5): 297–304.
72. Kobori, T. 1990. *Technology development and forecast of dynamical intelligent building (D.I.B.).*
73. Kobori, T., et al. 1993. Seismic response controlled structure with active variable stiffness system. *Earthquake Engineering and Structural Dynamics* 22 (11): 925–941.
74. Riche, P.J., D.K. Nims, J.M. Kelly, and R.M. Kallenbach. 1990. *The EDR-energy dissipating restraint. A new device for mitigating seismic effective.* Lake Tahoe: Structure Engineering Association of California.
75. Nasu, T., et al. 1995. Analytical study on the active variable stiffness system applied to a high-rise building. *Journal of Structural Engineering B* 41: 33–39.
76. Liu, J., and M. Li. 1999. Semiactive structural control with variable stiffness. *Journal of Vibration Engineering* 02: 19–25.
77. Takahashi, M., et al. 1998. Active response control of buildings for large earthquakes—seismic response control system with variable structural characteristics. *Smart Materials and Structures* 7 (4): 522–529.
78. Hrovat, D., P. Barak, and M. Rabins. 1983. Semiactive versus passive or active tuned mass dampers for structural control. *Journal of Engineering Mechanics* 109 (3): 691–705.
79. Motoichi Takahashi, T.K., T. Nasu, et al. 1998. Active response control of buildings for large earthquake-seismic response control system with variable structural characteristics. *Smart Material Structure* 7: 522–529.
80. Zhou, F., P. Tan, and W. Yan. 2002. Theoretical and experimental research on new system of semi-active control of structure. *Journal of Guangzhou University (Natural Science Edition)* (01): 69–74.
81. He, Y., Y. He, and J. Huang. 2002. Shaking-table test on structural semi-active control. *Journal of Building Structures* 04: 10–15.

Chapter 2
Origination, Development and Applications of Particle Damping Technology

2.1 Basic Concept and Development of Particle Damping

The concept of particle damping could be traced back to 1937, when Paget [1] was studying the vibration attenuation problem of the turbine blades, during which he invented the impact damper. The impact damper only involves a single particle, resulting in high noise levels and significant impact forces during the impact process, and it tends to become sensitive to the change in certain parameters (such as the excitation amplitude and the restitution coefficient). Later on, in 1945, Lieber and Jensen [2] proposed the concept of using a mass moving between two walls of a container to eliminate the vibration of mechanical systems, which evolved to the form of the impact damper. Because of a variety of flaws in a single-particle impact damper, its further application and development in more fields are limited. Hence, subsequent researchers have replaced the single particle with many smaller particles of equivalent total mass, thus resulting in the particle damper.

2.1.1 Basic Traditional Particle Dampers

Based on the number of units and particles per unit, the traditional particle dampers can be grouped into four major types: the impact damper [2], the multi-unit impact damper [3, 4], the particle damper [5, 6] and the multi-unit particle damper [7, 8], as shown in Fig. 2.1. The development and evolution of these four fundamental types of particle dampers have broken the traditional single design concept, laying a solid foundation for the development of the subsequent multiform particle dampers.

The multi-unit impact damper is the descendant of impact dampers. In 1969, Masri [4] obtained the exact solution for the steady-state motion of a multiple-unit impact damper attached to a sinusoidally excited primary system, verified by numerical simulation and experiments. He found that, compared to the impact damper of equivalent

© China Machine Press and Springer Nature Singapore Pte Ltd. 2020
Z. Lu et al., *Particle Damping Technology Based Structural Control*,
Springer Tracts in Civil Engineering,
https://doi.org/10.1007/978-981-15-3499-7_2

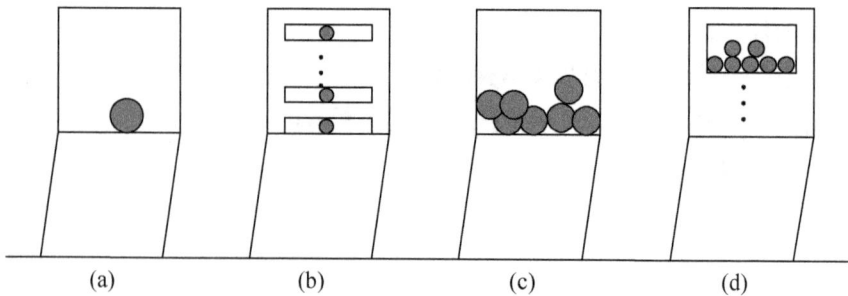

Fig. 2.1 Schematic diagram of the traditional particle dampers: **a** Impact damper; **b** Multi-unit impact damper; **c** Particle damper; and **d** Multi-unit particle damper

total mass, as to reducing vibration, especially reducing the noise generated during the damper's impact process, properly designed multi-unit impact dampers are more effective. The increasing interest in the particle damper dates from 1985, when Araki [5] replaced the shock unit of the traditional impact damper containing a single particle with a bed of granular materials. He studied the effectiveness of this kind of impact damper for reducing the vibration of a single-degree-of-freedom (SDOF) system under simple harmonic excitation, and the effect of mass ratio and clearance on the performance of the damper was determined. The non-obstructive particle damper (NOPD), which is one of the most common types of particle dampers, was proposed by Panossian [9] in 1991 through the experiment of an aluminum beam. Such technology consists of making small diameter cavities at appropriate locations inside the vibrating structure, which is capable of yielding the maximum damping effect for any desired mode. The multi-unit particle damper was first put forward by Saeki [8] in 2005. He found that the damping performance of the particle damper depends on the size of the cavity. If the optimum size of the cavity is too large, from the standpoint of practical design, it will lead to excessive clearance and decrease the number of effective collisions between the particles and the wall, which has negative effects on the damping performance. Hence, Saeki divided the cavity into multiple small cavities with an appropriate number and the influence of the number and the size of the cavities on damping performance was studied.

2.1.2   Development of Particle Dampers

Based on the traditional particle dampers, several variants have evolved. Each variant has its own features so that it can be selected depending on the engineering requirements. According to the different aspects of the improvements made in the traditional particle dampers, these variants can be classified into three main categories: the configuration improved type, the material improved type and the combination type, as shown in Fig. 2.2.

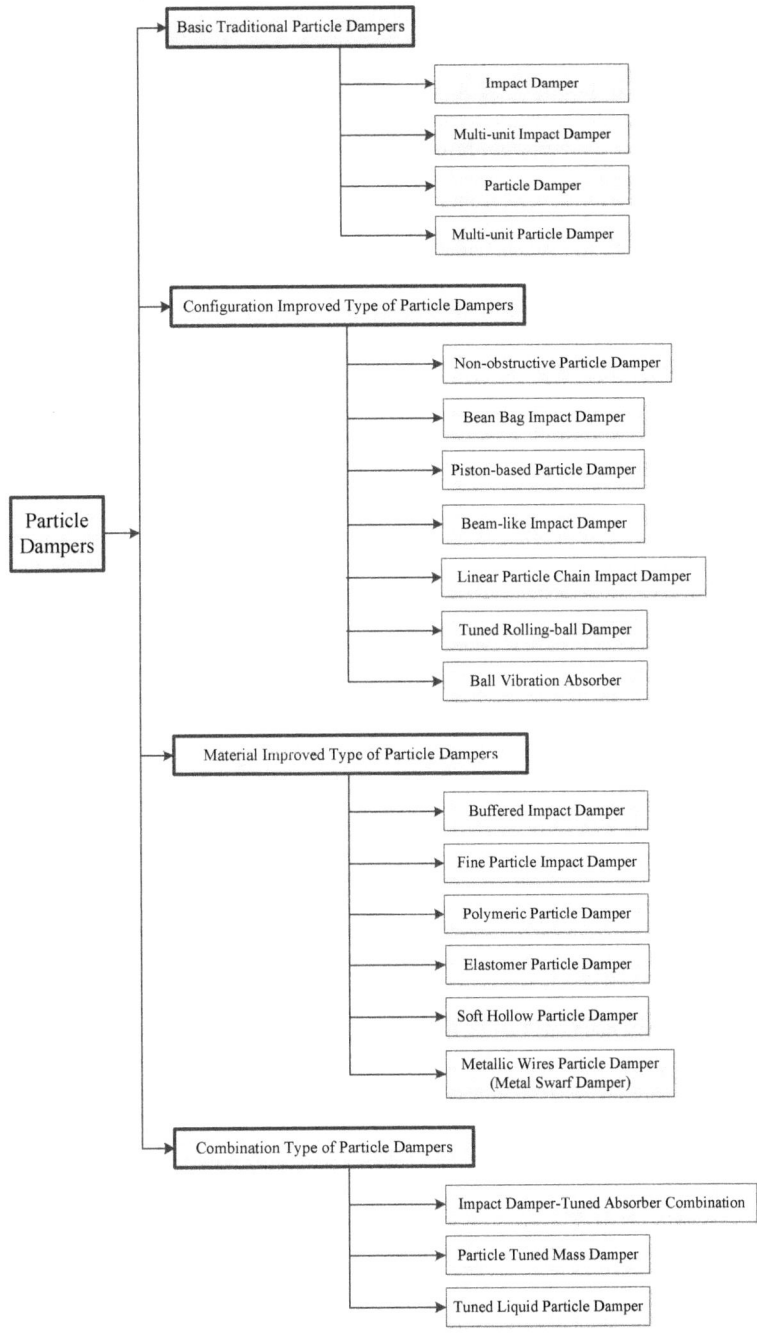

Fig. 2.2 Particle dampers categorization

2.1.2.1 The Configuration Improved Type of Particle Dampers

The configuration improved type improves the traditional particle dampers by changing structural configurations, e.g., particles and cavities. It can optimize the mechanism of vibration attenuation and improve its characteristics in order to apply it to a wider range of working conditions. This type of variants mainly includes the bean bag impact damper [10], the piston-based particle damper [11, 12], the beam-like impact damper [13], the linear particle chain impact damper [14], the tuned rolling-ball damper [15], the ball vibration absorber [16], etc. The bean bag impact damper (BBD) is also called the flexible restraint particle impact damper, which originated in the early 1980s. Popplewell et al. [10] adopted this kind of damping technology when he was studying the vibration attenuation problem of the boring bar. The bean bag impact damper wraps a certain amount of metal or non-metal micro particles with a soft bag which has a certain elastic resilience, putting it into a specific structural cavity instead of the rigid mass block of the traditional impact damper, as shown in Fig. 2.3. Compared with the traditional impact damper, the bean bag impact damper has the feature of no impulsive noise and a wide frequency band of vibration attenuation.

The piston-based particle damper introduces damping pole on the basis of the non-obstructive particle damper, shown in Fig. 2.4a. When the main structure is

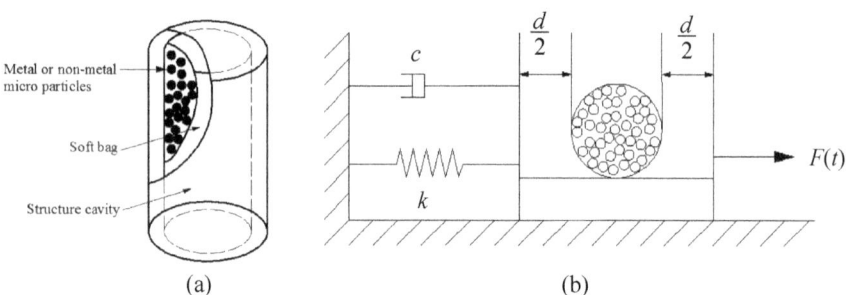

(a)   (b)

Fig. 2.3 **a** Schematic diagram of BBD; **b** mechanical model of BBD

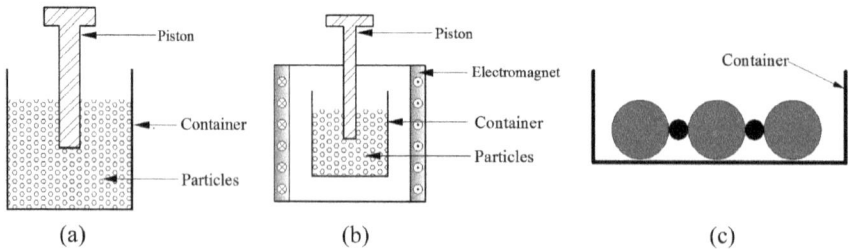

(a)   (b)   (c)

Fig. 2.4 Schematic diagram of **a** the piston-based particle damper under passive control; **b** the piston-based particle damper under semi-active control by magnetic field; and **c** LPC impact damper

vibrating, the particles and the damping pole collide, press and rub together, converting mechanical energy into thermal energy and acoustic energy, which generates the damping effect. Through experiments, Binoy et al. [17] found that the damping effect of such piston-based particle dampers is entirely unrelated to the temperature. Besides, it can still exert damping effect under the condition of small external excitation, whereas the impact damper usually cease to be effective in that case. Afterwards, on the basis of the piston-based particle damper under passive control, Binoy et al. [11] proposed to achieve semi-active control by means of magnetic field, shown in Fig. 2.4b. Regarding some specific structures, such as flexible mechanical arm and cantilever beam, they will generate a large instantaneous movement because of their flexibility when they are stopped suddenly during the process of fast running. In order to alleviate the damage into the structure caused by impulsive instantaneous vibration, Chen et al. [13] put forward the beam-like impact damper. Compared with the traditional impact damper, the beam-like impact damper can overcome the dependence of its damping performance on the direction. To enhance the dissipation of kinetic energy during the impact process, Gharib et al. [14] put forward a novel impact damper, called the linear particle chain impact damper (LPC impact damper), which was formed by a small-sized ball between each two large-sized balls, shown in Fig. 2.4c.

2.1.2.2 The Material Improved Type of Particle Dampers

The material improved type improves the materials of particles and cavities by employing novel materials or using traditional materials skillfully, in order to reduce noise levels, impact forces and increase plastic deformation, thus increasing the energy absorption. This type of variant mainly includes the buffered impact damper [18], the fine particle impact damper [19], the polymeric particle damper [20, 21], the elastomer particle damper [22], the soft hollow particle damper [23], the metal swarf damper [24], the metallic wires particle damper [25–27] etc. Li et al. [18] covered the inner wall of the particle damper with thin rubber material, which formed the buffered impact damper. According to the experimental study, it was shown that rubber can alter the stiffness of the inner wall, leading to considerable improvements on the working performance of the dampers over a wide range of frequencies, especially decreasing the impact force and noise levels. Du [19] first proposed the concept of the fine particle impact damper in which a small quantity of fine particles are enrolled as damping agent, coupled with some steel spheres enrolled as impact partners. When the main structure is vibrating, the impact of steel spheres brings about plastic deformation of fine particles which are usually surrounded and covered around steel spheres, resulting in consuming the vibration energy of the main structure permanently. Darabi et al. [20] substituted relatively large particles with significant viscoelasticity for traditional rigid particles, thus forming the polymeric particle damper. He found that at relatively higher amplitude, the particles get into a convection region, in which the damping levels due to the combined effects of both friction and viscous are irrespective of the material properties.

Bustamante et al. [22] put forward the elastomer particle damper, which was formed by using elastomer as the material for the particles. The experimental results revealed that when the movement of particles reaches the fluidization point, the elastomer particles become optimally excited, thus leading to maximum damping. Michon et al. [23] replaced traditional rigid particles by soft hollow particles, in order to strengthen the properties of honeycomb structures in aerospace applications and alleviate the mass impact on the main structure, which formed the soft hollow particle damper. Experimental research indicated that this kind of particle dampers present good performance over a large frequency range with low impact on the main structure. Hussain et al. [24] proposed to substitute metal swarfs with rubber spheres and studied the damping characteristics of three metals (namely aluminum, stainless steel and mild steel) through experiments. The experiments found that metal swarfs can yield considerable damping in the main structure, across a wide range of excitation amplitudes and frequencies.

2.1.2.3   The Combination Type of Particle Dampers

The combination type combines the traditional particle damping technology with some existing technologies, which improves the damping performance in certain conditions and breaks through the bottleneck of particle dampers in engineering applications. This type of variant mainly includes the impact damper-tuned absorber combination [28–30], the particle tuned mass damper [31–33], the tuned liquid particle damper [34], etc.

The concept of integrating the impact damper with the tuned absorber was introduced first by Masri [35] in 1971 who proposed the use of the Dynamic Vibration Neutralizer (i.e., Tuned Mass Damper) with motion-limiting stops that reflect the physics of an impact damper interacting with the primary system when a motion threshold is exceeded. The study in Ref. [35] provided an exact analytical solution under steady-state conditions, as well as experimental verification demonstrating the potential advantages of such devices to provide effective damping at relatively low as well as high levels of the primary system response. Semercigil et al. [29] found that the addition of the impact damper can improve the vibration attenuation effect of the tuned vibration absorber significantly through experimental study. Chen et al. [28] put forward the tuned particle damper, which was achieved by a particle damper attached to the main structure with flexible supports, based on the operational principles of the classical dynamic vibration absorber, as shown in Fig. 2.5. Compared with the particle damper with rigid supports and traditional dynamic vibration absorber, the tuned particle damper solves the ineffectiveness of conventional particle dampers at low vibration acceleration, especially, less than the acceleration of gravity (1 g).

Researchers have introduced particle damping technology into tuned mass dampers (TMDs), in order to take advantage of each other's advantages and combine multiple damping mechanisms, for instance, tuning, impact, friction, etc. Under this combined energy-consuming mechanism, non-linear energy consuming system is formed, which can widen the frequency band of vibration attenuation, enhance

Fig. 2.5 Mechanical model of **a** the classical dynamic vibration absorber; **b** the tuned particle damper

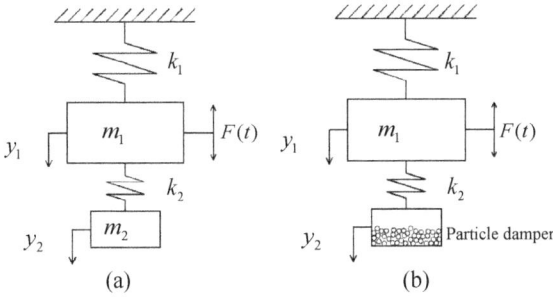

(a)   (b)

robustness and heighten the control effects. This new device is called the particle tuned mass damper (PTMD). Lu et al. [31, 32, 36] introduced the PTMD to the vibration control field of high-rise buildings, and a series of shaking table tests and wind tunnel tests have been conducted, during which the PTMD is suspended on the top of the primary structure by four strands with the same length. It has been shown that the PTMD has significant effect on the suppression of wind-induced and earthquake-induced vibration of high-rise buildings. Yan et al. [33] carried out the experimental research of shaking table tests for a 1/10-scale model bridge with this technology, and found that the damping effect and the robustness can be enhanced.

2.2 Damping Mechanism and Theoretical Analysis of Particle Damping

2.2.1 Damping Mechanism of Particle Damping

The damping mechanism of particle damping mainly includes the energy consumption among particles and the impact energy dissipation between particles and the main structure. The method of using energy consumption among particles to reduce the oscillations of vibrating body has been used for several centuries, such as putting a bag filled with sands on the vibrating body, placing the particle material around the vibrating body, applying the sand-closed damping structure to the metal-cutting machine tool, by which the damping can be increased by 8–11 times [37], etc. The explanations about the mechanism of the damping among particles mainly include as follows: Kerwin [38] argued that particle material consumes the energy of the main structure by three approaches, which were: (i) the friction among the particles, (ii) nonlinear deformation of the contacting points among the particles, and (iii) the resonance of the particle material. Lenzi [39] stipulated that dry friction among the particles is the main mechanism of damping. Sun et al. [40] thought that damping is generated during the process in which acoustic energy radiated by structures is being consumed by sand particles. Single impact damper, with the rigid mass block as the impact body, is the typical representative of the impact damping theory, which is also the origin of the particle damper. The explanations about the mechanism of impact damping mainly include the following interpretations: Some scholars held the opinion that the mechanism of impact damping is a certain kind of energy loss which is produced based on the non-perfect elastic collision [41]. Popplewell [10] considered that the impact damping is mainly achieved by momentum exchange during the process of impact.

More specifically, regarding the particle dampers in different forms, many scholars have studied the balance between the different approaches of energy consumption, and the relationship between the damping mechanism and the design parameters. For example, energy dissipation of particle dampers was predicted by Wong et al. [42] via the discrete element method. The results revealed that despite the fact that friction plays a leading role in energy dissipation, the actual friction coefficient has a little influence on energy consumption. Besides, Masmoudi et al. [43] discovered that the loss factor is unrelated to both the number of particles and the material properties, whereas it is only closely associated with the total mass of particles and the excitation amplitude. Chen et al. [44] elaborated the new damping mechanism of non-obstructive particle dampers based on the rheological behavior of damping particles. It was shown that the optimal damping performance of NOPDs comes from two aspects, and the vibration energy of the main system is dissipated through thermal energy and potential energy.

2.2.2 Nonlinear Energy Sink (NES)

Despite the fact that researchers around the world have done a lot of work, with regard to the damping mechanism of particle damping, no common explanation has been agreed on. From the broad theory of view, actually, particle damping technology belongs to the category of Nonlinear Energy Sink (NES) which is a vibration absorption technology studied enthusiastically in many fields, such as machinery [45], aerospace [46] and civil engineering [47]. There are resemblances between this type of nonlinear absorbers and tuned mass dampers (TMDs). Both of them reduce the response of the main structure by movement of the additional mass and should be placed at the position with large displacement (as for building structures, it usually is the top storey). The major difference between the two is that the resilience generated by TMD is linear, whereas the resilience generated by NES is nonlinear. Consequently, NES is different from the general linear oscillators which have single inherent frequency. The inherent frequencies of NES are variable, which tends to increase with the increase of vibration energy. Therefore, NES can resonate with multiple modes of the main structure within a wider frequency band. In other words, the goal of reducing the vibration for a wide range of frequencies can be attained by NES [48–50]. Based on the traditional tuned mass damper, Li and Song et al. [51–53] added the impact energy dissipation and put forward the pounding tuned mass damper. In the design, the vibration energy can be dissipated through heat generated by the impact or pounding between a tuned mass block and a delimiter covered with

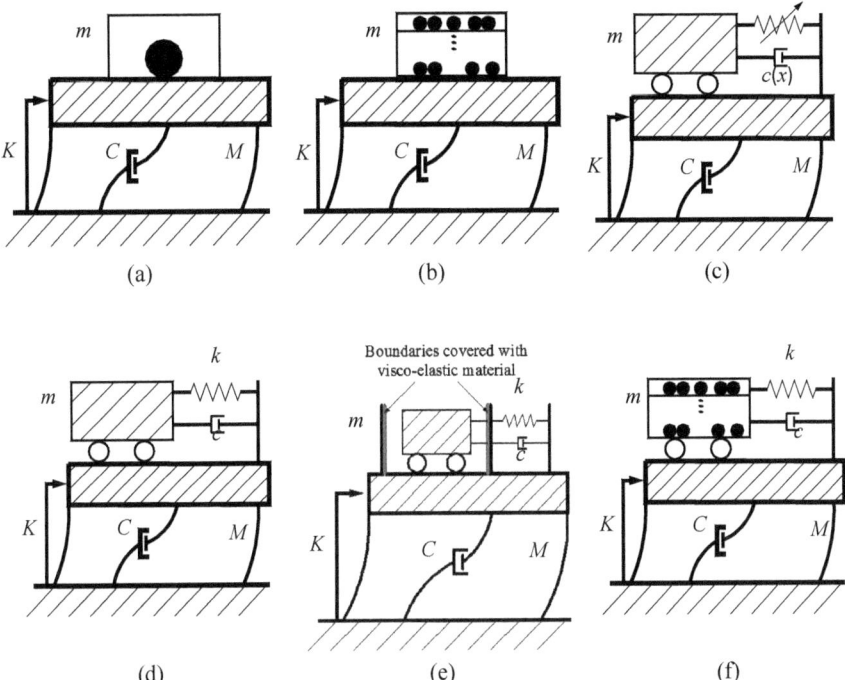

Fig. 2.6 Mechanical model of **a** impact damper; **b** particle damper; **c** nonlinear energy sink; **d** tuned mass damper; **e** pounding tuned mass damper; and **f** particle tuned mass damper

viscous-elastic material. Consequently, pounding TMD (i.e., a Dynamic Vibration Neutralizer with motion-limiting stops [35, 54]) is also regarded as a kind of NES with good robustness. Besides, it has been successfully employed to the vibration control of subsea jumpers [51, 53] and power transmission tower [52]. The mechanical models of different types of dampers are shown in Fig. 2.6.

2.2.3 Theoretical Analysis of Particle Damping

Regarding the theoretical analysis of particle damping, a lot of researchers took a SDOF system as the research object. Since the particle-particle and particle-wall collisions will cause abrupt changes of the motion parameters (such as velocity), its dynamic behavior shows a strong non-linearity. Therefore, only under certain conditions, can analytical solution be obtained, e.g., the main structure with single-particle dampers (impact dampers) under simple excitation (harmonic excitation or steady-state vibration) and assuming symmetrical collisions as twice per cycle.

2.2.3.1 Theoretical Analysis of Single-Particle Dampers (Impact Dampers)

Theoretical analysis of impact dampers first began with the work of Lieber et al. [2]. They considered each collision as a completely plastic collision. Grubin [55] introduced the elastic restitution coefficient of collision into analysis. He took account of the energy loss during the collision and established the theoretical model of the SDOF system attached with an impact damper under harmonic excitation. Masri [56] extended this assumption to the case of symmetric collisions twice per cycle. Furthermore, the non-linear control equations were adopted by Bapat [57] to analyze the vibration state of a SDOF system in case of colliding N times per cycle under harmonic excitation. Masri [4, 56, 58] deduced the analytical solution for the steady-state motion of the impact damper and the multi-unit impact damper attached to the sinusoidally excited main structure. Moreover, their stabilities were also analyzed. Bapat and Sankar [59, 60] analyzed the influence of the coulomb friction. They plotted a table that shows the optimum gaps and the corresponding amplitude reduction when the impact damper is subjected to forced vibration [59]. Ema and Marui's research [61] showed that the additional damping provided by the impact damper is due to the collisions of the impact mass with the primary system. In addition, the optimal damping is affected by the mass ratio and the clearance. The damping characteristics of a vertical impact damper under a wide range of frequencies and multiple amplitudes was studied by Duncan et al. [62] via numerical simulation.

2.2.3.2 Theoretical Analysis of Multi-particle Dampers

Regarding multi-particle dampers, involving interactions between particles, it is difficult to acquire the analytical solution for such a system. One example of an exact analytical solution, under the assumption of steady-state response under harmonic excitation was derived and its stability boundaries were established by Masri [63]. Consequently, theoretical analysis of the nonlinear damping mechanism are very limited, and have to rely on experimental studies and numerical simulation. Mao et al. [64] simulated particle damping via the discrete element method. The simulation results demonstrated that the particle damper can provide considerable additional damping generated by the combined effects of impact damping and friction damping, which can produce a rapid decay in the amplitude of the main structure over a finite period of time. The characteristics of energy dissipation as for the vibrating particles were studied by Saluena et al. [65] through molecular dynamics simulations. They concluded that under different amplitudes and frequencies of external excitation, rates of energy loss correspond to three different regimes: namely solid, convective and gas-like regimes. Sánchez et al. [66] simulated the complicated dynamic interaction between the particles by the discrete element method. It was shown that with the increase of the excitation frequency, particle motion switches from a periodic motion to a chaotic motion. Ben et al. [67] extended the application of the Fourier-based Power Flow Method proposed by Yang [68] and developed a new experimental method, through which the loss factor of NOPDs can be measured quantitatively. Xiao et al. [69] presented the model of frictional energy dissipation and the model of collision energy dissipation based on the powder mechanics and the collision theory, respectively. The concept of 'Effective Momentum Exchange' that can be used to interpret the physical properties of the particles' movement was proposed by Lu et al. [70–72]. This concept is also regarded as a comprehensive index to indicate the optimal damping performance.

2.3 Numerical Simulation of Particle Damping

As theoretical studies of multi-particle dampers are constantly enriched and evolving, researchers have worked out a series of simplified methods and numerical methods in order to simulate the non-linear effects generated by the particle-particle and particle-wall interactions. Papalou and Masri [6, 73, 74] simplified a multi-unit particle damper to an equivalent single-unit impact damper. Analogously, Friend and Kinra [75] wrapped all the mechanisms of energy dissipation into a new concept named 'effective coefficient of restitution' by means of modeling the multi-particle as a cohesive mass block. This coefficient was obtained by fitting the test data, and thereby a novel analytical method was formed. Moreover, Liu et al. [76] adopted equivalent viscous damping to simulate the non-linear characteristics of particle dampers, based on summarizing the corresponding experimental results. Xu et al. [77] put forward an empirical method for designing particle dampers, by which the relationship between the damping effect and each parameter was obtained through fitting the test data. Wu et al. [78] introduced multiphase flow theory of gas particles to the analysis of particle dampers and the corresponding theoretical model was presented. On the basis of that, Fang and Tang [79] further improved and completed the theoretical model to reduce analysis complexity and its computational cost. Recently, Wu et al. [80, 81] utilized multiphase flow theory to the numerical analysis of vibration and acoustic radiation for rectangular plate with particle dampers by the aid of COMSOL multiphysics software package. Furthermore, the simulation methods, utilized to simulate the particle damping, also include restoring force method [82, 83], neural network method [84], molecular dynamics method [65], turbulence theory [85], etc. Despite the fact that these simplified models and experiment-based studies have gained substantial achievements, most of them belong to phenomena-based methods, which means that the conclusions drawn from these methods are difficult to extrapolate to other cases.

2.3.1 Discrete/Distinct Element Method (DEM)

For the past few years, the discrete/distinct element method (DEM) has been widely introduced to the analysis of particle dampers [8, 86–89]. DEM, proposed by Cundall [90] in 1971, is an effective method which is specially utilized to solve the non-continuum problems, such as the mechanical behavior of rock. Because DEM can account for particle-particle and particle-wall interactions, this method is quite suitable to analyze particle dampers quantitatively [8, 86–89].

The DEM is a numerical analysis method that can analyze the mechanical behavior of discrete bodies via iterative solutions of step by step. In this method, discrete bodies are divided into a collection of numerous discrete elements whose movements are described by local contact laws and Newton's equation of motion. It should be emphasized that DEM is based on the hypothesis that if the time step chosen is extremely small, during a single time step, disturbances cannot propagate from any particle further than its immediate neighbors. Hence, at all times, the forces acting on any particle are determined exclusively by its interaction with the adjacent particles, and then the behavior of the entire system can be captured in detail.

It is worth mentioning that even though the DEM possesses the advantages of simple principle and can simulate the non-linear effect of particle dampers with high fidelity, many problems still exist mainly owing to its high requirements in computational resources. These problems mainly include how to build the contact force models that can describe the physical behaviors accurately, how to determine the calculation parameters reasonably, such as stiffness coefficient in normal and tangential direction, damping coefficient and time step, and how to search and judge the contact status of particles effectively. Moreover, it is quite time-consuming to adopt DEM to simulate when the particle number is extremely large. All of these problems have a significant influence on the accuracy of the results [91].

2.3.2 Simplified Analytical Method

Considering the challenges in large calculations involving the DEM, Lu et al. [31] proposed a simplified analytical method, based on the correlational studies conducted by Papalou and Masri [6, 73, 74]. The essence of this method is to make all particles being equivalent to a single one, based on certain equivalence principles, to simulate the vibration effects of particle dampers. During the process of equivalent simulation, the collisions among the particles are neglected. However, the impact between particles and the wall tends to be the dominant factor that has a significant influence on the global response of the main structure. It should be noted that this simplified analytical method captures the most principal control force of particle dampers, so that it possesses a satisfactory degree of accuracy in practical applications, and can provide a useful reference for realistic applications in engineering projects.

Actually, this equivalent simplified method has a sound theoretical foundation, which is validated by scholars through numerical simulation and experimental studies. Sánchez et al. [92] has pointed out that when the number of particles is sufficiently large and the particles get into a state of dense lumped mass, the damping performance of particle dampers is insensitive to the material properties of the particles as well as the particle–particle interaction. This phenomenon was called as 'universal response', which not only explains the suitability of particle dampers working in harsh environments without having a specialized maintenance, but also greatly simplifies the design schemes of particle dampers. Moreover, Sánchez et al. [66] also simulated the complex dynamic interaction among the particles by DEM, and found that when the optimum damping is obtained, the particles tend to be dense. Interestingly, the dynamic characteristics of particle dampers in that case are similar to the impact damper with single particle. Furthermore, Bannerman et al. [93] attached a particle damper to the free end of a spring blade and conducted a vibration control experiment under micro-gravity environment. It was shown that the effective energy dissipation is carried out under the collect-and-collide regime. Additionally, Sack et al. [94] performed a similar experiment with regard to steady state systems and drew the same conclusion.

2.3.3 The Coupled Algorithm of FEM and DEM

When particle dampers are employed in conjunction with MDOF structures, the simulation of particle dampers can be conducted by DEM, while the analysis of the main structure is usually performed via the finite element method (FEM). The contact force between the particles and the wall eventually acts on the main structure, and then the damping effect can be calculated. Hence, it is a natural trend to introduce the combination of FEM and DEM to the design of particle dampers applied in MDOF structures. Xia and Liu et al. [95] utilized particle damping to a rotating brake drum of the vehicle to control its noise and vibration. They implemented the design and calculation of particle dampers based on the coupled algorithm of FEM and DEM. In addition, Liu et al. [96] adopted the coupled algorithm to study the influence of distributed damping on the torsional vibration of a plate. Furthermore, Xia et al. [97] put forward the coupled algorithm for the rotating plate structure with particle damping. Although the combined algorithm of FEM and DEM has been applied in simple members, they lay a theoretic foundation for the extension of this coupled algorithm to much complex host structures.

2.4 Experimental Study of Particle Damping

Due to the highly-nonlinear characteristic of particle damping, the theoretical analysis and numerical simulation have some limitations; consequently, it is difficult to establish a universal theoretical analysis method. On the other hand, many scholars carried out extensive experimental studies on particle damping. It can be summarized as following three broad approaches.

Approach #1:
The first is to carry out theoretical analysis of particle damping by virtue of experiments so as to explore the essence of the damping mechanism through experimental phenomenon. For example, Sadek et al. [98, 99] studied the influence of gravity on the impact damper and found that the damper has better effect in zero gravity environment. Moreover, in the vicinity of resonance, asymmetric collisions twice per cycle is dominant. Cempel and Lotz [100] studied the vibration damping through filling particles in a container and found that the energy dissipation of the impact particle not only relies on the collision of internal particles, but also relates to the external collision (referring to the collision between the particles and the wall). In addition, the friction is also a factor that affects the energy dissipation of the impact particle. Hollkamp and Gordon [101] used metal and ceramic particles as the impacting bodies and put them into the structural cavities. It was found that when the main structure vibrates, the vibrational energy can be consumed by the collisions among the particles. The research of Xu et al. [102] focused on the contribution of shear friction induced by strain gradient along the length of the structure to damping. The experimental results indicated that particle dampers can provide significant additional damping over a broad frequency range, and the optimal damping might be obtained by utilizing multi-particles as the impact bodies given that the impact, friction and shear mechanisms are accounted for.

Approach #2:
The second is to validate the numerical results and investigate the damping performance of various forms of particle dampers utilized in different vibration objects as well as external excitations. For example, an experimental study of a vertical multi-particle impact damper under free excitation was performed by Trigui et al. [103], by which the influence of clearance and excitation intensity was studied. The damping effectiveness of a particle damper under centrifugal loads was analyzed by Daniel [104] via the experiment of a rotating cantilever beam. Jadhav et al. [105] compared the damping performance of the single-cell and multi-cell particle dampers by lots of experiments. Moreover, they established the models of the single-cell and multi-cell particle dampers using dimensional analysis method and validated their correctness by experiments. Zhang et al. [106] simplified the resilient bag in BBD (bean bag impact damper) to a particle-spring model based on DEM, which was verified by test data. The experiment of an L-shaped cantilever beam with a particle damper in the top free end was conducted by Wang et al. [107], through which the damping performance under horizontal-vertical excitations in the context of free decay was

investigated. It was shown that the damping characteristics of the particle damper under horizontal-vertical excitations is similar to those under only vertical excitation. Additionally, Rongong et al. [108–110] carried out several experimental studies on the amplitude dependent behavior of particle dampers.

Approach #3:
The third is to study the influence of various design parameters on the vibration attenuation effect of the system, and provide guidance for the practical applications in engineering by means of the parametric analysis. For example, Veluswami et al. [111, 112] used three types of material as the coating of strike plate inside the particle damper. It was concluded that soft materials with small restitution coefficient provide light damping during the process of resonance. Yokomichi et al. [113, 114] and Saeki [89] studied the response of the particle damper under harmonic excitation and found that the impact body with greater mass tends to provide more additional damping to the main structure, whereas the impact body with lighter mass tends to exert damping effect rapidly at the beginning of the vibration. Furthermore, the optimal clearance was also determined. Yang [68] summarized a series of design curves to predict the damping characteristics of the particle damper. Li [115] conducted a series of experiments to study the properties of the impact damper attached to a MDOF structure. It was indicated that increasing the mass of the particles does not necessarily enhance the damping of the primary structure for all modes. Regarding the various design parameters of particle dampers, such as particle type, particle placement, packing ratio, etc., a series of tests were carried out by Hollkamp et al. [101]. Shaking table tests of a three-storey steel frame with a buffered particle damper [116, 117], and a five-storey steel frame with a particle tuned mass damper [31] were carried out by Lu et al. Several design parameters, such as the auxiliary mass ratio, gap clearance, etc., were analyzed. Sathishkumar et al. [118] filled a boring bar with different metal particles and investigated the effects of certain particle parameters, such as particle size, density and hardness, on the surface roughness of machined surface.

In summary, through experimental study, the limitations of theoretical research owing to the nonlinear characteristics of particle damping have been overcome to a great extent. It is noteworthy that many theoretical methods were proposed based on the experiments, and through experimental phenomenon, the law of particle damping can be intuitively explored. Thus, the factors affecting the attenuation effect of particle damping can be summarized into two aspects, namely external and internal causes. To be specific, the external causes mainly include excitation amplitude and its frequency characterization. The internal causes mainly include the number, size and material of particles, container size, mass ratio of total particles to primary structure, group effect between particles, restitution coefficient of the particles, particle placement, the damping of primary structure, etc. However, at the present stage, the published experimental studies are mainly concentrated in mechanical and aerospace fields, whereas the experiments based on civil engineering are relatively scarce.

2.5 Application of Particle Damping Technology

With the continuous development and evolution of particle damping technology, faced with a variety of vibration reduction problems in engineering fields, particle damping technology has taken full advantage of its favorable characteristics, which has led to many types of successful application. These engineering applications mainly focus on the aerospace field, machinery field, lifeline engineering, etc. For example, the protection for antenna structures [119, 120] and printed circuit board [121] under shock-loading conditions, reducing the wind-induced vibration of lamp pole, chimney and some flexible structures [122], the suppression on the vibration of tennis rackets [123], electro-mechanical relays [124], aircraft [2], engine turbine system of space shuttle [9, 125], fan and compressor blisks [126] and metal-cutting machine tools [127].

2.5.1 Particle Damping Applications in Aerospace Field

There are hundreds of applications in aerospace field lasting for more than 70 years. Lieber and Jensen [2] adopted impact damper to control the vibration of aircraft. They considered twice collision in each cycle and found that when the phase angle between impact mass and major structure is 180 degrees, the damping effect is best. Rocke and Masri [120] used an impact damper to attenuate the shock response of an antenna structure. Oledzki [128] utilized the impact damper to inhibit the vibration of a long tube on the light-weight spacecraft and adopted a rheological model for numerical simulation. The numerical results showed that resonance amplitude was weakened, which fitted well to the experimental results. Moore et al. [125] used the impact damper for high-speed rotor in the rocket engine turbine system in low temperature working condition, such as the main engine of the space shuttle. Experimental results revealed that it is feasible for the impact damper to inhibit the vibration of rotor-bearing system at a low temperature. Gibson [129], Torvik and Gibson [130] used the particle damper for spatial application, and found that the attenuation rate of the system response, and the minimum effective amplitude, are two major parameters in the damper design. Simonian [131] conducted experimental studies on employing particle dampers to reduce excessive vibration of cantilever beam type space structural subsystems. Veeramuthuvel et al. [121] proposed the novel use of particle damper capsule on a printed circuit board so as to suppress the vibration of printed circuit boards.

Honeycomb composite structures, commonly used in spacecraft, have great superiority in the installation of particle dampers. The damping particles with smaller mass can be directly placed in the cells of honeycomb without adding excessive weight. Liu et al. [132] filled the hollow glass microspheres into the honeycomb lattices and sealed it with face sheets, which formed the honeycomb composite beam. According to an experimental study, it was found that honeycomb composite beams

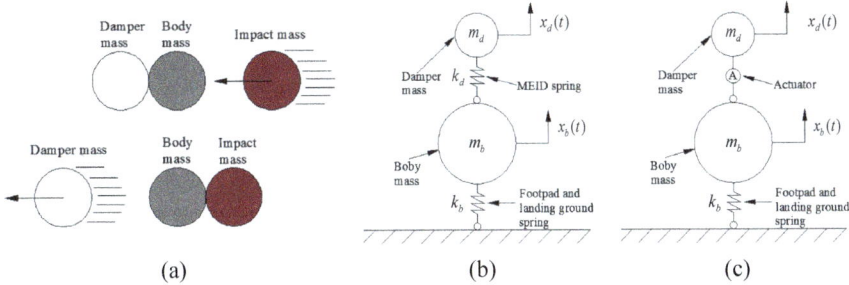

Fig. 2.7 a Conceptual diagram of a MEID; **b** landing gear system with PMEID; and **c** landing gear system with AMEID

can significantly increase damping, without causing significant modifications in the mass. Moreover, Panossian [133] introduced the NOPD into honeycomb composite beams and found that the beam with the optimal NOPD treatment has the lowest response amplitude through modal vibration tests. Furthermore, it is common for a spacecraft to suffer from large shock loads during its landing process, accompanied by rebound, swing and sideslip of the spacecraft. To reduce that adverse effects on spacecraft upon landing, Hara et al. [134] proposed the momentum exchange impact dampers (MEIDs) which can be classified into two types: the passive (PMEID) and active (AMEID) momentum exchange impact dampers, respectively, as shown in Fig. 2.7. It was shown that the active momentum exchange impact damper can effectively reduce the responses of spacecraft upon landing. Additionally, based on using actuators in combination with passive elements, Kushida et al. [135] put forward the Hybrid Momentum Exchange Impact Damper (HMEID) and applied it into the robust landing gear system.

2.5.2 Particle Damping Applications in Machinery Field

A great deal of applications in the machinery field have appeared since 1967, when Park [136] studied the response of mass-spring damper exposed to repeated impact effect. Those devices, like electric and pneumatic hammers, bin vibrators as well as pile drivers, are all driven by the repeated impact force. Thus, these devices usually generate strong vibration when they are working, whereas mass-spring damper makes it very convenient to control the vibration. Furthermore, Sims et al. [137] proposed that the chatter stability during a machining process can be improved by applying particle dampers. Fuse [138] utilized the impact damper, which generates the mutual collisions between the main system and the additional vibration system in opposite phases, to eliminate the adverse effects exerted by mechanical system resonance.

There are some applications used for mitigating the vibration and noise during the running course of office equipment. For example, Skipor [139] utilized the impact damper to the carrying cylinder of a web-fed printing press. Sato et al. [140] employed particle dampers to reduce the vibration in pantograph-support systems. Xu et al. [141] utilized the particle damper to substantially reduce the vibration and noise generated by the banknote processing machine. To control the vibration produced by metal processing, Aiba et al. [127] proposed that the variable-attractive-force impact damper can weaken the vibration produced in the process of face milling for low-rigidity workpieces. It is a quite effective way to restrain the vibration during the process of metal cutting by this kind of impact damper. Moreover, Sathishkumar et al. [118] employed particle damping to control the vibration of stationary boring bars, which are used for conducting boring operations on rotating parts, by filling them with various metal particles.

It has also been demonstrated that it is feasible for particle damping to reduce the vibration and noise of automobile [131]. For example, Hu et al. [142] first studied acoustic pressure reduction at a target point inside an enclosed cavity using particle dampers. In addition, they utilized it to control the acoustic radiation and vibration of the thin panel in the auto body. Xia et al. [95] applied the particle damper to attenuate the vibration of rotating brake drum. Regarding the vibration control towards some special structures of mechanical equipment, Chan et al. [143] conducted experimental studies on particle damping applied to a lightly damped bond arm in a die bonding machine. Additionally, Li et al. [144] introduced bean bag dampers to plate structure. Furthermore, Xiao et al. [145] pointed out that the tooth surfaces of transmission gears produce significant vibration and noise under centrifugal force, which causes negative influence on the service life of gears. Therefore, they filled the lightening holes inside the gears with different particle packing ratios and achieved beneficial damping effects.

2.5.3 Particle Damping Applications in Lifeline Engineering

Lifeline engineering is the basic engineering of facilities and systems that can maintain the survival of cities, and has significant influence on the national economy and people's livelihood. Hence, lifeline engineering should have the ability to cope with harsh environments, such as earthquake and windstorm. Particle damping applications in lifeline engineering mainly include the vibration control of wind turbines [15], power transmission towers [52, 146], subsea jumpers [51, 53], oil-well or gas-well drilling [147], the vibration and noise reduction of high-speed rail wheels [148], etc.

To attenuate the dynamic response and prevent the fatigue failure of wind turbines, the tuned rolling-ball damper, which contains single or multiple balls rolling in a spherical container and is mounted on the top of wind turbines, was proposed by Chen et al. [15]. Furthermore, Zhang et al. [16] put forward a ball vibration absorber (BVA) to prevent offshore wind turbines from the destruction of vibration caused by earthquakes or combined wind-wave loads. The schematic diagram of BVA is shown in Fig. 2.8a. As an increasing number of longer-span and taller-span power transmission towers have been constructed recently, they are susceptible to dynamic

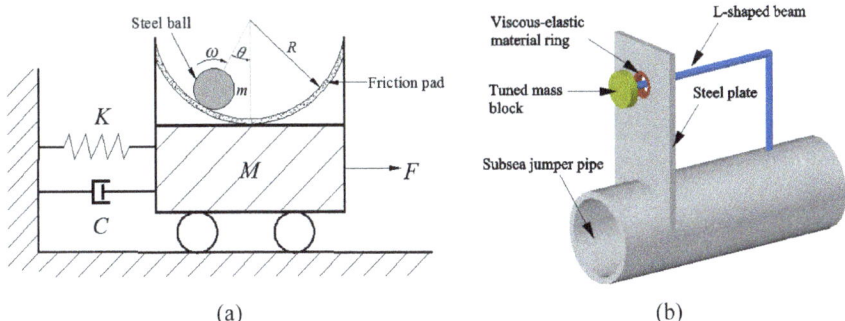

Fig. 2.8 a The schematic diagram of the ball vibration absorber; **b** L-shaped pounding tuned mass damper designed for a subsea jumper

excitations, such as wind, earthquake, and ice shedding. Pounding tuned mass damper was applied to mitigate the dynamic response of a power transmission tower under earthquake excitation [52] and wind load [146].

Particle damping technology also has impressive applications in the oil and gas production. Subsea jumper is a flexible pipeline structure commonly used to transport the production fluids, which are usually a mixture of oil, gas and water, from the reservoir to the topside processing facilities. However, subsea jumper is susceptible to flow-induced vibration, which may cause pipeline break, production loss and even resonance risk [149]. To solve this problem, Song et al. [51, 53] introduced an L-shaped pounding tuned mass damper installed at the middle of subsea jumper, as shown in Fig. 2.8b. The vibration energy can be dissipated as heat energy by pounding between a tuned mass block and a ring covered with viscous-elastic material. It was found that unlike TMD, the pounding tuned mass damper is maintenance-free and has better robustness. Soil-well or gas-well drilling may have detrimental effects on the life of the drill string elements and the surface equipment owing to the accompanying significant vibrations. Hence, Velichkovich et al. [147] applied a vibration-impact damper to change the dynamic conditions of the object by transporting the vibration energy from the object to the damper and enhancing the consumption of vibration energy.

Nowadays, high-speed rail plays an important role in people's daily life as it brings significant convenience and efficiency to our society. However, high-speed rail may generate noise radiation owing to the excessive vibration caused by the roughness of wheel and rail running surface [150]. Therefore, Guo [148] introduced the particle damping rings into the high-speed rail wheels and found that compared to the existing wheel product with solid rings, it is more effective to attenuate the radial and axial vibration of wheels with particle damping rings.

2.5.4 Particle Damping Applications in Civil Engineering

In order to ensure various kinds of structures staying safe and comfortable under wind load and earthquake action, numerous structural vibration control systems have been proposed over the years. The concept of structural control was first proposed by John Milne [151] over a hundred years ago, who built a house by wood and placed it on ball bearings to demonstrate that the structure could be isolated from earthquake shaking. Later, Yao [152] applied the modern control theory to civil structures in 1972, which marks the beginning of the intensive study of vibration control in civil engineering. During the same period, Kelly and Skinner et al. [153, 154] presented an idea that the vibration energy of a structure can be dissipated through additional energy dissipation devices. Since then, energy dissipation technology has been widely applied in civil and architectural engineering, and shows great promise for realistic applications.

By now, passive control technology has already been developed and is comparatively mature in the field of civil engineering. As a type of passive control technology, it is a new tendency to apply particle damping to civil engineering and its development is just at the beginning. Consequently, there have been relatively rare practical engineering projects in the civil engineering field that incorporate particle dampers, which mainly include the following: Ogawa [155] utilized the impact damper for the pylon of the cable-stayed Fuchu Lake Bridge in Japan so as to attenuate wind-induced vibration of the pylon. Furthermore, another typical application is a tall building named the Parque Araucano building in Santiago, Chile [156]. A particle damper was installed on the 21st floor, and it performed very well during the 2010 offshore Maule, Chile earthquake.

2.5.4.1 Research and Application Status of Particle Damping in Civil Engineering

In recent years, the study of particle damping based on disaster prevention and reduction in civil engineering is still at the stage of experimental study and theoretical analysis. The application of particle damping to several basic structural elements, such as cantilever beams, stiffened plates, etc., provides worthy guidelines to the practical application in the full-scale building-like structures. For example, Masri et al. [157, 158] utilized the impact damper to reduce the forced vibration of continuous beams and plates. Xia et al. [132] took a cantilever beam filled with particles as the research object and investigated the nonlinear variation of structural damping with the change of the particle parameters. In addition, the influence of distributed damping on the torsional vibration of a plate was studied by Liu et al. [96] through the coupled algorithm of FEM and DEM. Zhao et al. [159] carried out a preliminary experimental study on the damping performance of NOPD columns.

With the purpose of exploring the application of particle dampers in civil engineering more deeply, some scholars started to study whole buildings and other structures as the research objects. Lu [160] carried out the theoretical and experimental studies of frame structure with particle dampers under the excitations of different earthquake waves. Yan et al. [33] applied the tuned particle damper to the seismic control of continuous viaducts. Additionally, Papalou et al. [161] proposed that particle damping technology can be employed to protect ancient monuments (such as columns) from seismic damage by using multi-drum particle dampers. The multi-drum columns are made up of several drum-shaped cylinders, among which the damaged and missing ones can be replaced by the new ones with a hollow part containing particles. It is also common for cable-stayed bridges to suffer from vibrations induced by wind or rain–wind, and some novel techniques to solve this problem have been proposed [162, 163]. Egger et al. [164] introduced an advanced form of the impact damper named distributed-mass impact damper, originated from the traditional impact damper. They applied this type of impact damper to a cable-stayed bridge in order to attenuate rain-wind induced vibrations.

2.5.4.2 The Advantages of Particle Damping Applied to Civil Engineering

Through theoretical and experimental studies performed by scholars all over the world, it has been demonstrated that the particle damping technology has great superiority in vibration attenuation control encountered in the civil engineering field.

First, particle damping possesses a wide frequency band of vibration attenuation, with effective vibration attenuation performance within the range of 0–6000 Hz [9], so it can be considered to be capable (if properly designed) to suppress earthquake, wind vibration and other low frequency vibrations faced by civil engineering structures. It can also mitigate the environmental vibrations caused by metro, high-speed rail and other working conditions.

Second, among the existing passive control technologies which are comparatively mature towards civil engineering structures, it is obvious that viscous damping is sensitive to temperature. Moreover, with regard to mechanical damping such as friction, its material properties tend to degrade and will exhibit fatigue effects under various kinds of dynamic action. Hence, the application of these passive control dampers in civil engineering has been restricted greatly. However, particle damping is almost unlimited by the ambient temperature. The problems such as material degradation and fatigue effect are also solved. Furthermore, as the damping performance won't be reduced over time, its resonance peak value could be effectively restrained. In consideration of such characteristics of particle material, this kind of technology is especially suitable for structures under extreme field conditions, such as the vibration control of power transmission tower.

Third, the layout locations of the particles are very flexible. They could be attached to the external periphery of civil engineering structures as a TMD, and could also be embedded into structural members. Any interlayer and internal cavity is acceptable, without influence on structural usage, and without causing significant modifications in the main structure.

Fourth, the material of the particles can be found easily with low prices. Especially, some ordinary building materials, such as steel balls, sands, stones, etc., could be used. Consequently, the applicability of particle damping technology in civil engineering has a great potential that has not yet been fully exploited.

2.5.4.3 The Challenges of Particle Damping Applied to Civil Engineering

At present, many studies of the main structure with particle dampers in the technical literature are mainly focused on SDOF systems, or other structures generally used in the mechanical field, such as cantilevers, stiffened plates, etc., while accurate analysis for MDOF structures with particle dampers has not been carried out yet. In addition, fewer theoretical and experimental studies on the control effect of civil engineering structures with particle dampers in the case of earthquake and wind vibration have been performed. Moreover, the researches on the standardized and normalized design of particle dampers which could guide the engineering application have not been conducted. Hence, there are several challenges for particle damping technology being used in civil engineering, some of which can be outlined as follows:

First, in the aerospace and machinery fields, both the frequency and amplitude of excitation are relatively high, whereas in civil engineering, those tend to be relatively low. According to the current research results, it was found that under certain conditions, the reduction in the response of the main structure may be enhanced through increasing the excitation intensity. Apparently, under weak excitation conditions in civil engineering, the efficiency of particle damping technology might be confined to some extent. Hence, improved particle dampers would be the focus in future studies.

Second, the application objects in civil engineering are different from that in the aerospace and machinery fields, which means the mass and scale of the main structure are usually quite huge in civil engineering. Therefore, despite the fact that a large mass ratio generally can be utilized in the aerospace and machinery fields, the mass ratio should be strictly controlled in civil engineering. In general, the inertial mass takes up 0.5–1.5% of the generalized mass of the main structure.

Third, Hollkamp et al. [101] proposed that the optimum arrangement form of the particles is that the particle group should be asymmetrically distributed at the maximum amplitude of each major mode of the structure (or members). However, from the comfort point of view, installing a particle damper at the top of a building can reduce the impact of the noise generated during the impact process. Consequently, it is still an open question as to which arrangement form of particle damping is most suitable to civil structures.

2.6 Discussion

While this survey paper has focused primarily, on the past and present of the field of particle damping, in the future, there will be some very promising avenues where the field will expand and develop to leverage ever improving capabilities in sensors, devices, microprocessors, and high-performance materials that open the door to incorporate more sophisticated approaches for enhancing the effectiveness of the class of nonlinear devices under discussion, when they are employed for vibration mitigation under rapidly changing dynamic loads (e.g., earthquakes). As known from basic physics, all passive devices have reduced effectiveness when used under transient (i.e., non-stationary) loads, compared to their performance under stationary loads whose spectral and amplitude characteristics are nearly constant. Consequently, there is a pressing need to develop in the future modified particle damping devises whose properties evolve (i.e., change with time) as the dynamic environment to which they are subject is changing, so as to continuously furnish the maximum amount of damping that they can provide. This is the motivation for considering the concept of semi-active particle dampers.

The idea of using semi-active control in the civil engineering field has been considered in the work of Housner et al. [151] since the use of full active control, which is very effective but infeasible for application to full-scale civil structures due to unrealistic energy demands. By contrast, semi-active control approaches use a very modest amount of energy (typically provided by a small battery) to actively control the critical parameters in a given damping device. This basic idea was evaluated in the work of Masri et al. [165] who developed a procedure to use nonlinear auxiliary dampers (operating as particle dampers) with adjustable motion-limiting stops. A mathematical model of the system to be controlled is not needed for implementing the control algorithm. The degree of the primary structure oscillation near each vibration damper determines the damper's actively-controlled gap size and activation time. By using a tiny amount of control energy to adjust the damper parameters instead of directly attenuating the motion of the primary system, a significant improvement is achieved in the level of vibration attenuation, in comparison to what an optimally-designed passive impact damper can provide. A proof-of-concept experimental study showed that this semi-active impact damper approach has a great promise for extension to other applications in which the primary system is subjected to no-stationary excitation.

References

1. Paget, A.L. 1937. Vibration in steam turbine buckets and damping by impacts. *Engineering* 143: 305–307.
2. Lieber, P., and D.P. Jensen. 1945. An acceleration damper: Development, design and some applications. *Transactions of the ASME* 67: 523–530.

3. Nayeri, R.D., S.F. Masri, and J.P. Caffrey. 2007. Studies of the performance of multi-unit impact dampers under stochastic excitation. *Journal of Vibration and Acoustics* 129 (2): 239–251.
4. Masri, S.F. 1969. Analytical and experimental studies of multiple-unit impact dampers. *Journal of the Acoustical Society of America* 45 (5): 1111–1117.
5. Araki, Y., I. Yokomichi, and Y. Jinnouchi. 1986. Impact damper with granular materials: 4th Report, frequency response in a horizontal system. *Transactions of the Japan Society of Mechanical Engineers C* 29 (258): 4334–4338.
6. Papalou, A., and S.F. Masri. 1996. Performance of particle dampers under random excitation. *Journal of Vibration and Acoustics, ASME* 118 (4): 614–621.
7. Masri, S.F. 1970. Periodic excitation of multiple-unit impact dampers. *Journal of the Engineering Mechanics Division* 96: 1195–1207.
8. Saeki, M. 2005. Analytical study of multi-particle damping. *Journal of Sound and Vibration* 281 (3–5): 1133–1144.
9. Panossian, H.V. 1992. Structural damping enhancement via non-obstructive particle damping technique. *Journal of Vibration and Acoustics* 114 (1): 101–105.
10. Popplewell, N., and S.E. Semercigil. 1989. Performance of the bean bag impact damper for a sinusoidal external force. *Journal of Sound and Vibration* 133 (2): 193–223.
11. Shah, B.M., et al. 2011. Semi-active particle-based damping systems controlled by magnetic fields. *Journal of Sound and Vibration* 330 (2): 182–193.
12. Bai, X., et al. 2009. Particle dynamics simulations of a piston-based particle damper. *Powder Technology* 189 (1): 115–125.
13. Chen, L.A., and S.E. Semercigil. 1993. A beam-like damper for attenuating transient vibrations of light structures. *Journal of Sound and Vibration* 164 (1): 53–65.
14. Gharib, M., and S. Ghani. 2013. Free vibration analysis of linear particle chain impact damper. *Journal of Sound and Vibration* 332 (24): 6254–6264.
15. Chen, J., and C.T. Georgakis. 2013. Tuned rolling-ball dampers for vibration control in wind turbines. *Journal of Sound and Vibration* 332 (21): 5271–5282.
16. Zhang, Z.L., J.B. Chen, and J. Li. 2014. Theoretical study and experimental verification of vibration control of offshore wind turbines by a ball vibration absorber. *Structure and Infrastructure Engineering* 10 (8): 1087–1100.
17. Shah, B.M., et al. 2009. Construction and characterization of a particle-based thrust damping system. *Journal of Sound and Vibration* 326 (3–5): 489–502.
18. Li, K., and A.P. Darby. 2008. A buffered impact damper for multi-degree-of-freedom structural control. *Earthquake Engineering and Structural Dynamics* 37 (13): 1491–1510.
19. Du, Y.C., and S.L. Wang. 2010. Modeling the fine particle impact damper. *International Journal of Mechanical Sciences* 52 (7): 1015–1022.
20. Darabi, B., and J.A. Rongong. 2012. Polymeric particle dampers under steady-state vertical vibrations. *Journal of Sound and Vibration* 331 (14): 3304–3316.
21. Darabi, B., J.A. Rongong, and T. Zhang. 2016. Viscoelastic granular dampers under low amplitude vibration. *Journal of Vibration and Control*. https://doi.org/10.1177/1077546316650098.
22. Bustamante, M., et al. 2013. Experimental study on some parameters that affect the performance of an elastomer particle damper. In *21st international congress on acoustics, ICA 2013—165th meeting of the acoustical society of America, Montreal, QC.*
23. Michon, G., A. Almajid, and G. Aridon. 2013. Soft hollow particle damping identification in honeycomb structures. *Journal of Sound and Vibration* 332 (3): 536–544.
24. Abbas, H., et al. 2014. Damping performance of metal swarfs in a horizontal hollow structure. *Journal of Mechanical Science and Technology* 28 (1): 9–13.
25. Lord, C., N. Tang, and J. Rongong. 2016. Damping of metallic wool with embedded rigid body motion amplifiers. In *Proceedings of 6th European conference on structural control, Sheffield.*
26. Hong, J., et al. 2013. Hysteretic properties of metal rubber particles. *Proceedings of the Institution of Mechanical Engineers, Part C: Journal of Mechanical Engineering Science* 227 (4): 693–702.

27. Tang, N., J.A. Rongong, and G.R. Tomlinson. 2015. Nonlinear behaviour of tangled metal wire particle dampers. In *International conference on structural engineering dynamics (ICEDYN2015)*, Lagos, Portugal.
28. Yao, B., et al. 2014. Experimental and theoretical investigation on dynamic properties of tuned particle damper. *International Journal of Mechanical Sciences* 80: 122–130.
29. Semercigil, S.E., D. Lammers, and Z. Ying. 1992. A new tuned vibration absorber for wide-band excitations. *Journal of Sound and Vibration* 156 (3): 445–459.
30. Semercigil, S.E., F. Collette, and D. Huynh. 2002. Experiments with tuned absorber-impact damper combination. *Journal of Sound and Vibration* 256 (1): 179–188.
31. Lu, Z., et al. 2017. Experimental and analytical study on the performance of particle tuned mass dampers under seismic excitation. *Earthquake Engineering and Structural Dynamics* 46 (5): 697–714.
32. Lu, Z., et al. 2016. An experimental study of vibration control of wind-excited high-rise buildings using particle tuned mass dampers. *Smart Structures and Systems* 18 (1): 93–115.
33. Yan, W., et al. 2014. Experimental research on the effects of a tuned particle damper on a viaduct system under seismic loads. *Journal of Bridge Engineering* 19 (3): 165–184.
34. Dai, K., et al. 2017. Experimental investigation on dynamic characterization and seismic control performance of a TLPD system. *The Structural Design of Tall and Special Buildings* 26 (7): e1350.
35. Masri, S.F. 1972. Theory of the dynamic vibration neutralizer with motion-limiting stops. *Journal of Applied Mechanics, ASME* 39 (2): 563–568.
36. Lu, Z., D. Wang, and Y. Zhou. 2017. Experimental parametric study on wind-induced vibration control of particle tuned mass damper on a benchmark high-rise building. *The Structural Design of Tall and Special Buildings*. https://doi.org/10.1002/tal.1359.
37. Dai, D. 1986. *Damping technology for vibration and noise control* (in Chinese). Xi'an: Xi'an Jiaotong University Press.
38. Kerwin, E.M. 1965. Macro-mechanisms of damping in composite structures. In *Internal friction damping and cyclic plasticity*. Baltimore, Md: ASTM-STP.
39. Lenzi, A. 1985. *The use of damping material in industrial machine*. England: Institute of Sound and Vibration Research, University of Southampton.
40. Sun, J.C., and H.B. Sun. 1986. Predictions of total loss factors of structures Part II: Loss factors of sand filled structure. *Journal of Sound and Vibration* 104 (2): 243–257.
41. Deng, W. 1964. Effects of impact damper and determination of its parameters. *Journal of Mechanical Engineering* 12 (4): 83–94. (in Chinese).
42. Wong, C.X., M.C. Daniel, and J.A. Rongong. 2009. Energy dissipation prediction of particle dampers. *Journal of Sound and Vibration* 319 (1–2): 91–118.
43. Masmoudi, M., et al. 2016. Experimental and numerical investigations of dissipation mechanisms in particle dampers. *Granular Matter* 18 (3).
44. Zhang, K., et al. 2016. Rheology behavior and optimal damping effect of granular particles in a non-obstructive particle damper. *Journal of Sound and Vibration* 364: 30–43.
45. Gourc, E., et al. 2015. Quenching chatter instability in turning process with a vibro-impact nonlinear energy sink. *Journal of Sound and Vibration* 355: 392–406.
46. Bergeot, B., S. Bellizzi, and B. Cochelin. 2017. Passive suppression of helicopter ground resonance using nonlinear energy sinks attached on the helicopter blades. *Journal of Sound and Vibration* 392: 41–55.
47. Lu, X., Z. Liu, and Z. Lu. 2017. Optimization design and experimental verification of track nonlinear energy sink for vibration control under seismic excitation. *Structural Control and Health Monitoring*. https://doi.org/10.1002/stc.2033.
48. Vakakis, A.F. 2001. Inducing passive nonlinear energy sinks in vibrating systems. *Journal of Vibration and Acoustics* 123 (3): 324–332.
49. Lee, Y.S., et al. 2008. Passive non-linear targeted energy transfer and its applications to vibration absorption: A review. *Proceedings of the Institution of Mechanical Engineers, Part K: Journal of Multi-body Dynamics* 222 (2): 77–134.

50. Roberson, R.E. 1952. Synthesis of a nonlinear dynamic vibration absorber. *Journal of the Franklin Institute* 254 (3): 205–220.
51. Li, H., et al. 2015. Robustness study of the pounding tuned mass damper for vibration control of subsea jumpers. *Smart Materials and Structures* 24 (9).
52. Zhang, P., et al. 2013. Seismic control of power transmission tower using pounding TMD. *Journal of Engineering Mechanics* 139 (10): 1395–1406.
53. Zhang, P., et al. 2016. Parametric study of pounding tuned mass damper for subsea jumpers. *Smart Materials and Structures* 25 (1): 015028.
54. Masri, S.F. 1972. Forced vibration of a class of non-linear two-degree-of-freedom oscillators. *International Journal of Non-Linear Mechanics* 7: 663–674.
55. Grubin, C. 1956. On the theory of the acceleration damper. *Journal of Applied Mechanics* 23 (3): 373–378.
56. Masri, S.F. 1970. General motion of impact dampers. *The Journal of the Acoustical Society of America* 47 (1B): 229–237.
57. Bapat, C.N. 1998. Periodic motion of an impact oscillator. *Journal of Sound and Vibration* 209 (1): 43–60.
58. Masri, S.F. 1967. Effectiveness of two-particle impact dampers. *Journal of the Acoustical Society of America* 41 (6): 1553–1554.
59. Bapat, C.N., and S. Sankar. 1985. Single unit impact damper in free and forced vibration. *Journal of Sound and Vibration* 99 (1): 85–94.
60. Bapat, C.N., and S. Sankar. 1985. Multiunit impact damper—Re-examined. *Journal of Sound and Vibration* 103 (4): 457–469.
61. Ema, S., and E. Marui. 1994. A fundamental study on impact dampers. *International Journal of Machine Tools and Manufacture* 34 (3): 407–421.
62. Duncan, M.R., C.R. Wassgren, and C.M. Krousgrill. 2005. The damping performance of a single particle impact damper. *Journal of Sound and Vibration* 286 (1–2): 123–144.
63. Masri, S.F. 1967. Motion and stability of two-particle, single-container impact dampers. *Journal of Applied Mechanics* 34 (2): 506–507.
64. Mao, K., et al. 2004. Simulation and characterization of particle damping in transient vibrations. *Journal of Vibration and Acoustics, ASME* 126 (2): 202–211.
65. Saluena, C., T. Poeschel, and S.E. Esipov. 1998. Dissipative properties of vibrated granular materials. *Physical Review E* 59 (59): 4422–4425.
66. Sanchez, M., and C.M. Carlevaro. 2013. Nonlinear dynamic analysis of an optimal particle damper. *Journal of Sound and Vibration* 332 (8): 2070–2080.
67. Ben Romdhane, M., et al. 2013. The loss factor experimental characterisation of the non-obstructive particles damping approach. *Mechanical Systems and Signal Processing* 38 (2): 585–600.
68. Yang, M.Y., et al. 2005. Development of a design curve for particle impact dampers. *Noise Control Engineering Journal* 53 (1): 5–13.
69. Xiao, W.Q., L.N. Jin, and B.Q. Chen. 2015. Theoretical analysis and experimental verification of particle damper-based energy dissipation with applications to reduce structural vibration. *Shock and Vibration*.
70. Lu, Z., S.F. Masri, and X. Lu. 2011. Studies of the performance of particle dampers attached to a two-degree-of-freedom system under random excitation. *Journal of Vibration and Control* 17 (10): 1454–1471.
71. Lu, Z., S.F. Masri, and X.L. Lu. 2011. Parametric studies of the performance of particle dampers under harmonic excitation. *Sturctural Control and Health Monitoring* 18 (1): 79–98.
72. Lu, Z., X.L. Lu, and S.F. Masri. 2010. Studies of the performance of particle dampers under dynamic loads. *Journal of Sound and Vibration* 329 (26): 5415–5433.
73. Papalou, A., and S.F. Masri. 1996. Response of impact dampers with granular materials under random excitation. *Earthquake Engineering and Structural Dynamics* 25 (3): 253–267.
74. Papalou, A., and S.F. Masri. 1998. An experimental investigation of particle dampers under harmonic excitation. *Journal of Vibration and Control* 4 (4): 361–379.

75. Friend, R.D., and V.K. Kinra. 2000. Particle impacting damping. *Journal of Sound and Vibration* 233 (1): 93–118.
76. Liu, W., G.R. Tomlinson, and J.A. Rongong. 2005. The dynamic characterisation of disk geometry particle dampers. *Journal of Sound and Vibration* 280 (3–5): 849–861.
77. Xu, Z.W., K.W. Chan, and W.H. Liao. 2004. An empirical method for particle damping design. *Shock and Vibration* 11 (5–6): 647–664.
78. Wu, C.J., W.H. Liao, and M.Y. Wang. 2004. Modeling of granular particle damping using multiphase flow theory of gas-particle. *Journal of Vibration and Acoustics* 126 (2): 196–201.
79. Fang, X., and J. Tang. 2006. Granular damping in forced vibration: Qualitative and quantitative analyses. *Journal of Vibration and Acoustics* 128 (4): 489–500.
80. Wang, D.Q., and C.J. Wu. 2015. Vibration response prediction of plate with particle dampers using cosimulation method. *Shock and Vibration*.
81. Wang, D.Q., and C.J. Wu. 2016. A novel prediction method of vibration and acoustic radiation for rectangular plate with particle dampers. *Journal of Mechanical Science and Technology* 30 (3): 1021–1035.
82. Liu, W., G. Tomlinson, and K. Worden. 2002. Nonlinearity study of particle dampers. In *Proceedings of the 2002 international conference on noise and vibration engineering, ISMA, Leuven*.
83. Chen, Q., K. Worden, and J. Rongong. 2005. Characterisation of particle dampers using restoring force surface technique. In *6th international conference on structural dynamics-EURODYN 2005, Paris, France*.
84. Tanrikulu, A.H. 2009. Application of ANN techniques for estimating modal damping of impact-damped flexible beams. *Advances in Engineering Software* 40 (10): 986–990.
85. Cui, Z., et al. 2011. A quantitative analysis on the energy dissipation mechanism of the non-obstructive particle damping technology. *Journal of Sound and Vibration* 330 (11): 2449–2456.
86. Mao, K.M., et al. 2004. DEM simulation of particle damping. *Powder Technology* 142 (2–3): 154–165.
87. Lu, Z., et al. 2014. Discrete element method simulation and experimental validation of particle damper system. *Engineering Computations* 31 (4): 810–823.
88. Wong, C., A. Spencer, and J. Rongong. 2009. Effects of enclosure geometry on particle damping performance. In *50th AIAA/ASME/ASCE/AHS/ASC structures, structural dynamics, and materials conference*. Palm Springs, California: American Institute of Aeronautics and Astronautics.
89. Saeki, M. 2002. Impact damping with granular materials in a horizontally vibrating system. *Journal of Sound and Vibration* 251 (1): 153–161.
90. Cundall, P.A. 1971. A computer model for simulating progressive large scale movements in blocky rock systems. In *Proceedings of the symposium of the international society of rock mechanics, Nancy, France*.
91. Zhou, X., et al. 2007. A review of distinct element method researching progress and application. *Rock and Soil Mechanics* S1: 408–416. (in Chinese).
92. Sanchez, M., G. Rosenthal, and L.A. Pugnaloni. 2012. Universal response of optimal granular damping devices. *Journal of Sound and Vibration* 331 (20): 4389–4394.
93. Bannerman, M.N., et al. 2011. Movers and shakers: Granular damping in microgravity. *Physical Review E—Statistical, Nonlinear, and Soft Matter Physics* 84 (1).
94. Sack, A., et al. 2013. Energy dissipation in driven granular matter in the absence of gravity. *Physical Review Letters* 111 (1).
95. Xia, Z., X. Liu, and Y. Shan. 2011. Application of particle damping for vibration attenuation in brake drum. *International Journal of Vehicle Noise and Vibration* 7 (2): 178–194.
96. Tan, D., X. Liu, and Y. Shan. 2011. Simulation of applying particle damper to torsion vibration of plate. *Journal of System Simulation* 23 (8): 1594–1597. (in Chinese).
97. Xia, Z., H. Wen, and X. Liu. 2014. Dynamic characteristics of a rotating plate structure with particle damping. *Journal of Vibration and Shock* 33 (9): 61–65, 88 (in Chinese).

98. Sadek, M.M., and B. Mills. 1970. Effect of gravity on the performance of an impact damper, part 1: Steady-state motion. *Journal of Mechanical Engineering Science* 12 (4): 268–277.
99. Sadek, M.M., C.J.H. Williams, and B. Mills. 1970. Effect of gravity on the performance of an impact damper, part 2: Stability of vibrational modes. *Journal of Mechanical Engineering Science* 12 (4): 278–287.
100. Cempel, C., and G. Lotz. 1993. Efficiency of vibrational energy dissipation by moving shot. *Journal of Structural Engineering* 119 (9): 2642–2652.
101. Hollkamp, J.J., and R.W. Gordon. 1998. Experiments with particle damping. In *Smart structures and materials 2002: Damping and Isolation, San Diego, CA*.
102. Xu, Z.W., M.Y. Wang, and T.N. Chen. 2005. Particle damping for passive vibration suppression: Numerical modelling and experimental investigation. *Journal of Sound and Vibration* 279 (3–5): 1097–1120.
103. Trigui, M., et al. 2009. An experimental study of a multi-particle impact damper. *Proceedings of the Institution of Mechanical Engineers, Part C: Journal of Mechanical Engineering Science* 223 (9): 2029–2038.
104. Els, D.N. 2011. Damping of rotating beams with particle dampers: experimental analysis. *AIAA Journal* 49 (10): 2228–2238.
105. Jadhav, T.A., and P.J. Awasare. 2016. Enhancement of particle damping effectiveness using multiple cell enclosure. *Journal of Vibration and Control* 22 (6): 1516–1525.
106. Zhang, C., et al. 2014. Discrete element method model and damping performance of bean bag dampers. *Journal of Sound and Vibration* 333 (23): 6024–6037.
107. Wang, Y.R., et al. 2016. Experimental and numerical investigations on the performance of particle dampers attached to a primary structure undergoing free vibration in the horizontal and vertical directions. *Journal of Sound and Vibration* 371: 35–55.
108. Rongong, J., and G. Tomlinson. 2005. Amplitude dependent behaviour in the application of particle dampers to vibrating structures. In *46th AIAA/ASME/ASCE/AHS/ASC structures, structural dynamics and materials conference*. Austin, Texas: American Institute of Aeronautics and Astronautics.
109. Wong, C., and J. Rongong. 2006. Macromodel characterisation and application of particle dampers to vibrating structures. In *47th AIAA/ASME/ASCE/AHS/ASC structures, structural dynamics, and materials conference*. Newport, RI, United States: American Institute of Aeronautics and Astronautics.
110. Wong, C.X., M.C. Daniel, and J.A. Rongong. 2007. Prediction of the amplitude dependent behaviour of particle dampers. In *48th AIAA/ASME/ASCE/AHS/ASC structures, structural dynamics and materials conference, Honolulu, Hawaii*.
111. Veluswami, M.A., and F.R.E. Crossley. 1975. Multiple impacts of a ball between two plates, part 1: Some experimental observations. *Journal of Engineering for Industry, ASME* 97 (3): 820–827.
112. Veluswami, M.A., F.R.E. Crossley, and G. Horvay. 1975. Multiple impacts of a ball between two plates. Part 2: Mathematical modeling. *Journal of Engineering for Industry, ASME* 97 (3): 835–838.
113. Yokomichi, I., et al. 1996. Impact damper with granular materials for multibody system. *Journal of Pressure Vessel Technology* 118 (1): 95–103.
114. Yokomichi, I., H. Muramatsu, and Y. Araki. 2001. On shot impact dampers applied to self-excited vibrations. *International Journal of Acoustics and Vibration* 6 (4): 193–199.
115. Li, K., and A.P. Darby. 2006. Experiments on the effect of an impact damper on a multiple-degree-of-freedom system. *Journal of Vibration and Control* 12 (5): 445–464.
116. Lu, Z., W.S. Lu, and S.F. Masri. 2012. Experimental studies of the effects of buffered particle dampers attached to a multi-degree-of-freedom system under dynamic loads. *Journal of Sound and Vibration* 331 (9): 2007–2022.
117. Lu, Z., et al. 2012. Shaking table test of the effects of multi-unit particle dampers attached to an MDOF system under earthquake excitation. *Earthquake Engineering and Structural Dynamics* 41 (5): 987–1000.

118. Sathishkumar, B., K.M. Mohanasundaram, and M.S. Kumar. 2014. Impact of particle damping parameters on surface roughness of bored surface. *Arabian Journal for Science and Engineering* 39 (10): 7327–7334.

119. Simonian, S.S. 1995. Particle beam damper. In *Proceedings of SPIE Conference on Passive Damping, San Diego.*

120. Rocke, R.D., and S.F. Masri. 1969. Application of a single-unit impact damper to an antenna structure. *Shock and Vibration Bulletin* 39 (1): 1–10.

121. Veeramuthuvel, P., K.K. Sairajan, and K. Shankar. 2016. Vibration suppression of printed circuit boards using an external particle damper. *Journal of Sound and Vibration* 366: 98–116.

122. Akl, F.A., and A.S. Butt. 1995. Application of impact dampers in vibration control of flexible structures. In *NASA Johnson Space Center, National Aeronautics and Space Administration (NASA) (American Society for Engineering Education (ASEE) Summer Faculty Fellowship Program, 1 15 P.* Washington, DC.

123. Ashley, S. 1995. New racket shakes up tennis. *Mechanical Engineering* 117 (8): 80–81.

124. Casciati, S., A.G. Chassiakos, and S.F. Masri. 2014. Toward a paradigm for civil structural control. *Smart Structures and Systems* 14 (5): 981–1004.

125. Moore, J.J., et al. 1995. Forced response analysis and application of impact dampers to rotordynamic vibration suppression in a cryogenic environment. *Journal of Vibration and Acoustics, ASME* 117 (3): 300–310.

126. Kielb, R., et al. 1998. Advanced damping systems for fan and compressor blisks. In *34th AIAA/ASME/SAE/ASEE Joint propulsion conference and exhibit.* Cleveland, United States: American Institute of Aeronautics and Astronautics Inc, AIAA.

127. Aiba, T., et al. 1995. An investigation on variable-attractive-force impact damper and application for controlling cutting vibration in milling process. *Journal of the Japan Society for Precision Engineering* 61 (1): 75–79.

128. Oledzki, A. 1981. New kind of impact damper—from simulation to real design. *Mechanism and Machine Theory* 16 (3): 247–253.

129. Gibson, B.W. 1983. *Usefulness of impact dampers for space applications.* Air Force Institute of Technology: Wright-Patterson AFB, OH.

130. Torvik, P.J., and W. Gibson. 1987. Design and effectiveness of impact dampers for space applications. *Design Engineering Division, ASME* 5: 65–74.

131. Simonian, S. 2004. Particle damping applications. In *45th AIAA/ASME/ASCE/AHS/ASC Structures, structural dynamics and materials conference.* Palm Springs, California: American Institute of Aeronautics and Astronautics.

132. Xia, Z.W., Y.C. Shan, and X.D. Liu. 2007. Experimental research on particle damping of cantilever beam. *Journal of Aerospace Power* 22 (10): 1737–1741. (in Chinese).

133. Panossian, H. 2006. Optimized Non-Obstructive Particle Damping (NOPD) treatment for composite honeycomb structures. In *47th AIAA/ASME/ASCE/AHS/ASC Structures, Structural Dynamics and Materials Conference, Newport, RI.*

134. Hara, S., et al. 2011. Momentum-exchange-impact-damper-based shock response control for planetary exploration spacecraft. *Journal of Guidance, Control and Dynamics* 34 (6): 1828–1838.

135. Kushida, Y., et al. 2013. Robust landing gear system based on a hybrid momentum exchange impact damper. *Journal of Guidance, Control and Dynamics* 36 (3): 776–789.

136. Park, W.H. 1967. Mass-spring-damper response to repetitive impact. *Journal of Manufacturing Science and Engineering* 89 (4): 587–596.

137. Sims, N.D., A. Amarasinghe, and K. Ridgway. 2005. Particle dampers for workpiece chatter mitigation. *Manufacturing Engineering Division, ASME* 16 (1): 825–832.

138. Fuse, T. 1989. Prevention of resonances by impact damper. In *Seismic engineering—1989: Design, analysis, testing, and qualification methods.* New York, NY, United States, Honolulu, HI, USA: Publ by ASME.

139. Skipor, E., and L.J. Bain. 1980. Application of impact damping to rotary printing equipment. *Journal of Mechanical Design* 102 (2): 338–343.

140. Sato, T., et al. 2002. Vibration reduction of pantograph-support system using an impact damper (influence of curve track). In *Proceeding of international conference on noise and vibration engineering, ISMA, Leuven, Belgium.*

141. Xu, Z.W., M.Y. Wang, and T.N. Chen. 2004. A particle damper for vibration and noise reduction. *Journal of Sound and Vibration* 270 (4–5): 1033–1040.

142. Hu, L., et al. 2016. Sound reduction at a target point inside an enclosed cavity using particle dampers. *Journal of Sound and Vibration* 384: 45–55.

143. Chan, K.W., et al. 2006. Experimental studies for particle damping on a bond arm. *Journal of Vibration and Control* 12 (3): 297–312.

144. Li, W., D. Zhu, and X. Huang. 1998. Study on the damping performance of flexible restraint particle impact damper applied to plate structure. *Noise and Vibration Control* 4 (1): 2–5. (in Chinese).

145. Xiao, W.Q., et al. 2016. Energy dissipation mechanism and experiment of particle dampers for gear transmission under centrifugal loads. *Particuology* 27: 40–50.

146. Tian, L., and X. Gai. 2015. Wind-induced vibration control of power transmission tower using pounding tuned mass damper. *Journal of Vibroengineering* 17 (7): 3693–3701.

147. Velichkovich, A.S., and S.V. Velichkovich. 2001. Vibration-impact damper for controlling the dynamic drillstring conditions. *Chemical and Petroleum Engineering* 37 (3–4): 213–215.

148. Guo, S. 2016. *Research on the characteristics of vibration and noise reduction of particle damping rings installed on high-speed rail wheels* (in Chinese). Harbin: Harbin Institute of Technology.

149. Lu, Y., et al. 2016. Flow-induced vibration in subsea jumper subject to downstream slug and ocean current. *Journal of Offshore Mechanics and Arctic Engineering* 138 (2): 021302-021302-10.

150. Yang, X., and G. Shi. 2014. The effect of slab track on wheel/rail rolling noise in high speed railway. *Intelligent Automation and Soft Computing* 20 (4): 575–585.

151. Housner, G.W., L.A. Bergman, T.K. Caughey, et al. 1997. Structural control: Past, present, and future. *Journal of Engineering Mechanics, ASCE* 123 (9): 897–971.

152. Yao, J.T.P. 1972. Concept of structural control. *Journal of the structural Division, ASCE* 98 (7): 1567–1574.

153. Kelly, J.M., R.L. Skinner, and A.J. Heine. 1972. Mechanics of energy absorption in special devices for use in earthquake-resistant structures. *Bulletin of New Zealand National Society for Earthquake Engineering* 5 (3): 63–88.

154. Skinner, R.I., J.M. Kelly, and A.J. Heine. 1974. Hysteretic dampers for earthquake-resistant structures. *Earthquake Engineering and Structural Dynamics* 3 (3): 287–296.

155. Ogawa, K., T. Ide, and T. Saitou. 1997. Application of impact mass damper to a cable-stayed bridge pylon. *Journal of Wind Engineering and Industrial Aerodynamics* 72 (1–3): 301–312.

156. Naeim, F., et al. 2011. Performance of tall buildings in Santiago, Chile during the 27 February 2010 offshore Maule, Chile earthquake. *The Structural Design of Tall and Special Buildings* 20 (1): 1–16.

157. Masri, S.F., and K. Kahyai. 1974. Steady-state motion of a plate with a discontinuous mass. *International Journal of Non-Linear Mechanics* 9: 451–462.

158. Masri, S.F. 1973. Forced vibration of a class of nonlinear dissipative beams. *Journal of the Engineering Mechanics Division* 99: 669–683.

159. Zhao, L., P. Liu, and Y.-Y. Lu. 2009. Experimental investigation on damping characteristics of NOPD columns. *Journal of Vibration and Shock* 28 (8): 1–5. (in Chinese).

160. Lu, Z. 2011. *Numerical Simulation and Performance Analysis of Particle Dampers*, in *College of Civil Engineering* (in Chinese), Tongji University: Shanghai.

161. Papalou, A., et al. 2015. Seismic protection of monuments using particle dampers in multi-drum columns. *Soil Dynamics and Earthquake Engineering* 77: 360–368.

162. Izzi, M., L. Caracoglia, and S. Noe. 2016. Investigating the use of Targeted-Energy-Transfer devices for stay-cable vibration mitigation. *Structural Control and Health Monitoring* 23 (2): 315–332.

163. Zhou, P., and H. Li. 2016. Modeling and control performance of a negative stiffness damper for suppressing stay cable vibrations. *Structural Control and Health Monitoring* 23 (4): 764–782.
164. Egger, P., L. Caracoglia, and J. Kollegger. 2015. Modeling and experimental validation of a multiple-mass-particle impact damper for controlling stay-cable oscillations. *Structural Control and Health Monitoring* 23 (6): 960–978.
165. Masri, S.F., et al. 1989. Active parameter control of nonlinear vibrating structures. *Journal of Applied Mechanics* 56 (3): 658–666.

Chapter 3
Theoretical Analysis and Numerical Simulation of Particle Impact Dampers

The numerical simulation of the behaviors of particles, which are placed in particle dampers under wind loads and seismic loads, has still not been well developed. Due to the sudden change of momentum caused by collisions of particles, the performance of particle dampers is highly nonlinear, and therefore, it is hard to obtain the analytical solution of the system. Hence, simplified analytical methods and numerical simulation methods have been carried out by many researchers. Papalou and Masri [1, 2] introduced an equivalent single-particle impact damper model to evaluate the performance of multi-particle dampers. Friend and Kinra [3] developed an analytical approach by treating multi particles as a lumped mass system and advanced a concept of effective coefficient of restitution, which was extracted from experimental results, to represent the energy dissipation caused by various mechanisms. Lu and Lv [4] established a numerical simulation method for particle dampers controlling vibration of multi-degree-of-freedom primary systems, based on the DEM (Discrete element method). Yan et al. [5] developed a simplified mechanical model to simulate the controlling effects of a tuned particle damper and established an analytical energy method to estimate the equivalent additional damping ratio of the damper. Moreover, the researchers presented numerical simulation methods for particle dampers such as the regression design method [6] and the restoring force surface method [7].

This section introduces the steady-state response of structures with particle impact dampers at first. Secondly, the simplified analytical method to simulate collisions of multi particles is elaborated, in which multi-particle damper is equivalent to single-particle damper based on the equivalent principles that the mass of the particles before and after the equivalent is equal and the void volume of the cavity before and after the equivalent is equal. An equivalent mechanical model for particle dampers is established, and the method to realize the model is introduced. Thirdly, a more sophisticated discrete element simulation method is introduced, which can accurately analyze the performance of particle dampers by fully considering particle-particle collisions, particle-wall collisions, and the motion of particles. Depending on the purpose of the actual analysis, either the simplified analytical method or the discrete element simulation method can be selected flexibly.

© China Machine Press and Springer Nature Singapore Pte Ltd. 2020
Z. Lu et al., *Particle Damping Technology Based Structural Control*,
Springer Tracts in Civil Engineering,
https://doi.org/10.1007/978-981-15-3499-7_3

3.1 Steady-State Response of Structures with Particle Impact Dampers

3.1.1 Steady-State Response of Structures with an Impact Damper

3.1.1.1 Steady-State Motion of a Plate

This part presents an "exact" solution for the steady-state motion of a viscously damped plate of arbitrary shape and with arbitrary boundary conditions, which is subjected to sinusoidally varying distributed as well as concentrated loads, and provided at some point within its domain with a discontinuous mass that is free to oscillate frictionlessly and unidirectionally with a given clearance.

The model of the system under consideration is shown in Fig. 3.1. In the absence of the discontinuous mass, the plate is governed by the partial differential equation

$$L[w(x, y, t)] + \frac{\partial}{\partial t} C[w(x, y, t)] + M(x, y)\frac{\partial^2 w(x, y, t)}{\partial t^2} = f(x, y, t) \qquad (3.1)$$

over the domain D of the plate, where

L is a linear homogeneous self-adjoint differential operator of order $2p$ with respect to spatial coordinates x and y that specifies the stiffness distribution of the plate
C is an operator that is a linear combination of operator L and function M, viz

$$C = \alpha M + \beta L \qquad (3.2)$$

Fig. 3.1 Model of system

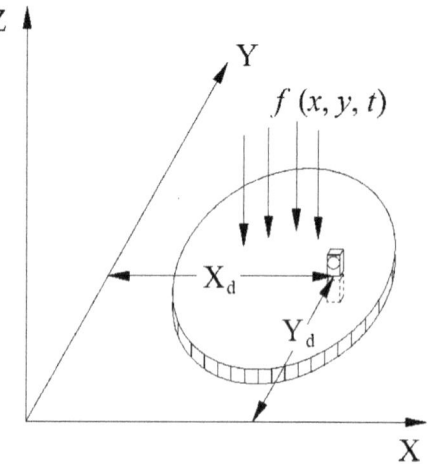

M is a function that specifies the mass distribution of the plate

$f(x, y, t)$ is a sinusoidally varying load equal to $f(x, y) \sin \Omega t$ with

$$f(x, y) = p(x, y) + \sum_{k=1}^{s} F_k \delta(x - x_k, y - y_k)$$

$p(x, y)$ is distributed excitation force

F_k is concentrated excitation force acting at point $y = y_k, x = x_k$

$\delta(x - x_k, y - y_k)$ is spatial Dirac's delta function defined by

$$\delta(x - x_k, y - y_k) = 0, \, x \neq x_k, \, y \neq y_k, \, \int_D \delta(x - x_k, y - y_k) dx = 1 \qquad (3.3)$$

At each of the boundary points there are p boundary conditions of the type

$$B_i[w(x, y, t)] = 0 \quad i = 1, 2, \ldots, p \qquad (3.4)$$

where B_i is linear homogeneous differential operators of order $(2p - 1)$.

Let $\varphi_i(x, y)$ be the ith eigenfunction associated with the homogeneous equation of the undamped system and assume that the eigenfunction satisfies the orthogonality condition

$$\int_D M(x, y)\varphi_i(x, y)\varphi_j(x, y) dD = \delta_{ij} M_i \qquad (3.5)$$

$$\int_D \varphi_i(x, y) L[\varphi_j(x, y)] dD = \delta_{ij} K_i \qquad (3.6)$$

where δ_{ij} is the Kronecker delta.

Using the normal mode approach

$$w(x, y, t) = \sum_{i=1}^{\infty} \varphi_i(x, y) q_i(t) \qquad (3.7)$$

Which upon substitution into (3.1) leads to

$$M_i \ddot{q}_i + C_i \dot{q}_i + K_i q_i = Q_i = \left[\int_D \varphi_i(x, y) f(x, y) dD \right] \sin(\Omega t + \alpha_0) \qquad (3.8)$$

where α_0 is a phase angle related to the time origin t_0 by $\alpha_0 = \Omega t_0$.

Consider the steady-state motion of the system shown in Fig. 3.1, and shift the origin of the time axis to coincide with the time of occurrence of an impact between the particle m_d and the plate at $\Omega t = \alpha_0$. Let the initial displacement and velocity at $t = 0_+$ be

$$w(x, y, 0) = w_0(x, y); \quad \dot{w}(x, y, 0_+) = \dot{w}_a(x, y) \tag{3.9}$$

Then in steady-state motion with two symmetric impacts per cycle of the excitation (a condition which predominates when steady-state motion is achieved) the following relations must be satisfied:

$$y(0) \equiv y_0 = z(0) - w(x_d, y_d, t) = \pm\tfrac{1}{2}d \tag{3.10}$$

$$\dot{z}(0_+) = -(2\Omega/\pi)[w(x_d, y_d, 0) + y_0] \tag{3.11}$$

$$w(x, y, t)|_{\Omega t = \pi} = -w(x, y, 0) \equiv -w(x, y)_0 \tag{3.12}$$

$$\dot{w}(x, y, t)|_{\Omega t = \pi_-} = -\dot{w}(x, y, 0) \equiv -\dot{w}(x, y)_b \tag{3.13}$$

where d is clearance where damper is free to oscillate.
   The solution of (3.8) is

$$
\begin{aligned}
q_i(t) = \exp\left(-\frac{\xi_i}{r_i}\Omega t\right) &\Big[\frac{1}{\eta_i}\left(\xi_i \sin\frac{\eta_i}{r_i}\Omega t + \eta_i \cos\frac{\eta_i}{r_i}\Omega t\right)q_{0_i} \\
&+ \frac{1}{\omega_i \eta_i}\left(\sin\frac{\eta_i}{r_i}\Omega t\right)\dot{q}_{a_i} - \frac{A_i}{\eta_i}\left(\xi_i \sin\frac{\eta_i}{r_i}\Omega t + \eta_i \cos\frac{\eta_i}{r_i}\Omega t\right)\sin\tau_i \\
&- \frac{A_i}{\eta_i}r_i\left(\sin\frac{\eta_i}{r_i}\Omega t\right)\cos\tau_i\Big] + A_i \sin(\Omega t + \tau_i) \qquad i = 1, 2, \ldots, n \tag{3.14}
\end{aligned}
$$

where
$\omega_i \equiv \sqrt{\frac{K_i}{M_i}},\ \xi_i \equiv 2\frac{C_i}{\sqrt{K_i M_i}},\ \eta_i \equiv \sqrt{1 - \xi_i^2},\ r_i \equiv \frac{\Omega}{\omega_i},\ f_i \equiv \int_D \varphi_i(x, y)f(x, y)dD,$
$A_i \equiv \frac{f_i/K_i}{\sqrt{(1-r_i^2)^2+(2\xi_i r_i)^2}},\ \Psi_i \equiv \tan^{-1}\frac{2\xi_i r_i}{1-r_i^2},\ \tau_i \equiv \alpha_0 - \Psi_i,\ q_{0_i} \equiv q_i(0),\ \dot{q}_{a_i} \equiv \dot{q}_i(0_+)$
where the $+$ subscript indicates conditions immediately after the specified time.
   Equation (3.14) can be expressed in concise form as

$$q_i(t) = E_i(t)\big[B_{2_i}(t)\dot{q}_{a_i} + B_{3_i}(t)q_{0_i} + B_{4_i}(t)S_{1_i} + B_{5_i}(t)S_{2_i} + S_{3_i}(t)\big] \tag{3.15}$$

Differentiating (3.15)

$$\dot{q}_i(t) = E_i(t)\big[B_{12_i}(t)\dot{q}_{a_i} + B_{13_i}(t)q_{0_i} + B_{14_i}(t)S_{1_i} + B_{15_i}(t)S_{2_i} + S_{4_i}(t)\big] \tag{3.16}$$

Using (3.7) and the orthogonality condition

$$q_i(t)M_i = \int_D \varphi_i(x, y)M(x, y)w(x, y, t)dxdy \tag{3.17}$$

$$\dot{q}_i(t)M_i = \int_D \varphi_i(x, y)M(x, y)\dot{w}(x, y, t)dxdy \tag{3.18}$$

Using (3.17) and (3.18) together with (3.9) in (3.15) and (3.16)

$$\int_D \varphi_i(x, y)M(x, y)w(x, y, t)dxdy = B_{21_i}(t)\int_D \varphi_i(x, y)M(x, y)\dot{w}_a(x, y)dxdy$$

$$+ B_{22_i}(t)\int_D \varphi_i(x, y)M(x, y)w_0(x, y)dxdy$$

$$+ B_{23_i}(t)S_{1_i} + B_{24_i}(t)S_{2_i} + M_i S_{3_i}(t) \quad (3.19)$$

and

$$\int_D \varphi_i(x, y)M(x, y)\dot{w}(x, y, t)dxdy = B_{31_i}(t)\int_D \varphi_i(x, y)M(x, y)\dot{w}_a(x, y)dxdy$$

$$+ B_{32_i}(t)\int_D \varphi_i(x, y)M(x, y)w_0(x, y)dxdy$$

$$+ B_{33_i}(t)S_{1_i} + B_{34_i}(t)S_{2_i} + M_i S_{4_i}(t) \quad (3.20)$$

Idealizing the impact to be a discontinuous process during which the displacement and velocity of the system remain the same except for the point of impact whose velocity changes discontinuously in accordance with the momentum equation and the definition of the coefficient of restitution, then

$$w(x, y, 0_-) = w(x, y, 0_+) \quad (3.21)$$

and

$$\dot{w}(x, y, 0_-) = [1 + (G - 1)\delta(x - x_d)\delta(y - y_d)]\dot{w}(x, y, 0_+) \quad (3.22)$$

Evaluating (3.19) and (3.20) at $\Omega t = \pi$, using conditions (3.10)–(3.13), noting that

$$S_{3_i}(\pi/\Omega) = -S_{1_i}, \quad S_{4_i}(\pi/\Omega) = -\Omega S_{2_i}$$

and employing (3.22), then

$$\int_D \varphi_i(x, y)M(x, y)w_0(x, y)dxdy = B_{2\,6_i}\int_D \varphi_i(x, y)M(x, y)\dot{w}_a(x, y)dxdy$$

$$+ B_{2\,7_i}S_{1_i} + B_{2\,8_i}S_{2_i} \quad (3.23)$$

$$\int_D \varphi_i(x, y)M(x, y)\dot{w}_a(x, y)dxdy = B_{36_i}\int_D \varphi_i(x, y)M(x, y)w_0(x, y)dxdy$$

$$+ B_{37_i}S_{1_i} + B_{38_i}S_{2_i} + D_{1_i}\dot{w}_a(x_d, y_d) \quad (3.24)$$

Substituting (3.24) into (3.23) to eliminate $\dot{w}_a(x, y)$

$$\int_D \varphi_i(x, y)M(x, y)w_0(x, y)dxdy = B_{44_i}S_{1_i} + B_{45_i}S_{2_i} + D_{3_i}\dot{w}_a(x_d, y_d) \quad (3.25)$$

Using (3.24) in conjunction with (3.25)

$$\int_D \varphi_i(x, y)M(x, y)\dot{w}_a(x, y)dxdy = B_{46_i}S_{1_i} + B_{47_i}S_{2_i} + D_{4_i}\dot{w}_a(x_d, y_d) \quad (3.26)$$

Equation (3.25) and (3.26) with the help of (3.17) and (3.18) can be expressed as

$$q_i(0) = B_{51_i}S_{1_i} + B_{52_i}S_{2_i} + D_{5_i}\dot{w}_a(x_d, y_d) \quad (3.27)$$

$$\dot{q}_i(0_+) = B_{61_i}S_{1_i} + B_{62_i}S_{2_i} + D_{6_i}\dot{w}_a(x_d, y_d) \quad (3.28)$$

Multiplying each equation of system (3.27) and (3.28) by $\phi_i(x_d, y_d)$ and summing over i:

$$w_0(x_d, y_d) = \sum_{i=1}^{\infty} \varphi_i(x_d, y_d)\left[B_{51_i}S_{1_i} + B_{52_i}S_{2_i}\right] + \dot{w}_a(x_d, y_d)\sum_{i=1}^{\infty} \varphi_i(x_d, y_d)D_{5_i}$$
$$(3.29)$$

and

$$\dot{w}_a(x_d, y_d) = \sum_{i=1}^{\infty} \varphi_i(x_d, y_d)\left[B_{61_i}S_{1_i} + B_{62_i}S_{2_i}\right] + \dot{w}_a(x_d, y_d)\sum_{i=1}^{\infty} \varphi_i(x_d, y_d)D_{6_i}$$
$$(3.30)$$

Expressing S_{1_i} and S_{2_i} in Eqs. (3.29) and (3.30) in terms of $\sin \alpha_0$ and $\cos \alpha_0$, and noting that

$$\dot{w}_a(x_d, y_d) = -2G_1[w_0(x_d, y_d) + y_0] \quad (3.31)$$

leads to

$$h_1 \sin \alpha_0 + h_2 \cos \alpha_0 = h_3 \quad (3.32)$$

whose solution is

$$\alpha_0 = \tan^{-1}\left[(h_1 h_3 \pm h_2 h_4)/(h_2 h_3 \mp h_1 h_4)\right] \quad (3.33)$$

Once α_0 is determined from (3.33), the remainder of the unknowns can be found by back-substitution.

Application:

(1) Uniform plates

For a uniform thin plate satisfying the assumptions of the Kirchhoff's hypotheses, the previously mentioned operators L and M are:

$$L = D\nabla^4, \, M = \mu_0$$

where D is the flexural rigidity of the plate, ∇^4 is the biharmonic operator, and μ_0 is the mass density.

(2) Base excitation of plates

If a plate is excited through support motion rather than directly applied load, then letting $S_0(t)$ be the rigid body translation and $w(x, y, t)$ be the elastic deformation measured relative to rigid body motion, the governing equation of motion (3.1) remains the same except that the applied load $f(x, y, t)$ is replaced by the inertia load $M(x, y)S_0(t)$.

Figure 3.2 shows the response of a typical uniform square cantilever plate provided with a discontinuous mass and subjected to sinusoidal base excitation. Since no exact analytical solution is available in a closed form for all modes of a cantilever plate, the natural frequencies and mode shapes used in the present analysis were obtained from a finite element solution using standard techniques [8].

The left-hand-side ordinate of Fig. 3.2 is the ratio of the maximum displacement at the free end, $w(L, 0)_{max}$, for the plate using the damper divided by the peak displacement, $w(L, 0)_p$, of the same point with no damper being used. In the absence of the damper, the solution curve is a straight line at $w(L, 0)_{max}/w(L, 0)_p = 1$. The abscissa in Fig. 3.2 is the clearance ratio, d/S_0, in which S_0 is the amplitude of the sinusoidal base motion.

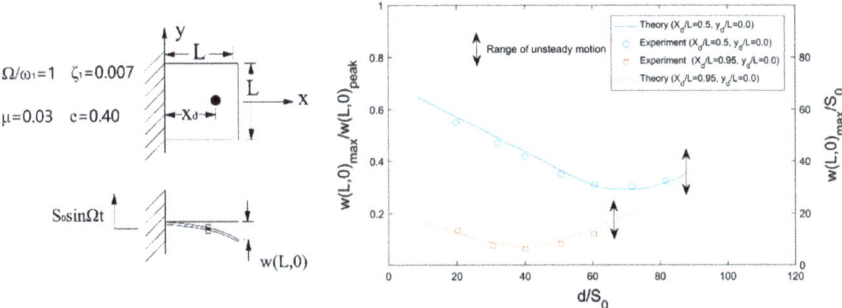

Fig. 3.2 Comparison of theoretical and experimental results

Experiments with a mechanical model were conducted to check the theoretical results and to determine which branch of the solution (3.33), if any, was asymptotically stable. The primary system consisted of a plate measuring $40 \times 40 \times 0.4$ cm. The auxiliary mass was a hardened-steel ball, of a type used in ball bearings, which was constrained to oscillate horizontally with a negligible amount of friction in a cylindrical container that was attached to the plate at a specified point. As indicated by the experimental results shown in Fig. 3.2, the solution curve associated with α_0^+ (the upper choice of signs in Eq. (3.33)) was found to be stable throughout most of its range.

The predominant type of steady-state motion encountered in the experimental studies had two symmetric impacts per cycle, and as expected, no subharmonic resonances were found for any combination of practical values of system parameters.

Comparison between the experimental and analytical results for a typical case is also shown in Fig. 3.3. It is clear that the experimental measurements completely corroborate the theoretical predictions. Figure 3.4 shows the influence of viscous damping on the response of a cantilever plate provided with damper.

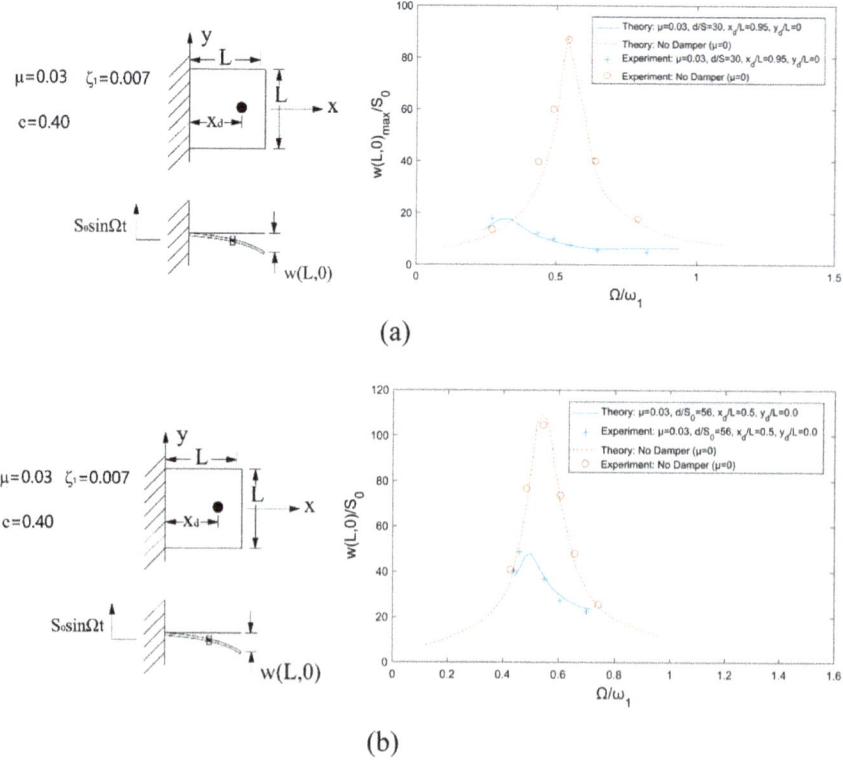

Fig. 3.3 Frequency-response of base-excited cantilever plate: **a** $x_d/L = 0.95$ and **b** $x_d/L = 0.5$

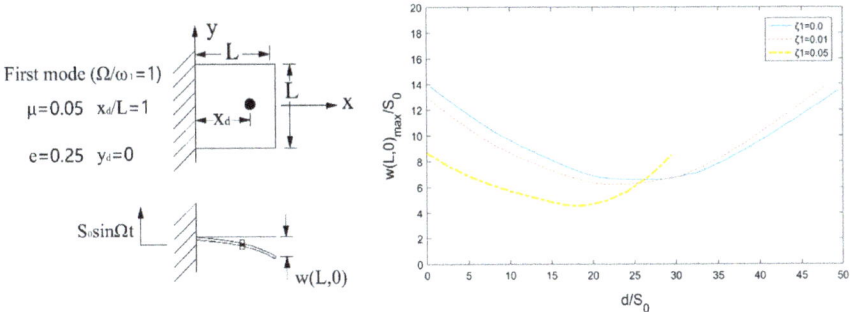

Fig. 3.4 Effect of viscous damping: First mode of cantilever plate

The effects of damper mass ratio, μ, and coefficient of restitution, e, on the response of a base-excited cantilever plate that is oscillating with a forcing frequency corresponding to the fundamental frequency, is shown in Fig. 3.5. It is seen from Fig. 3.5a that, even with a $\mu = 0.01$, a reduction of $\approx 80\%$ can be achieved if the clearance ratio is adjusted to its optimum value.

(3) Discrete force excitation of a simply-supported plate

The response of a simply-supported square subjected to a discrete sinusoidal force is shown in Fig. 3.6 where the effects of mode shape, force location, and damper location are exhibited. It is found that a properly designed damper is effective at any selected frequency. In general, the optimum location of the damper (regarding amplitude attenuation) coincides with the point of maximum deflection.

3.1.1.2 Steady-State Motion of a Beam

The next part presents the "exact" solution for the steady-state motion of a class of uniform Bernoulii-Euler beams with proportional-type viscous damping subjected to sinusoidally varying distributed, as well as concentrated, loads and provided with an impact vibration damper that is attached to the beam at some point along its length.

Consider the class of beams that, in the absence of the impact damper, is governed by the partial differential equation:

$$L[w(x,t)] + \frac{\partial}{\partial t}C[w(x,t)] + M(x)\frac{\partial^2 w(x,t)}{\partial t^2} = f(x,t) \qquad (3.34)$$

over the length, l, of the beam, in which $w(x,t)$ is transverse displacement of beam; t is time; L is a linear homogeneous self-adjoint differential operator of order $2p$ with respect to spatial coordinate x, that specifies the stiffness distribution of the beam; C is an operator that is a linear combination of operator L and function M, i.e.

$$C = \alpha M + \beta L \qquad (3.35)$$

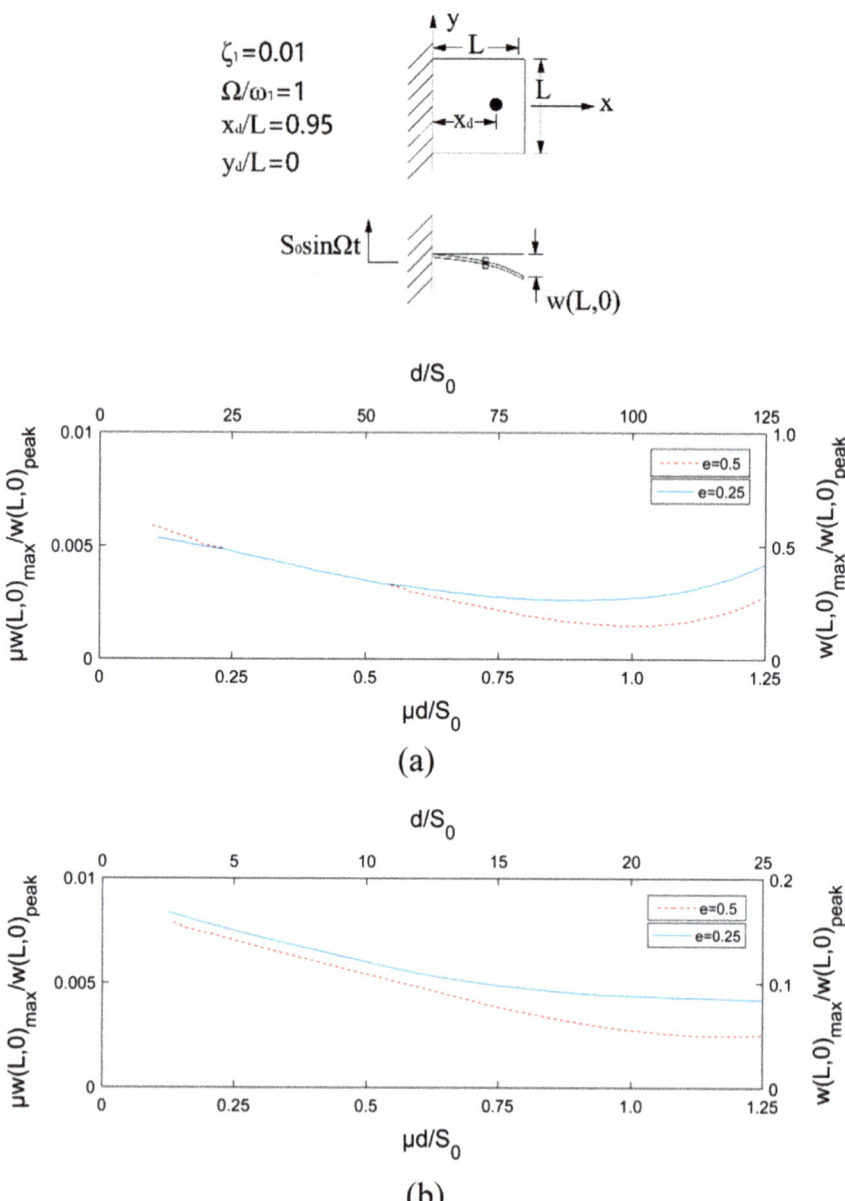

Fig. 3.5 Effects of mass ratio and coefficient of restitution: **a** $\mu = 0.01$ and **b** $\mu = 0.05$

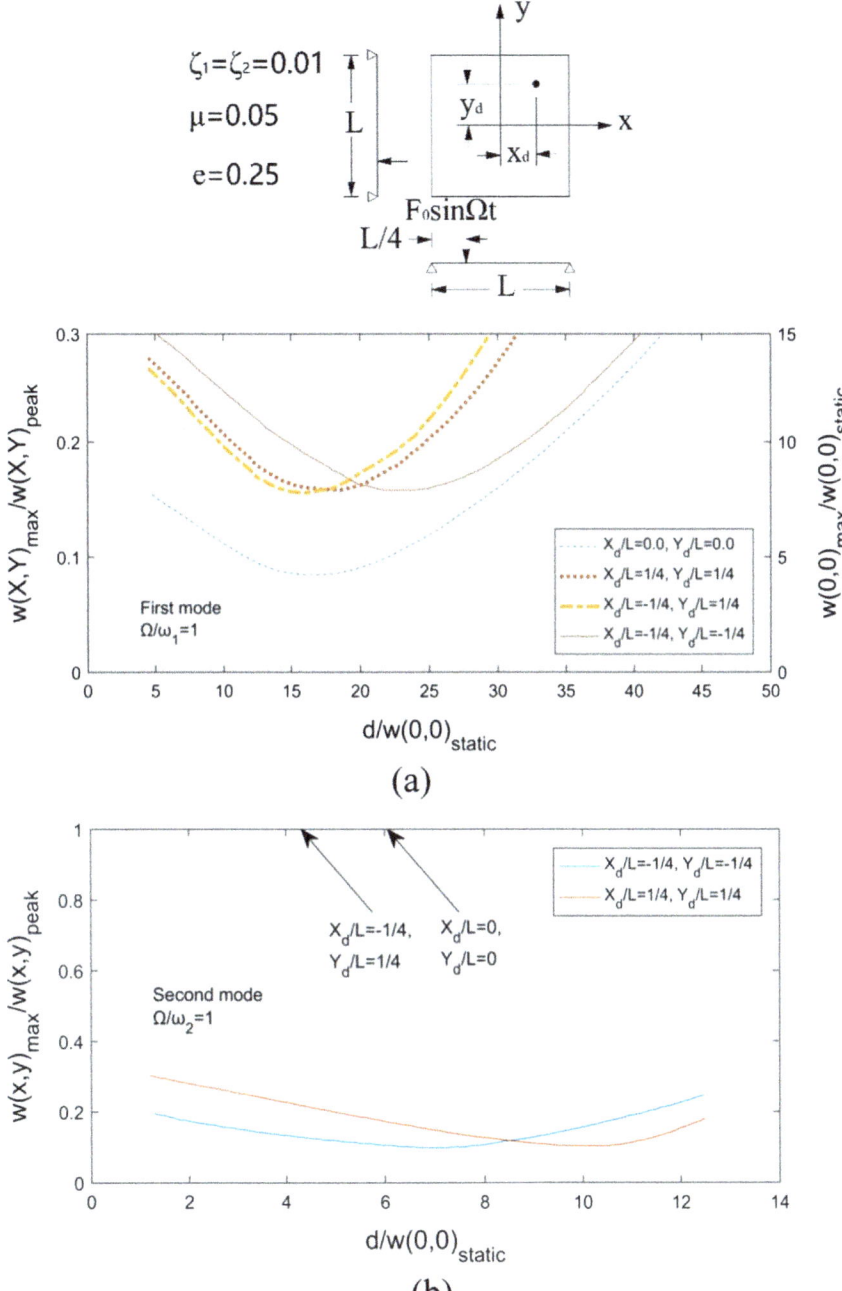

Fig. 3.6 Effect of damper location and mode shapes: discrete force excitation of a simply-supported plate: **a** First mode; and **b** Second mode

in which α and β are constant coefficients; M is a function that specifies the mass distribution of the beam; $f(x, t)$ is a sinusoidally varying load equal to $f(x)\sin\Omega t$ with

$$f(x) = p(x) + \sum_{k=1}^{s} F_k \delta(x - x_k);$$

$p(x)$ is distributed excitation force; F_k is concentrated excitation force acting at point $x = x_k$; and $\delta(x - x_k)$ is spatial Dirac's delta function.

Let $\varphi_i(x)$ be the ith eigenfunction associated with the homogeneous equation of the undamped system and assume that the eigenfunctions satisfy the orthogonality condition:

$$\int_0^l M(x)\,\varphi_i(x)\,\varphi_j(x)\,dx = \delta_{ij}M_i \tag{3.36}$$

$$\int_0^l \varphi_i(x)\,L[\varphi_j(x)]\,dx = \delta_{ij}K_i \tag{3.37}$$

in which δ_{ij} is the Kronecker delta.

Using a Galerkin-type approach, consider an approximate solution of the form:

$$w_n(x, t) = \sum_{j=1}^{n} \varphi_j(x)q_j(t) \tag{3.38}$$

which leads to (3.34):

$$[m]\,\ddot{q}(t) + [c]\,\dot{q}(t) + [k]\,q(t) = Q_{ex}(t) \tag{3.39}$$

in which $q(t) = \mathrm{col}\{q_1(t), q_2(t), \ldots, q_n(t)\}$

$$\left.\begin{array}{l} m_{rj} \equiv \int_0^l \varphi_r(x)M(x)\varphi_j(x)dx \\ k_{rj} \equiv \int_0^l \varphi_r(x)L[\varphi_j(x)]dx; \quad r, j = 1, 2, \ldots, n \\ Q_{ex,r}(t) \equiv \int_0^l \varphi_r(x)f(x,t)dx \end{array}\right\} \tag{3.40}$$

As a result of using the actual eigenfunctions of the system rather than just some comparison functions for shape functions in Eq. (3.38), matrices $[m]$, $[c]$, and $[k]$ will be diagonal:

$$[M]\,\ddot{q} + [C]\,\dot{q} + [K]\,q = Q_{ex}(t) \tag{3.41}$$

in which the diagonal matrices $[M]$, $[C]$, and $[K]$ correspond to the generalized mass, damping, and stiffness matrices, respectively.

Introducing the following notations:

$$\left.\begin{array}{l} x_i \equiv \frac{2i-1}{2n}l, \; \varphi_{ij} = \varphi_j(x_i), \; w_i(t) = w_n(x_i, t) \\ x = \mathrm{col}.\{x_1, x_2, \ldots, x_n\}; \quad i, j = 1, 2, \ldots, n \\ w(x, t) = \mathrm{col}.\{w_1(t), w_2(t), \ldots, w_n(t)\} \end{array}\right\} \tag{3.42a}$$

then from Eq. (3.38)

$$w(x, t) = [\varphi]\, q(t) \tag{3.42b}$$

in which square matrix $[\varphi]$ is the modal matrix in which the n components of the jth eigenvector are found by evaluating the jth eigenfunction, $\varphi_j(x)$, at the n discrete spatial points, x_i, $i = 1, \ldots, n$.

Consider the steady-state motion of the system shown in Fig. 3.7, and shift the origin of the time axis to coincide with the time of occurrence of an impact at $\Omega t = \alpha_0$. In the new time scale, the ith equation of system, Eq. (3.41), is

$$M_i \ddot{q}_i + C_i \dot{q}_i + K_i q_i = Q_{ex,i} = \left[\int_0^l \varphi_i(x) f(x) dx \right] \sin(\Omega t + \alpha_0) \tag{3.43}$$

Let the initial displacement and velocity at $t = 0_+$, be

$$w(x, 0) \equiv w_0; \quad \dot{w}(x, 0_+) \equiv \dot{w}_a \tag{3.44}$$

Then in steady-state motion with two symmetric impacts per cycle the following conditions must be satisfied:

$$y(0) \equiv y_0 = z(0) - w(x_j, t) = \pm\frac{d}{2} \tag{3.45}$$

$$\dot{z}(0)_+ = -\frac{2\Omega}{\pi}\left[w(x_j, 0) + y_0\right] \tag{3.46}$$

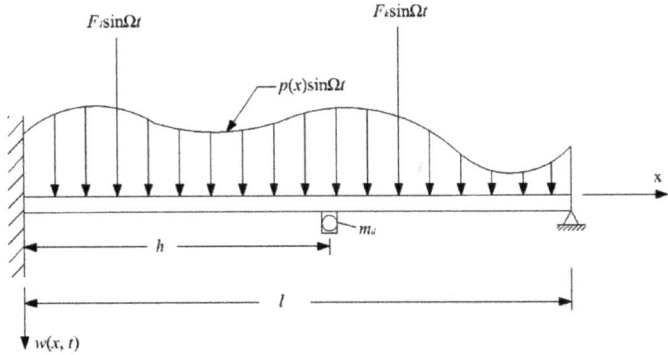

Fig. 3.7 Model of system

$$w(x, t)|_{\Omega t=\pi} = -w(x, 0) \equiv -w_0 \tag{3.47}$$

$$\dot{w}(x, t)|_{\Omega t=\pi_-} = -\dot{w}(x, 0_-) \equiv -\dot{w}_b \tag{3.48}$$

in which d is clearance where damper is free to oscillate; $y(t)$ is relative displacement of damper with respect to beam at $x = h$; $z(t)$ is absolute displacement of damper.

Treating the impact as a discontinuous process governed by the momentum equation, and by using the definition of the coefficient of restitution, the velocity discontinuity introduced at point x_j due to an impact can be expressed as

$$\frac{\partial w}{\partial t}(x_j, 0_-) = \left[1 + \left(\frac{G_2}{G_1} - 1\right)\delta(x - x_j)\right]\frac{\partial w}{\partial t}(x_j, 0_+) \tag{3.49}$$

Following steps similar to those in reference [9], the solution of Eq. (3.43) in conjunction with Eqs. (3.42b)–(3.49) leads to

$$\alpha_0 = \tan^{-1}\frac{h_1 h_3 \pm h_2\sqrt{h_1^2 + h_2^2 - h_3^2}}{h_2 h_3 \mp h_1\sqrt{h_1^2 + h_2^2 - h_3^2}} \tag{3.50}$$

Once α_0, is determined from Eq. (3.50), the rest of the unknowns can be found by back substitution

Application:

(1) Uniform Bernoulli-Euler beams.

For a Bernoulli-Euler beam, the previously mentioned operators, L and M, are:

$$\left.\begin{array}{l} L = \frac{d^2}{dx^2}\left[EI\frac{d^2}{dx^2}\right]; \quad p = 2 \\ M = \mu_0 \end{array}\right\} \tag{3.51}$$

in which EI is beam stiffness, $\mu_0 =$ uniform mass/unit length of beam.

The boundary conditions and the corresponding eigenvalues and eigenfunctions for the standard cases are well known and given, for example, in Ref. [10].

(2) Base excitation of beams.

If a beam is excited through support motion rather than direct applied load, then letting $s_0(t)$ be the rigid body translation and $w(x, t)$ be the elastic deformation measured, relative to rigid body motion, the governing equation of motion, Eq. (3.34), becomes

$$L[w(x, t)] + \frac{\partial}{\partial t}C[w(x, t)] + M(x)\frac{\partial^2 w(x, t)}{\partial t^2} = -M(x)\frac{\partial^2 s_0(t)}{\partial t^2} \tag{3.52}$$

Therefore, base excitation and force excitation are equivalent if the applied load, $f(x, t)$, is replaced by $-M(x)\ddot{s}_0(t)$.

(3) Base excitation of uniform cantilever beam.

Figure 3.8 shows the response of a typical uniform cantilever beam provided with an impact damper and subjected to sinusoidal base excitation. The approximate solution employed 10 modes. The left-hand side ordinate of Fig. 3.8 is the ratio of the maximum displacement at point x, $w(x)_{max}$, for the beam using the damper divided by the peak displacement, $w(x)_p$, of the same point with no damper being used. This ratio is a measure of the effectiveness of the damper (without the damper the solution curve is a straight line at $w(x)_{max}/w(x)_p = 1$). The lower abscissa in Fig. 3.8 is the clearance ratio, d/s_0, in which s_0 is the amplitude of the sinusoidal base motion.

If for a given set of parameters the radical in Eq. (3.50) is not negative, there are two possible solutions to this equation labeled α_0^+ and α_0^-, which correspond to two distinct steady-state solutions. The two solutions will coalesce at $d = 0$ and $d = d^*$ such that the radical in Eq. (3.50) becomes zero. Note that h_3 is directly proportional to d. The nonexistence of solutions beyond $d = d^*$ is due to the fact that for these conditions the auxiliary mass does not have enough energy to travel from one restraining stop to the other twice per cycle in steady-state motion.

Experiments with a mechanical model were conducted to check the theoretical results and to determine which branch of the solution, if any, was asymptotically stable. As indicated by the experimental results shown in Fig. 3.8, the solution curve associated with α_0^+ (which, in general, leads to lower response amplitude than solution curve α_0^-) was found to be stable throughout most of its range.

In the case of clearance ratios that were too small, either excessive impacts per cycle (more than two) or unsymmetric impacts (at unequal time intervals) occurred. However, even for these low ratios the measured response of the primary system was approximately the same as the steady-state solution based on the assumption of two symmetric impacts per cycle.

For clearance ratios that are much greater than the peak-to-peak displacement of the point $x = h$ to which the damper is attached (i.e., $2w(h)_p$) steady-state motion with two impacts per cycle could not be sustained. This is shown in Fig. 3.8 (using the upper abscissa) as the initiation of unstable motion. However, even in this case, the range of unsteady motion was still considerably less than the corresponding response without the damper.

It should be noted that the left-hand side ordinate in Fig. 3.8 applies to any point $0 \le x \le l$, i.e., all points along the beam undergo the same percentage reduction in amplitude.

Comparison of the experimental and analytical results for a typical case is also shown in Fig. 3.9. It is clear that the experimental measurements completely corroborate the theoretical predictions. The validity of the analytical results were further compared to, and were found to be in excellent agreement with, published analytical and experimental results pertaining to single-degree-of-freedom systems provided with impact dampers [11].

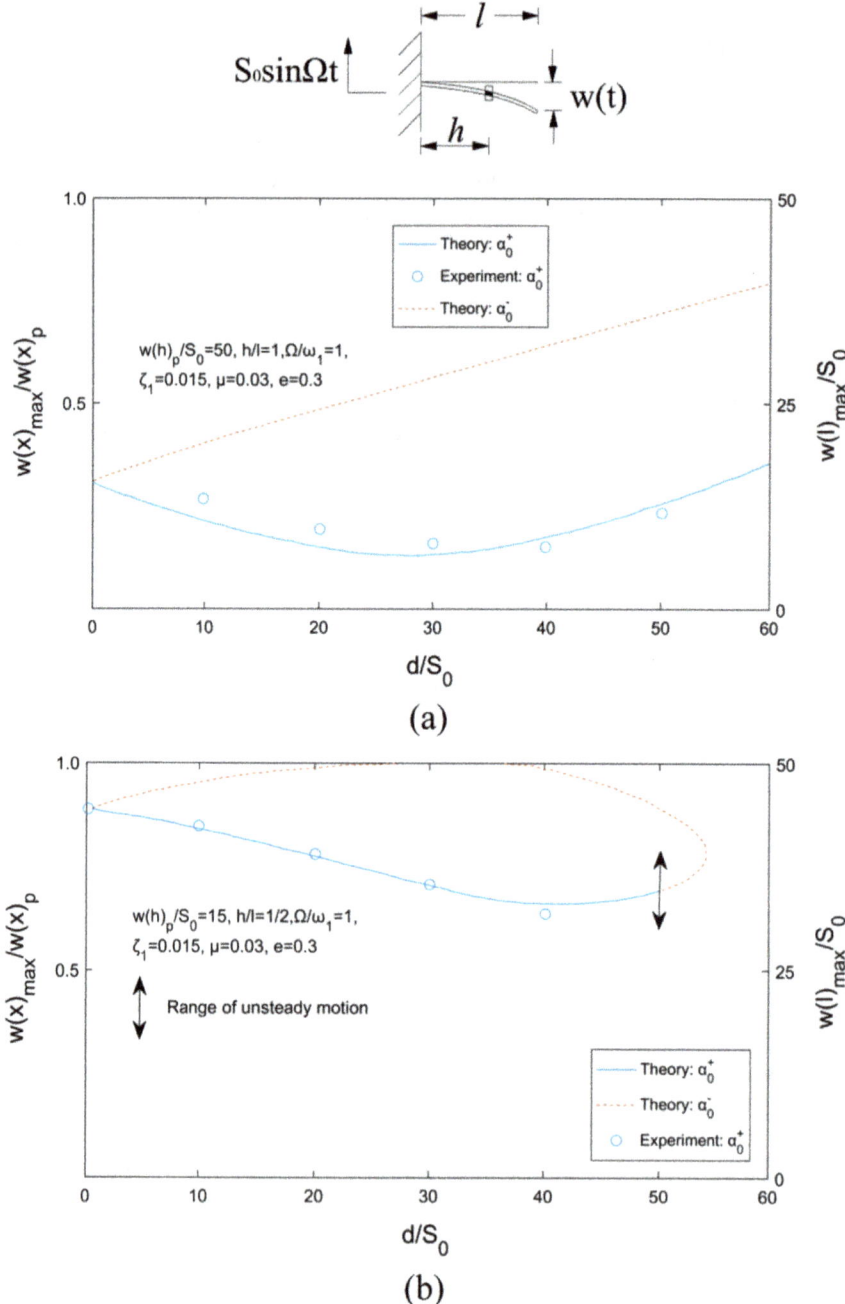

Fig. 3.8 Comparison of theoretical and experimental results: base-excitation of cantilever beam: **a** $h/l = 1$; and **b** $h/l = 1/2$

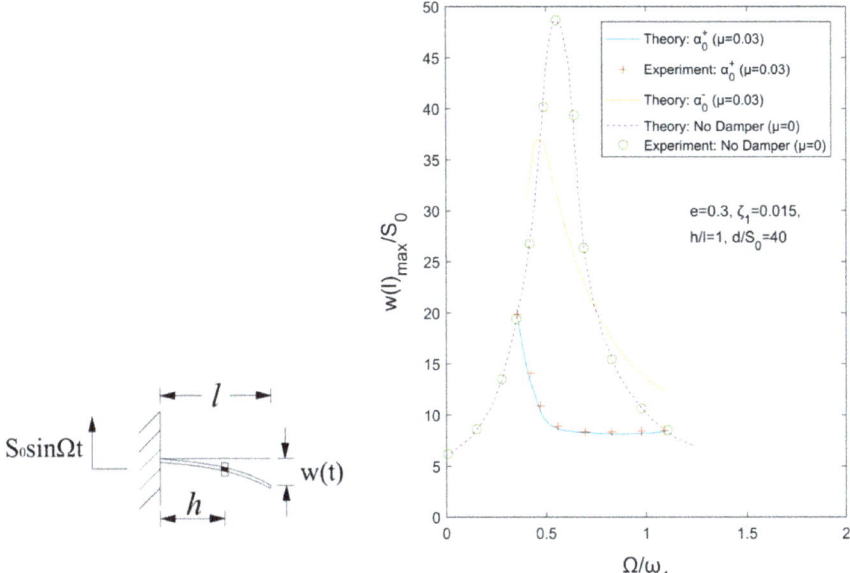

Fig. 3.9 Frequency-response of base-excited cantilever beam

Figure 3.10 shows the influence of viscous damping on the response of a uniformly loaded cantilever beam provided with a damper.

(4) Discrete force excitation of simply-supported beam.

The response of a simply-supported (s-s) beam subjected to a discrete force is shown in Figs. 3.11 and 3.12 where the effects of mode shape, force location, and damper location are exhibited.

(5) Effects of various system parameters.

A detailed investigation of the influence of the independent parameters on the response of the system was undertaken, and the results can be summarized as follows.

Clearance ratio $d/w(h)_{st}$. For a given set of parameters there is an optimum, d, that will lead to the maximum attenuation of the response of the primary system. The sensitivity of the system response to d ('tuning' effect) is increased as the energy dissipation of the system is increased. In general, the stability boundaries enclose the optimum, d.

Coefficient of restitution e. Since e measures the elasticity of the collisions between m_d (damper mass), and the stops, the more elastic the impact, the more effective will the damper be. However, a low e is more suitable if the parameters (particularly the excitation frequency) are varied considerably.

Mass ratio $\mu = m_d/(\mu_0 l)$. The system under analysis is effective even with mass ratios on the order of 1%. Note from Fig. 3.13a that even with a mass ratio $\mu = 0.01$ the response amplitude can be reduced by as much as 80%. An increased mass ratio

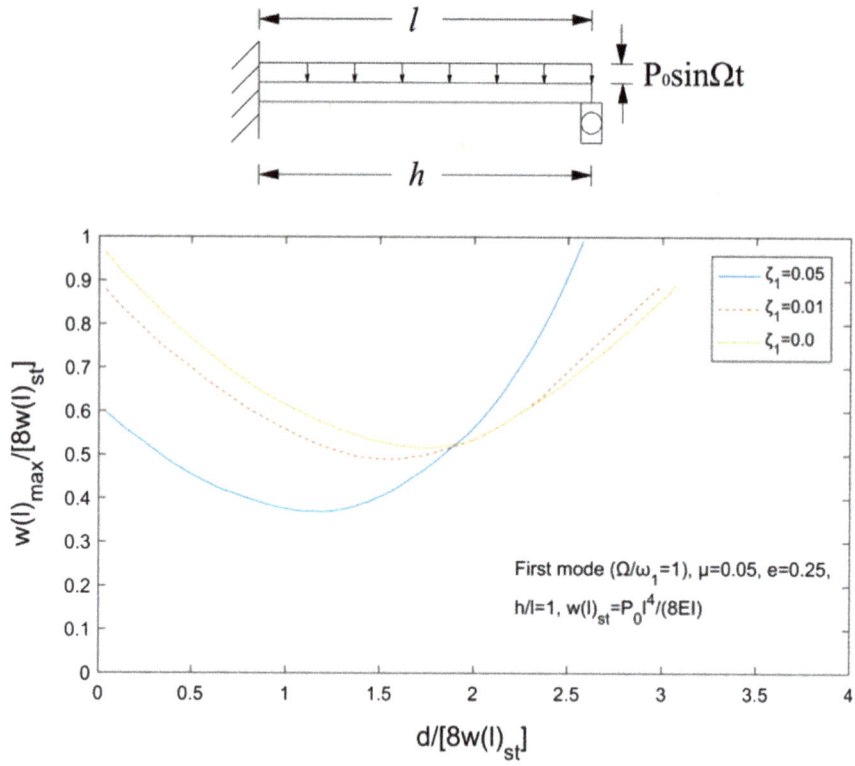

Fig. 3.10 Effects of viscous damping: first mode of cantilever beam

simplifies the optimization procedure, especially in regard to the "flatness" of the frequency response curve over a wide frequency band.

The impact damper employs both momentum transfer and mechanical energy dissipation during impact to attenuate the response of the primary system. Since the lower abscissa, $\mu d/w(h)_{st}$, in Fig. 3.13 is proportional to the momentum of the damper, it can be seen from Fig. 3.13a, b, and c that, in lightly damped primary systems, approximately the same results can be obtained with two different μ values if the following relationship is maintained: $\mu_1 d_1 = \mu_2 d_2$, i.e., if the product of the mass ratio and the clearance ratio are kept constant.

For cantilever beams operating in the vicinity of their first mode, similar to the one being considered herein, Fig. 3.13 shows that the optimum clearance ratio in regard to vibration attenuation satisfies the condition, $\mu d/w(h)_{st} \approx 0.7$, and results in a reduced amplitude which satisfies the relationship, $\mu w(x)_{max}/w(x)_p \approx 2 \times 10^{-3}$, $(0 \leq x \leq l)$ for relatively high values of e.

Note that the effectiveness of the damper per unit μ decreases as μ increases.

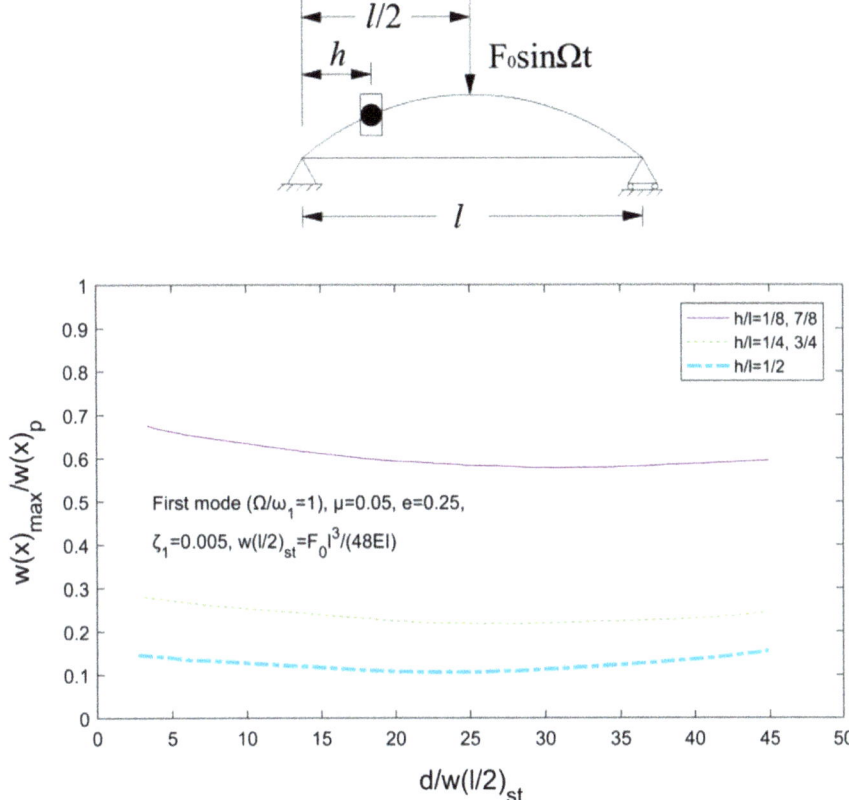

Fig. 3.11   Effects of damper location: discrete force excitation of simply-supported beam

(6)   Natural mode shapes.

The response of a simply-supported beam vibrating in each of its first two modes is given in Figs. 3.11 and 3.12. It can be seen that, with a given damper mass, the maximum percentage reduction in the response of the structure is achieved in the first mode. The maximum reduction decreases with mode number from $\approx 90\%$ in the first, to $\approx 70\%$ in the second. One of the reasons for this reduction in the efficiency of the damper is that ξ_i, the ratio of critical damping in the ith mode, increases with mode number from $\xi_1 \approx 0.005$ to $\xi_2 \approx 0.02$. It appears that the main factor that influences the efficiency of the damper in different modes is the magnitude of the displacement of point $x = h$ where the damper is attached.

Damper location h/l
It is seen from Fig. 3.8a that with the damper attached to the free end of the cantilever beam $(h/l = 1)$, the maximum reduction in amplitude is $\approx 80\%$ while if the damper is attached midway $(h/l = 1/2)$ along the beam (Fig. 3.8b) the maximum reduction is

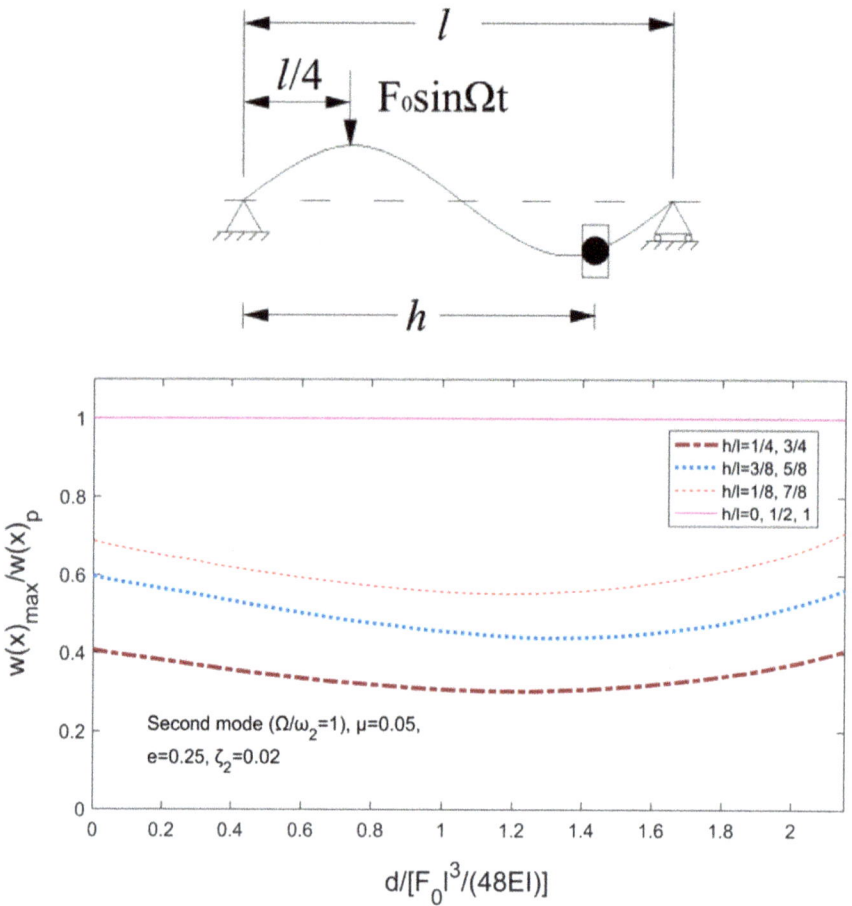

Fig. 3.12 Effects of mode shape: Discrete force excitation of simply-supported beam

$\approx 35\%$. This progressive increase in damper efficiency (regarding amplitude attenuation) with distance from the beam support points is to be expected since, in the first mode, the 'moment arm' of the damper is maximum for a cantilever beam at $h/l = 1$ and for a s-s beam at $h/l = 1/2$.

In modes other than the first, the relative locations of the damper and the nodal points must be considered (see Fig. 3.14). For example, although $h/l = 1/2$ is the optimum location for a s-s beam vibrating in its first mode, this is the worst location if the same beam is oscillating in the second mode, since $x/l = 1/2$ is a nodal point.

Force location k

The response curves for an impact damped beam are virtually independent of the point of application of the exciting force, provided that the results are expressed in terms of $d/w(h)_p$, and $w(x)_{max}/w(x)_p$.

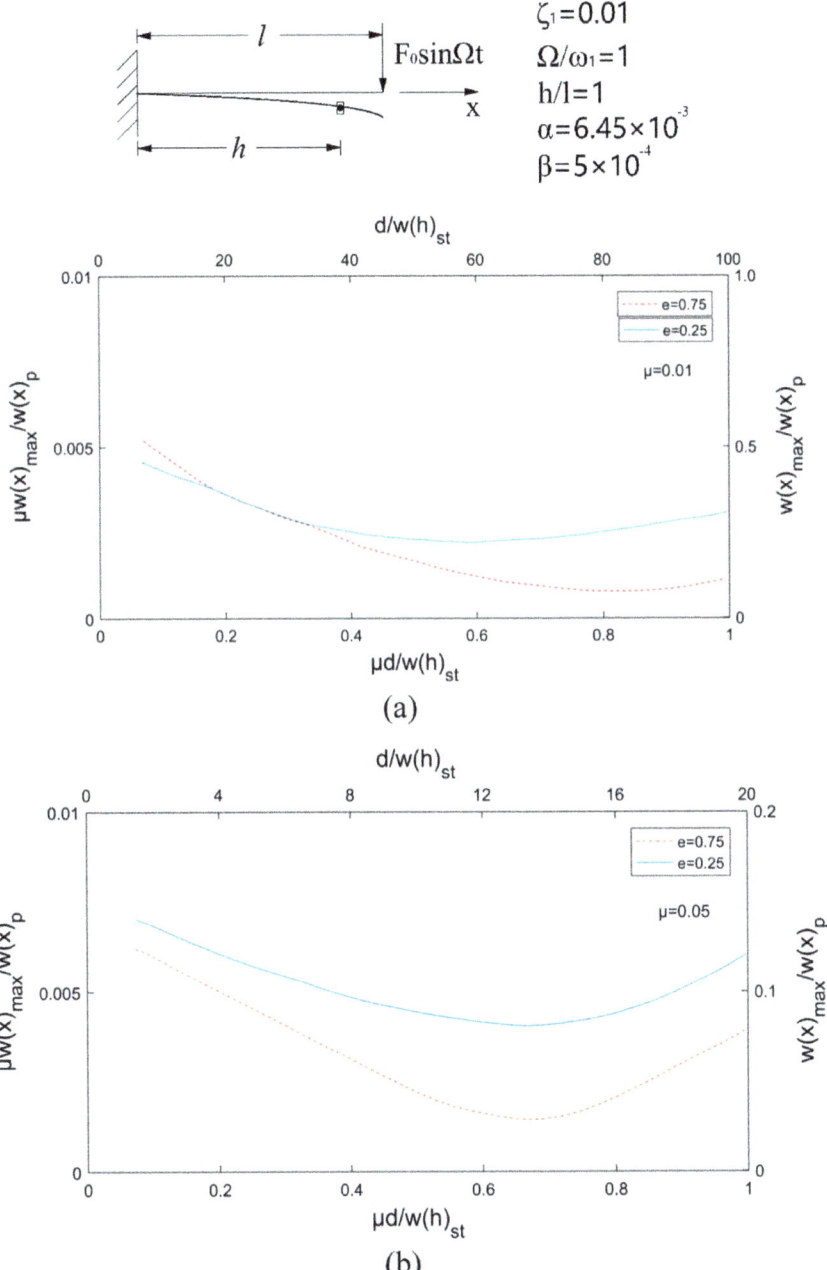

Fig. 3.13 Effects of mass ratio and coefficient of restitution: discrete force excitation of cantilever beam: **a** $\mu = 0.01$; **b** $\mu = 0.05$; and **c** $\mu = 0.10$

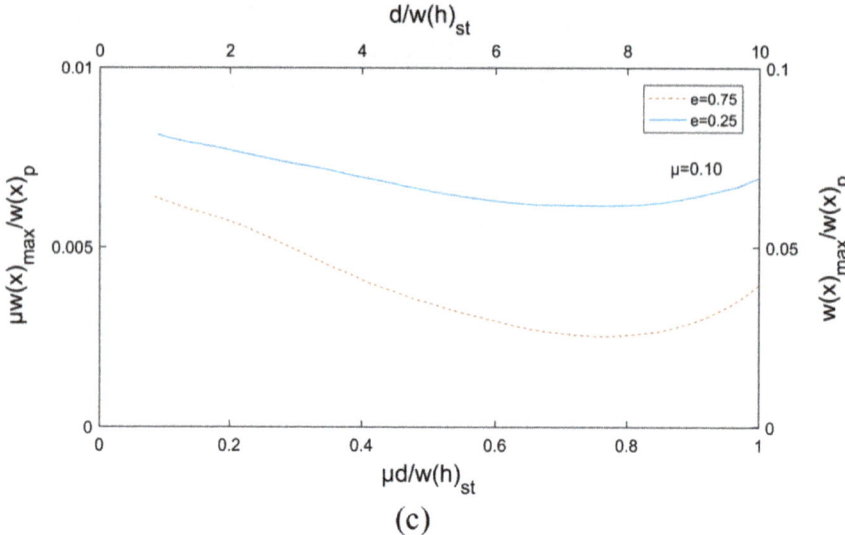

Fig. 3.13 (continued)

Damping in primary system (α, β). Some of the effects of damping in the primary system on the performance of the system are exhibited in Fig. 3.10 where the damping ratio in the first mode of a uniformly loaded cantilever beam is increased from zero to 5% of critical. As evidenced by the curves of Fig. 3.10, an increase in viscous damping is detrimental to the efficiency of the impact damper. With $\xi_1 = 0.01$ the maximum reduction in amplitude is $\left[w(x)_{max}/w(x)_p\right]_{min} \approx 0.075$, while with $\xi_1 = 0.05$ the corresponding value is ≈ 0.30.

In addition to reducing efficiency, an increase in energy dissipation in the form of lower e or higher α and β increases the sensitivity to "tuning". The addition of damping to the structure will tend to extend the stable ranges of operation of the system.

Excitation frequency Ω

An increase in μ and damping or a decrease in e will improve the performance of the damper if the response is to be controlled effectively over a relatively wide frequency band.

The typical response curves in Fig. 3.9 show the effects of vibrating a cantilever beam within a frequency band of $\approx \pm 10\%$ of the first natural frequency. With a mass ratio of 3% the response is held to only $\approx 40\%$ of that gotten without the damper. Although relatively high e values will lead to better reduction at $\Omega/\omega_1 = 1$, they will also result in much worse response (even amplifications, in some cases) at frequency ratios $\Omega/\omega_1 \approx 1/(1 + \mu)$. Consequently, for a broad suppression band, a low e is overall more advantageous, especially since it also reduces the attendant noise and transmitted impact force, both of which increase with e.

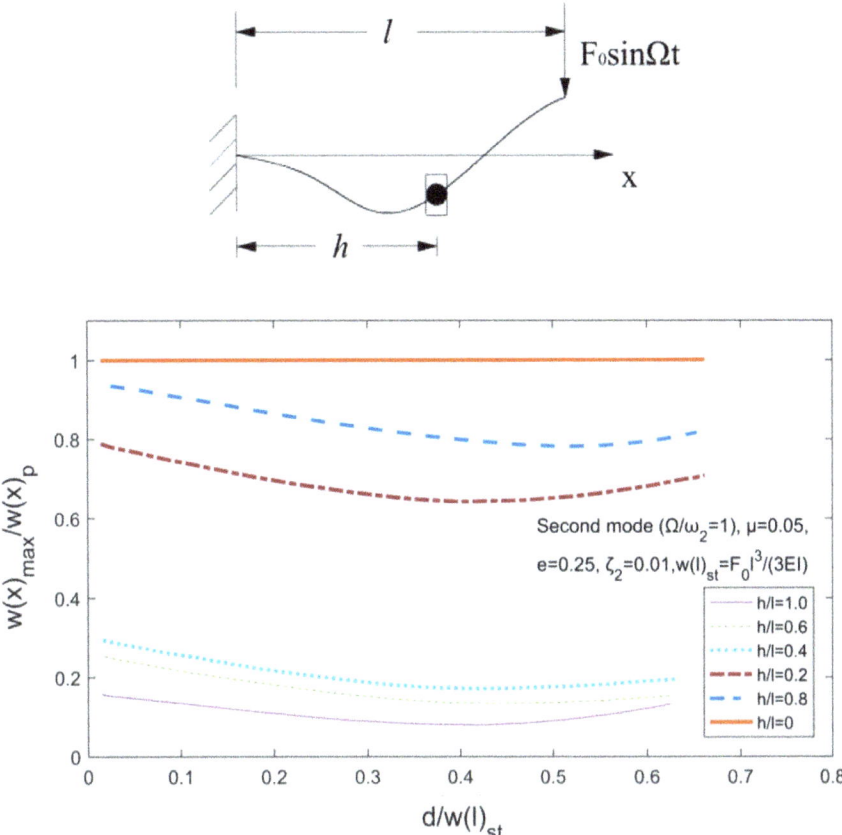

Fig. 3.14 Effect of damper position: second mode of cantilever beam (discrete force applied at free end)

If a damper is designed to operate at $\Omega/\omega_1 \approx 1$, then a frequency increase to $\Omega/\omega_2 \approx 1$, for example, will probably result in an unstable damper motion, since the clearance ratio $d/w(h)_p$ changes considerably from mode to mode. However, from the practical application point of view, this is not a major drawback since, for large clearance ratios, few or no impacts will occur and the response of the primary system will not be significantly changed. However, since the amplification ratios of a beam (without the damper) goes down drastically with mode number, it is clear that response attenuation in the higher modes is relatively unimportant.

(7)   Nonuniform beams.

The only complication that nonuniform beams add to the analysis is that, in general, the actual mode shapes are not known or not easy to determine. If the shape functions, $\varphi_i(x)$, used in Eq. (3.38) are just some comparison functions instead of being the actual eigenfunctions, the system Eq. (3.39) will be coupled since $[m]$, $[c]$, and $[k]$ are not diagonal. Thus the extra step of uncoupling system Eq. (3.39) and transforming it to system Eq. (3.41) by using standard procedures [10] is needed. The rest of the analysis will still apply.

3.1.1.3 Steady-State Motion of a Series-Type Multidegree-of-Freedom System

The following part presents the exact solution for the steady-state motion of a series-type multidegree-of-freedom system equipped with an impact damper and subjected to a sinusoidal force input. The detail about the model is illustrated in Sect. 5.1.1.

Consider the 10-degree-of-freedom lumped parameter system shown in Fig. 3.15c whose natural frequencies, mode shapes, and damping characteristics approximate those obtained experimentally from full-scale dynamic tests of a modern 10-story building [12].

To investigate the effects of various system parameters, the models shown in Fig. 3.15a and b were also studied. The hypothetical systems in Fig. 3.15a and b consist of the first 4 and first 7 stories, respectively, of the system in Fig. 3.15c. Figure 3.15d shows the first three mode shapes and the corresponding frequencies for the lumped parameter representation of the 10-story building in Fig. 3.15c.

Figure 3.16 shows a typical solution curve for the 7-story building in Fig. 3.15b. The left-hand-side (LHS) ordinate in Fig. 3.16 is the ratio of the maximum displacement of the ith floor, x_{maxi}; for the structure using the damper divided by the peak displacement, x_{pi}, of the same floor with no damper being used. This ratio is, therefore, a measure of the effectiveness of the damper (without a damper the solution curve is a straight line at $x_{maxi}/x_{pi} = 1$). The abscissa in Fig. 3.16 is the clearance ratio d/x_{stn}, where x_{stn} is the static deflection of the top floor with the excitation force and the damper both applied at the top.

For a given set of parameters there are two possible solutions to Eq. (3.70) labeled α_0^+ and α_0^-, which correspond to two distinct steady-state solutions. Experiments with an electronic analog computer were conducted to check the theoretical results and to determine which branch of the solution, if any, was asymptotically stable. As indicated by the experimental results shown in Fig. 3.16, the solution curve associated with α_0^+ (which, in general, leads to lower response amplitude than solution curve α_0^-) was found to be stable throughout most of its range.

The validity of the analytical results were further compared to, and were found to be in excellent agreement with, published analytical and experimental results pertaining to single-degree-of-freedom systems provided with impact dampers [11].

A detailed investigation of the independent parameters on the response of the system was undertaken, and the results can be summarized as follows:

Mass ratio μ_T (mass ratio of damper to total mass of structure)
The system under discussion is effective even with mass ratios on the order of 1%. Note from Fig. 3.16 that even with a mass ratio $\mu_T \approx 0.005$, the response amplitude can be reduced by as much as 80%.

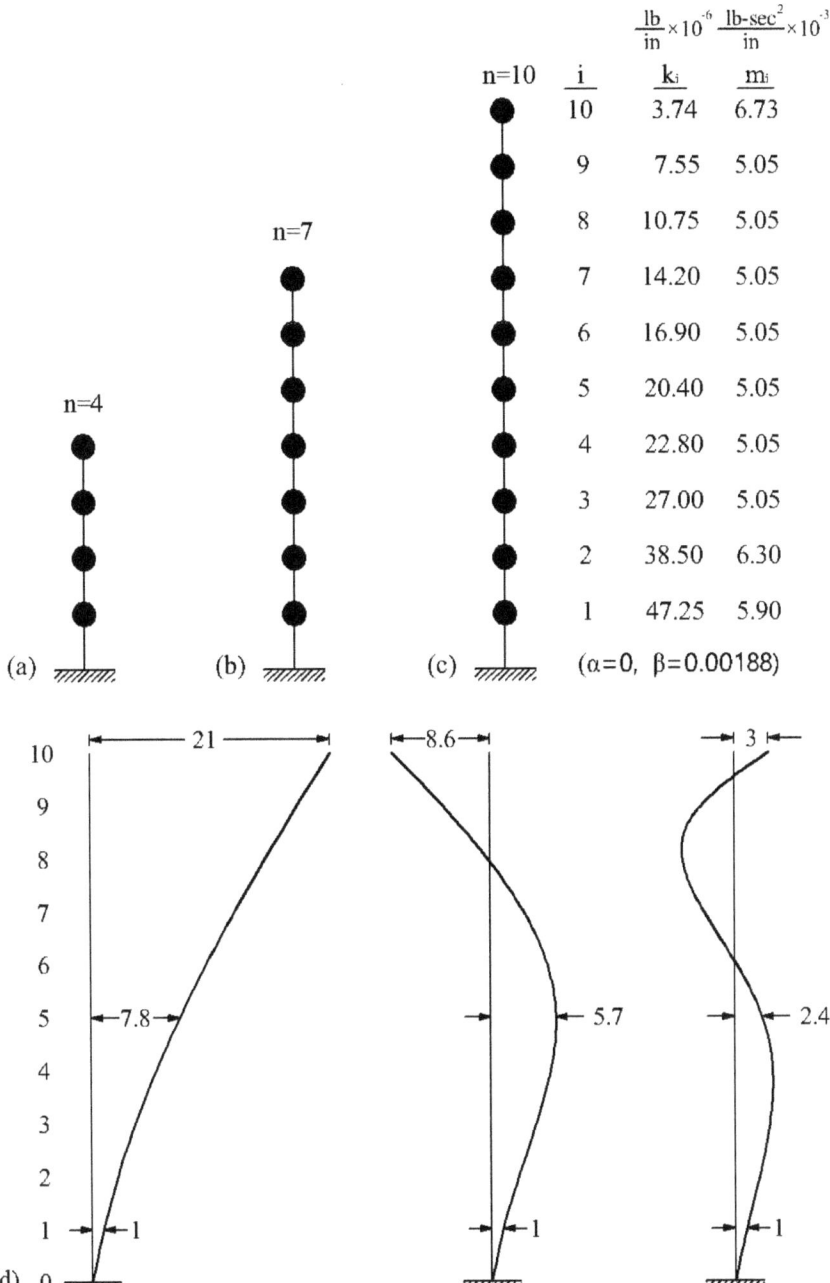

n=10	i	k_i	m_i
		$\frac{lb}{in} \times 10^{-6}$	$\frac{lb\text{-}sec^2}{in} \times 10^{-3}$
	10	3.74	6.73
	9	7.55	5.05
	8	10.75	5.05
	7	14.20	5.05
	6	16.90	5.05
	5	20.40	5.05
	4	22.80	5.05
	3	27.00	5.05
	2	38.50	6.30
	1	47.25	5.90

$(\alpha=0, \quad \beta=0.00188)$

Fig. 3.15 Example structure: **a** Four-degree-of-freedom system; **b** Seven-degree-of-freedom system; **c** 10-degree-of-freedom system; and **d** First three modes of 10-degree-of-freedom system

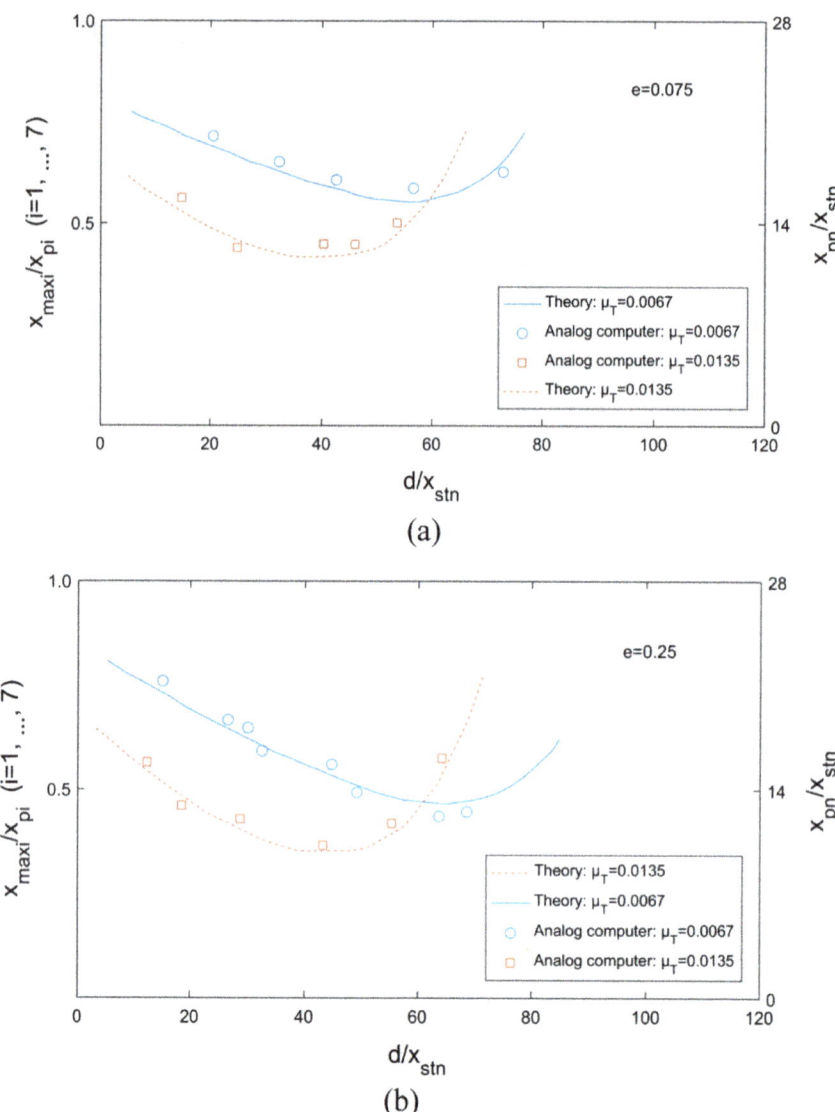

Fig. 3.16 Comparison of analytical and experimental results, $j = k = n = 7$, $\Omega/\omega_1 = 1$, $\alpha = 0$, $\beta = 0.0019$: **a** $e = 0.075$; **b** $e = 0.25$; and **c** $e = 0.75$

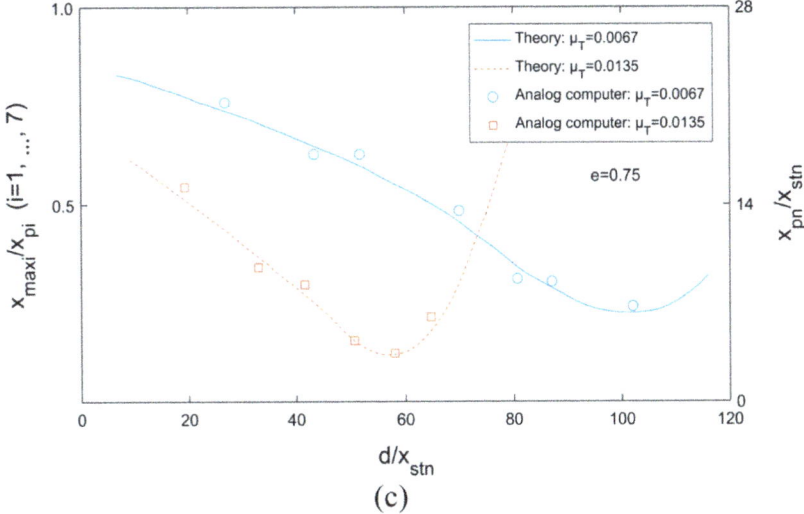

Fig. 3.16 (continued)

For "tall" structures operating in the vicinity of their first mode, similar to the one under discussion, the optimum clearance ratio (in regard to vibration attenuation) satisfies the condition $\mu_T d / x_{stn} \approx 0.6$ and results in a reduced amplitude which satisfies the relationship $\mu_T x_{maxi} / x_{pi} \approx 0.001$, $(i = 1, 2, \ldots, n)$ for relatively high values of e.

Natural mode shapes
The response of the 10-story building under discussion vibrating in each of its first three modes is given in Fig. 3.17. It can be seen that, with a given damper mass, the maximum percentage reduction in the response of the structure is achieved in the first mode. The maximum reduction decreases with mode number from 90% in the first, to 85% in the second, and 60% in the third. One of the reasons for this reduction in the efficiency of the damper is that ξ_i, the ratio of critical damping in the ith mode, increases with mode number from $\xi_1 \approx 0.01$ to $\xi_2 \approx 0.02$ and $\xi_3 \approx 0.03$. Figure 3.17a, b, and c also show the corresponding response of an equivalent single-degree-of-freedom (SDOF) system whose ratio of critical damping is the same as the 10-story building in a given mode. It is clear from Fig. 3.17 that factors in addition to damping are involved in modifying the performance of the damper. Since the "moment arm" of the damper with respect to the base of the building is not affected by the mode shape, it appears that the main factor that influences the efficiency of the damper is the magnitude of the displacement of m_j to which the damper is attached.

Damper location i. Figure 3.18 illustrates some of the possible damper locations in the 10-story ($n = 10$) building that is vibrating in its first mode.

With the force imposed on mass m_5 (i.e., $k = 5$), then, for the parameters in Fig. 3.19, the maximum reduction in amplitude will be about 2% if the damper is

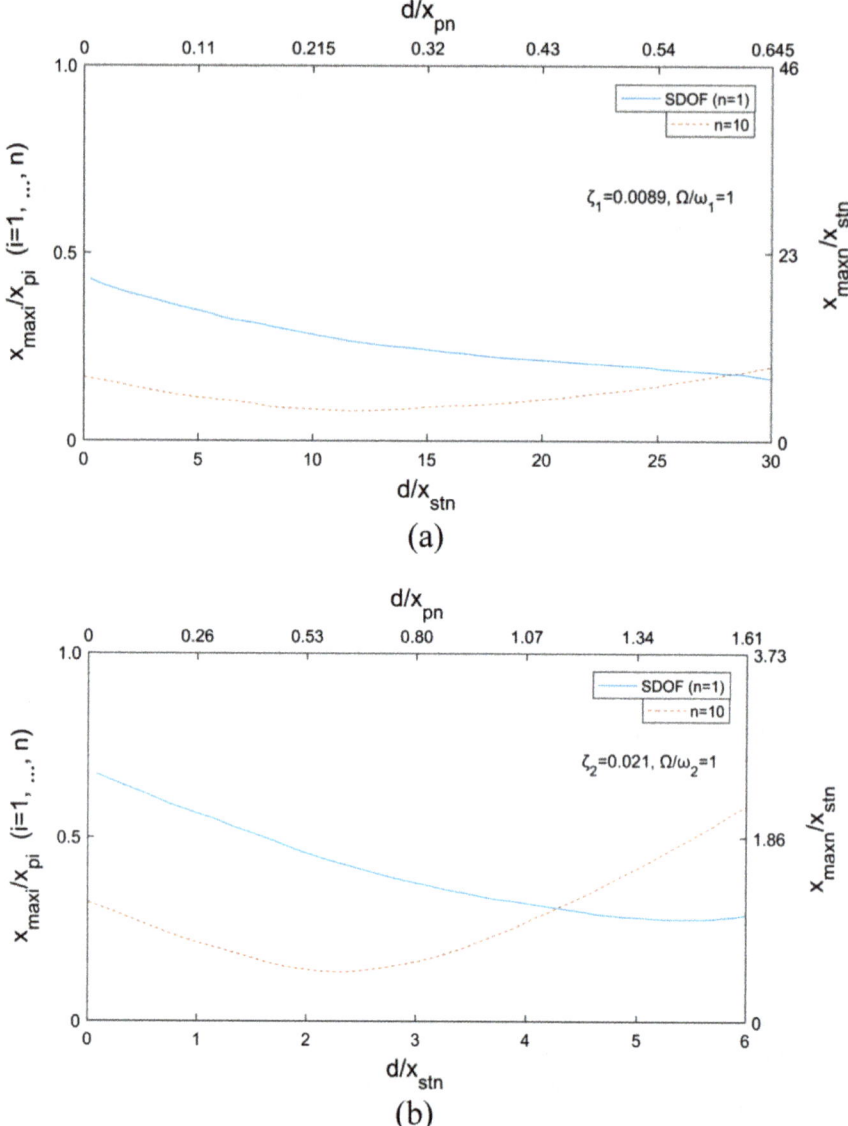

Fig. 3.17 Effects of mode shape, $j = k = n = 10$, $\mu_T = 0.05$, $e = 0.25$, $\alpha = 0$, $\beta = 0.0019$: **a** $\Omega/(\omega_1 = 1)$; **b** $\Omega/(\omega_2 = 1)$; and **c** $\Omega/(\omega_3 = 1)$

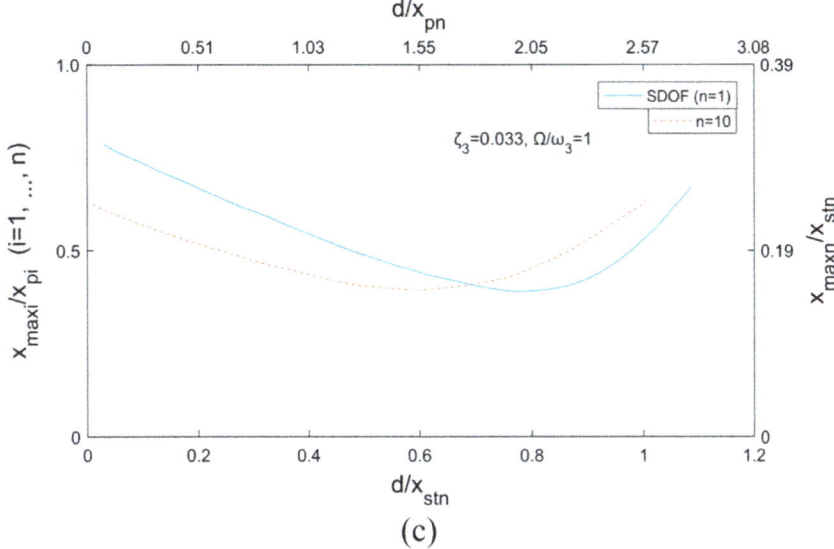

Fig. 3.17 (continued)

attached to m_1, Fig. 3.19a, about 50% if the damper is at m_5, Fig. 3.19b, and more than 90% if the damper is placed at the top floor, Fig. 3.19c. This progressive increase in efficiency with the damper "elevation" above the ground floor is to be expected since the moment arm increases with increasing j to a maximum at the top floor (i.e., $j = n$).

For modes other than the first, increase in moment arm must be modified to account for the relative locations of the damper and the nodal points.

Force location k

Although the application of an exciting force of a given amplitude at different points along the structure will affect the magnitude of the response of the structure, it was found that the response curves for an impact damped structure are virtually independent of the point of application of the exciting force, providing the results are expressed in terms of d/x_{pi}, and x_{maxi}/x_{pi}.

Telescopic effect

Figure 3.20 shows the relation between the amplification ratio x_{maxi}/x_{st_i}, $(i = 1, \ldots, n)$ and the clearance ratio d/x_{st_n} for the three buildings shown in Fig. 3.15, where each one is vibrating in its first mode and with both force and damper acting on the top floor so that $j = k = n$.

It is clear that if the damper mass ratio μ_T is kept constant, the effectiveness of the damper is proportional to the height of the building. This is due to the fact that as the height of the building is increased, the moment arm of the impact damper is simultaneously increased to amplify the dampening effects. The deviation of the

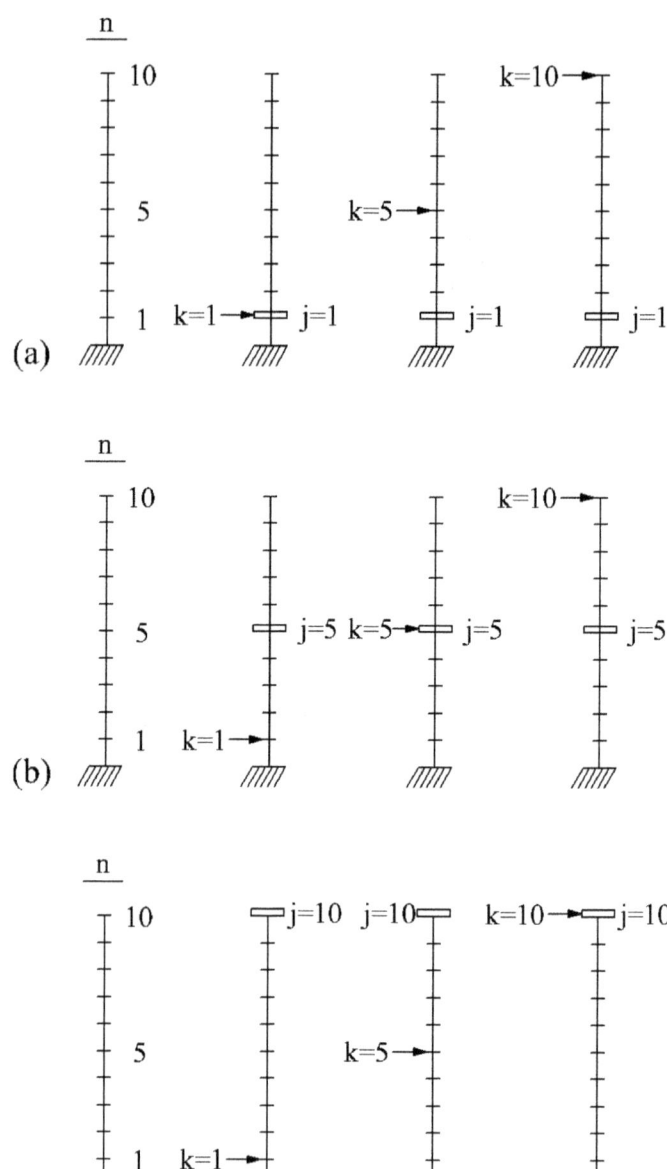

Fig. 3.18 Different combinations of force and damper location for 10-story building: **a** $j = 1$; **b** $j = 5$; and **c** $j = 10$

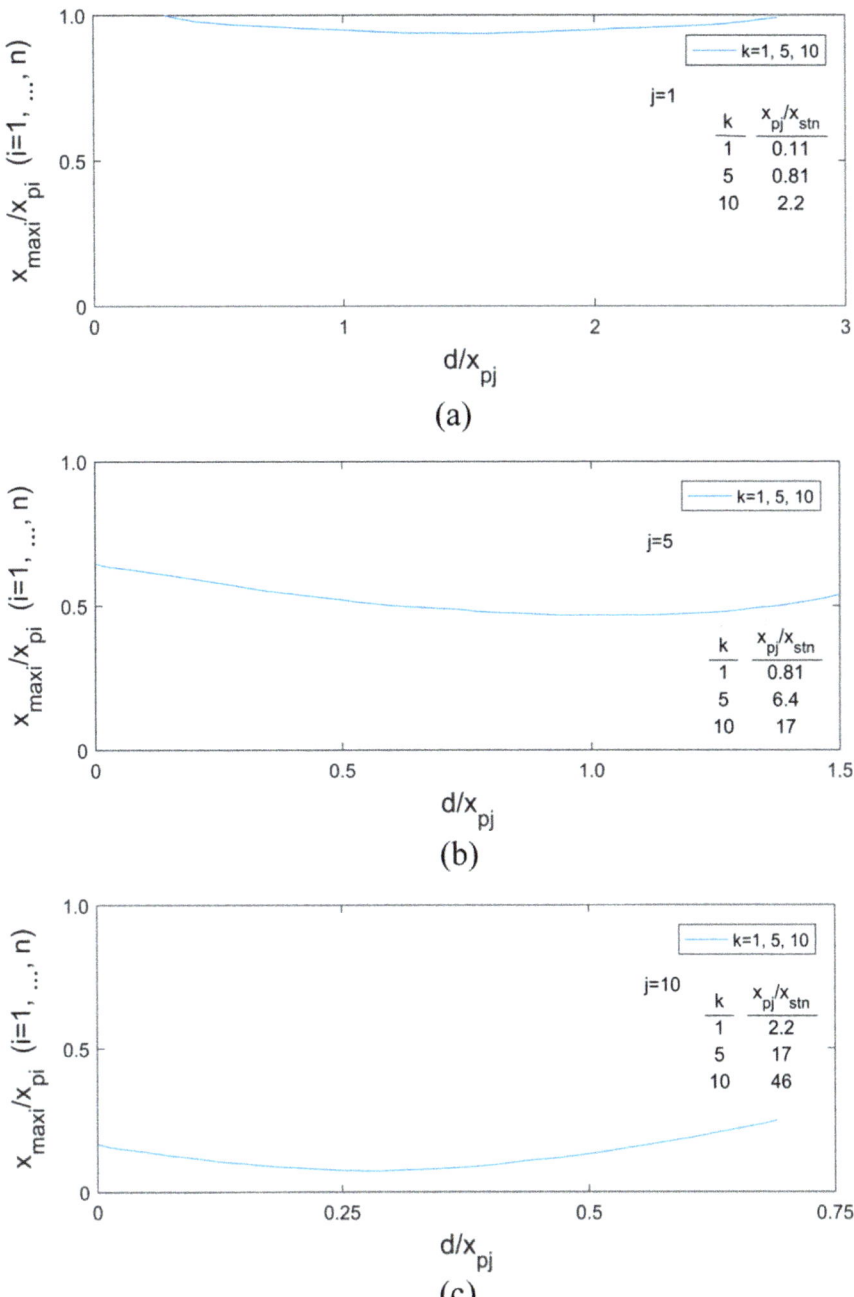

Fig. 3.19 Effects of force and damper location, $n = 10$, $\Omega/(\omega_1 = 1)$, $\mu_T = 0.05$, $e = 0.25$, $\alpha = 0$, $\beta = 0.0019$: **a** $j = 1$; **b** $j = 5$; and **c** $j = 10$

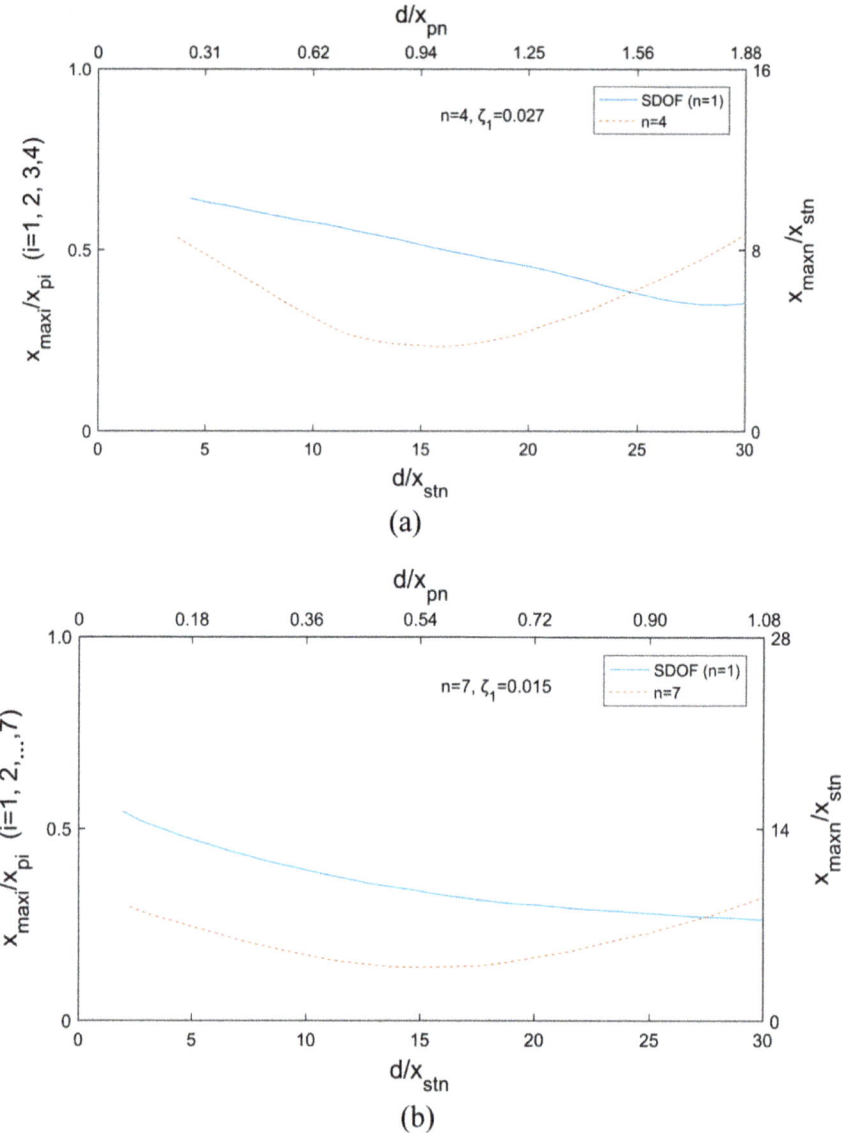

Fig. 3.20 'Telescoping' effect, $\Omega/(\omega_1 = 1)$, $\mu_T = 0.05$, $e = 0.25$, $\alpha = 0$, $\beta = 0.0019$, $j = k = n$: **a** $n = 4$; **b** $n = 7$; and **c** $n = 10$

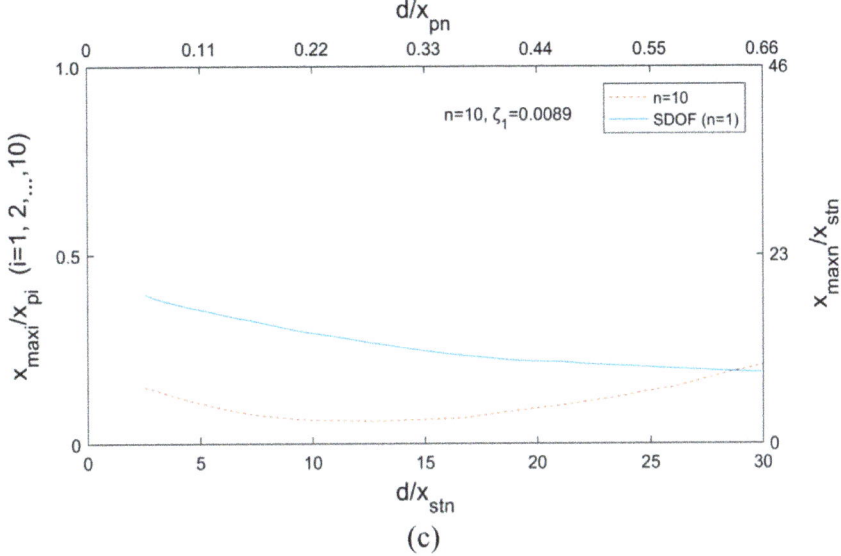

Fig. 3.20 (continued)

results from the corresponding ones for an equivalent SDOF system increases as the number of stories increases.

***Damping in the primary system* (α, β)**
An increase in the viscous damping of the primary system is detrimental to the efficiency of the impact damper.

***Excitation frequency* Ω**
If the structure is excited at a fixed frequency, the design parameter can be optimized according to the aforementioned criteria. However, if the response is to be controlled effectively over a relatively wide frequency band, an increase in μ_T and damping or a decrease in e will improve the performance of the damper.

Base excitation
The present theory can also analyze cases where the excitation is supplied through support displacement rather than force excitation. The base-excited two-degree-of-freedom system in Fig. 3.21a is equivalent to the force-excited three-degree-of-freedom system in Fig. 3.21b where m_1 replaces the support and the excitation amplitude is adjusted so that $S_0/(F_0/k_1) = 1$.

The theoretical predictions and the corresponding experimental measurements obtained by using an actual mechanical model are shown in Fig. 3.21c. As mentioned earlier, of the two possible steady-state solutions, the one associated with α_0^+ is the stable one. This is true in spite of the fact that, for the case in the frequency range $\Omega/\omega_1 \leq 0.95$, the solution α_0^+ leads to larger amplitudes than α_0^-. Note also the existence of a peak in the response curve at $\Omega/\omega_1 \approx 1/(1 + \mu_T)$.

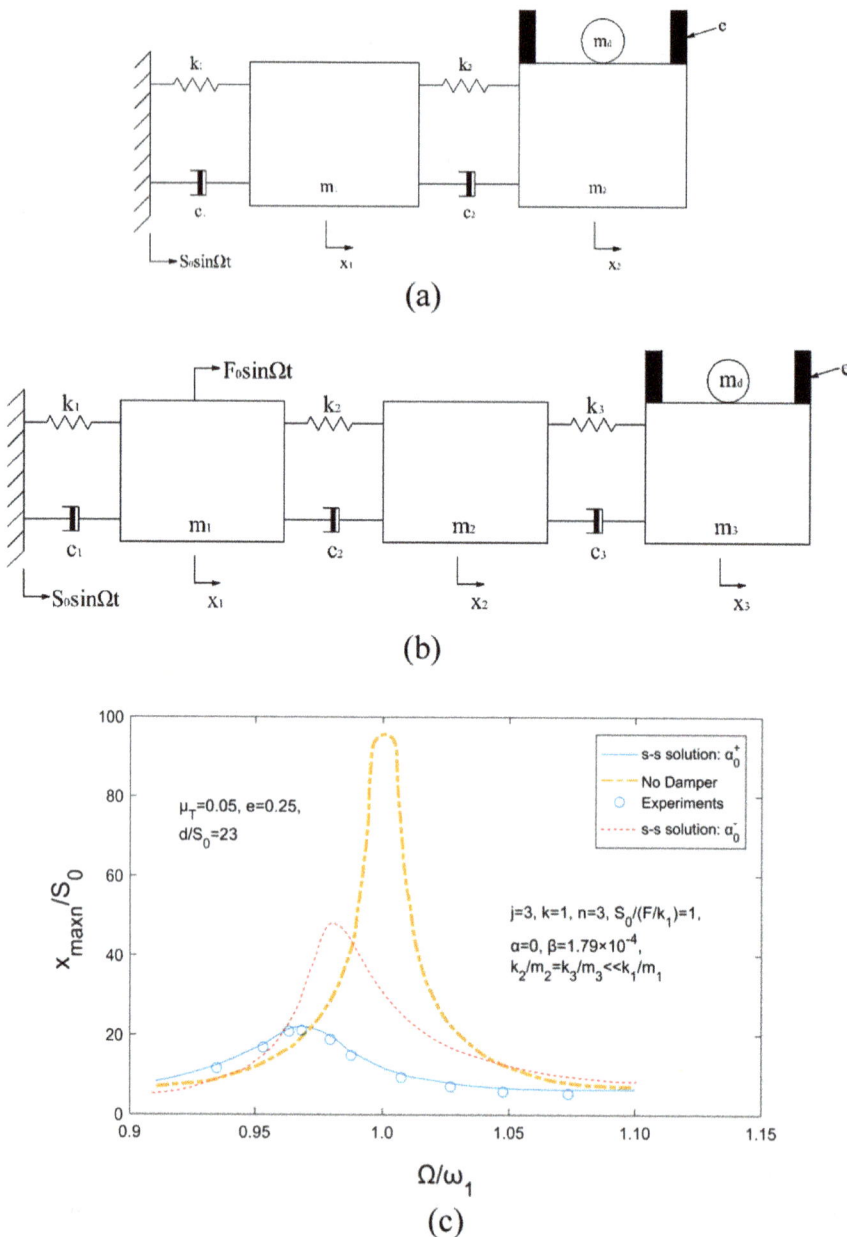

(a)

(b)

(c)

Fig. 3.21 Base excitation: **a** Base-excited two-degree-of-freedom model; **b** Equivalent force excited model; and **c** Frequency response of a base-excited two-degree-of-freedom system

Some of the design problems inherent in the application of practical impact dampers are (a) large accelerations developed during impact; (b) the accompanying noise; (c) design of the concentrated impacting mass (in massive structures, such as buildings, even a damper mass ratio of $\approx 1\%$ is still a large weight to have suspended in the structure). Analytical and experimental studies have shown that the foregoing problems can be alleviated to some extent by the use of multiple-unit impact dampers operating in parallel, instead of one single 'particle'.

3.1.2 Steady-State Motion of Multi-unit Impact Dampers

This part presents the exact (for the mathematical model) solution for the steady state motion of a multiple-unit impact damper attached to a periodically excited primary system.

A model of the subject system is shown in Fig. 3.22. Primary mass M is equipped with n impact dampers each of mass m_i in a container with clearance d_i and coefficient of restitution e_i. Period of exciting force $F(t)$ is $2\pi/\Omega$.

Suppose that after a certain number of impacts, e.g., k impacts, the motion of M has reached steady state conditions. Let the time of this kth impact be denoted by T_p. The complete motion of M is determined once the variables of the system have been evaluated for one period, i.e., from T_p to $T_p + T_0$, in which T_0 is the period of the impacts (not necessarily as same as the period of $F(t)$).

Without any loss of generality, the origin of the time axis can be shifted so as to coincide with the occurrence of the kth impact at time T_p. Then from $t = t_{i+1}$ immediately after the ith impact, until $t = t_{(i+1)_-}$, immediately preceding the next impact, the governing equation of motion of M is

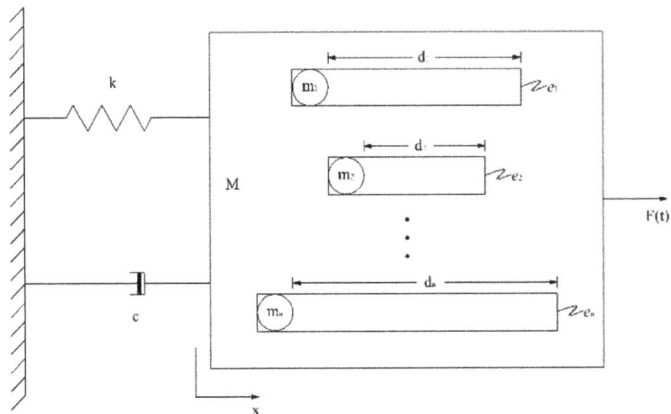

Fig. 3.22 System model

$$\left.\begin{array}{l} \ddot{x}(t) + 2\xi\omega\,\dot{x}(t) + \omega^2 x(t) = \frac{1}{M}F\big(t + T_p\big) \\ t_{i+} \leq t \leq t_{(i+1)-} \end{array}\right\} \qquad (3.53)$$

in which $t_1 = 0$; $\xi \equiv c/\big(2\sqrt{kM}\big)$; and $\omega \equiv \sqrt{k/M}$, with the initial conditions $x(t_i) = x_i$ and $\dot{x}(t_{i+}) = \dot{x}_{ia}$. The integral solution of Eq. (3.53) is

$$x(t) = \frac{1}{\eta}e^{-\xi\omega(t-t_i)}\left[\left(\xi\,x_i + \frac{1}{\omega}\dot{x}_{ia}\right)\sin\omega\eta(t-t_i) + \eta\,x_i\cos\omega\eta(t-t_i)\right] + I(t_i,t) \qquad (3.54)$$

and

$$\dot{x}(t) = e^{-\xi\omega(t-t_i)}\left[\dot{x}_{ia}\cos\omega\eta(t-t_i) - \frac{1}{\eta}(\xi\,\dot{x}_{ia} + \omega\,x_i)\sin\omega\eta(t-t_i)\right] + \dot{I}(t_i,t) \qquad (3.55)$$

in which $I(t_i,t) \equiv [1/(\omega\eta M)]\int_{t_i}^{t} F\big(t_v + T_p\big)\,e^{-\xi\omega(t-t_v)}\sin\omega\eta(t-t_v)dt_v$, and $\dot{I}(t_i,t) \equiv$ derivative of $I(t_i,t)$ with respect to t. Evaluating Eqs. (3.54) and (3.55) at $t = t_{(i+1)}$

$$x(t_{i+1}) \equiv x_{i+1} = S_{1_i}x_i + S_{2_i}\dot{x}_{ia} + I_i \qquad (3.56)$$

$$\dot{x}(t_{i+1})_- \equiv \dot{x}_{(i+1)b} = S_{3_i}x_i + S_{4_i}\dot{x}_{ia} + \dot{I}_i \qquad (3.57)$$

in which $\alpha_i \equiv \Omega t_i$; $I_i \equiv I(t_i, t_{i+1})$; $\dot{I}_i \equiv \dot{I}(t_i, t_{i+1})$.

Although, as previously stated, T need not be $2\pi/\Omega$, experimental and analytical studies of impact dampers [13] indicate that in the overwhelming majority of cases where M attains steady state motion with constant amplitude, $\Omega T_0 = 2\pi$. This is characterized by each particle undergoing two impacts per cycle of the forcing function, on opposite sides of its container.

Now consider the impacts occurring during period $0 \leq t \leq T_0$. Designate the impact occurring at $t = 0$ as impact number 1, and let t_i, $(i = 2, 3, \ldots)$ represent the time of the succeeding impacts to which M is subjected (note that $t_1 = 0$). Then, without any loss of generality, the n particles can be relabeled so that $x(t_i) \equiv x_i$ is the position of M at the time of its first collision with m_i. By this time M would have been subjected to i collisions, one with each of the i particles $m_1, m_2, \ldots, m_i$.

Taking the origin of each y_i, the relative displacement coordinate of the ith particle, to coincide with the center of the respective container, then the impact conditions require that

$$y_i(t_i) = z_i(t_i) - x(t_i) = \pm\frac{d_i}{2} \qquad (3.58)$$

The choice of the sign in Eq. (3.58) is immaterial. Its only effect is to renumber the stretches that make up the solution.

At t_n then the nth particle will collide with M, at t_{n+1}, the first particle will undergo its second impact on the other side of its container, and so on, so that

$$y_i(t_{n+i}) = z_i(t_{n+i}) - x(t_{n+i}) = -y_i(t_i) \tag{3.59}$$

By using the definition of the coefficient of restitution and the momentum equation, the following relationships can be obtained for impacts between M and m_i:

$$\dot{x}_- = k_{7_i}\dot{z}_{i_-} + k_{8_i}\dot{z}_{i_+} \tag{3.60}$$

$$\dot{x}_+ = k_{9_i}\dot{z}_{i_-} + k_{10_i}\dot{z}_{i_+} \tag{3.61}$$

Since there are no forces acting on the n particles between their impacts, the velocity of the ith particle during interval $\alpha_{i_+} \leq \Omega t \leq \alpha_{n+i_-}$ remains constant and is equal to

$$\dot{z}_i(t_i)_+ = \frac{\Omega}{\alpha_{n+i} - \alpha_i}\left[\left(x_{n+i} + y_{i_{n+i}}\right) - \left(x_i + y_{i_i}\right)\right] \tag{3.62}$$

Similarly, for $\alpha_{n+i_+} \leq \Omega t \leq \alpha_{2n+i_-}$

$$\dot{z}_i(t_{n+i})_+ = \frac{\Omega}{\alpha_{2n+i} - \alpha_{n+i}}\left[\left(x_{2n+i} + y_{i_{2n+i}}\right) - \left(x_{n+i} + y_{i_{n+i}}\right)\right] \tag{3.63}$$

in which $y_{i_j} \equiv y_i(t_j)$.

Note that

$$\left.\begin{aligned}
\dot{z}_i(t_i)_- &= \dot{z}_i(t_{n+i})_+ \\
\alpha_{2n+i} &= \alpha_i + 2\pi \\
x_{2n+i} &= x_i \\
y_{i_{2n+i}} &= y_{i_i} = -y_{i_{n+i}}
\end{aligned}\right\} \tag{3.64}$$

then from Eqs. (3.60), (3.62), (3.63), and (3.64)

$$\dot{x}_{ib} = G_{2_i}\left(-2y_{i_i} - x_i + x_{n+i}\right) \quad (i = 1, 2, \ldots, 2n) \tag{3.65}$$

Similarly, Eq. (3.61) results in

$$\dot{x}_{ia} = G_{1_i}\left(-2y_{i_i} - x_i + x_{n+i}\right) \quad (i = 1, 2, \ldots, 2n) \tag{3.66}$$

Using Eq. (3.66) to eliminate $\dot{x}_{i_a}$ from Eq. (3.56)

$$x_{i+1} = -2S_{2_i}G_{1_i}y_{i_i} + \left(S_{1_i} - S_{2_i}G_{1_i}\right)x_i + S_{2_i}G_{1_i}x_{n+i} + I_i \qquad (3.67)$$

Similarly, applying Eq. (3.65) at $i + 1$ and using it with Eq. (3.66) in Eq. (3.57)

$$\left(2G_{2_{i+1}}y_{i+1_{i+1}} - 2S_{4_i}G_{1_i}y_{i_i}\right) + \left(S_{3_i} - S_{4_i}G_{1_i}\right)x_i + G_{2_{i+1}}x_{i+1} + S_{4_i}G_{1_i}x_{n+i}$$
$$- G_{2_{i+1}}x_{n+i+1} + \dot{I}_i = 0 \qquad (3.68)$$

The unknowns in Eqs. (3.67) and (3.68) are T_p and the values of $(2n - 1)\alpha$ (since $\alpha_1 = 0$) and $2nx$. Applying Eqs. (3.67) and (3.68) for $i = (1, 2, \dots , 2n)$ furnishes the required $4n$ equations to determine all the unknowns. Then using Eqs. (3.65) and (3.66), the $2n$ values of $\dot{x}_{ia}$ and $\dot{x}_{ib}$ can be determined. With that, the $2n$ paths of the solution over one complete period can be constructed from Eqs. (3.54) and (3.55), thus completely determining the motion of the primary system.

As for the motion of the n particles, their constant velocities are given by Eqs. (3.62) and (3.63). The position of the ith particle as a function of time can be found by integrating Eqs. (3.62) and (3.63) and using the initial conditions

$$\left. \begin{array}{l} z_{i_i} = x_i + y_{i_i} \\ z_{i_{n+i}} = x_{n+i} + y_{i_{n+i}} \end{array} \right\} \qquad (3.69)$$

in which $z_{i_j} \equiv z_i\left(t_j\right)$.

Experimental evidence indicates that when the impact damper undergoes periodic motion with 2 impacts per particle per cycle, in most of the cases this type of motion simplifies to the case of symmetric 2 impacts per particle per cycle, with the impacts occurring at equal time intervals of $\Omega T_0 = \pi$ so that

$$\alpha_{n+i} = \alpha_i + \pi;\, x_{n+i} = -x_i;\, \dot{x}_{(n+i)a} = -\dot{x}_{ia};\, \dot{x}_{(n+i)b} = -\dot{x}_{ib};\, z_{i_{n+i}}$$
$$= -z_{i_i};\, \dot{z}_i\left(t_{n+i}\right)_+ = -\dot{z}_i\left(t_i\right)_+ \qquad (3.70)$$

Substituting Eq. (3.67) into Eq. (3.68), making use of Eq. (3.70), and solving for x_i

$$x_i = C_{40_i} - \frac{2G_{2_{i+1}}I_i + \dot{I}_i}{G_{6_i}} \qquad (3.71)$$

Similarly, from Eqs. (3.67), (3.70), and (3.71)

$$x_{i+1} = C_{35_i} + \left(1 - 2G_{2_{i+1}}\frac{G_{10_i}}{G_{6_i}}\right)I_i - \frac{G_{10_i}}{G_{6_i}}\dot{I}_i \qquad (3.72)$$

Evaluating Eq. (3.71) at $i + 1$ and equating it to Eq. (3.72)

$$\left(1 - 2G_{2_{i+1}}\frac{G_{10_i}}{G_{6_i}}\right)I_i - \frac{G_{10_i}}{G_{6_i}}\dot{I}_i + \frac{2G_{2_{i+1}}}{G_{6_{i+1}}}I_{i+1} + \frac{1}{G_{6_{i+1}}}\dot{I}_{i+1} = C_{40_{i+1}} - C_{35_i}$$

$$i = 1, 2, \ldots, n \quad (3.73)$$

The solution of the n coupled nonlinear algebraic Eq. (3.73) furnishes the values of T_p and $n - 1$ α $(\alpha_2, \alpha_3, \ldots, \alpha_n)$. Furthermore, Eqs. (3.71) or (3.72) can be used to determine the n x_i. Making use of the simplifying conditions Eq. (3.70) the n $\dot{x}_{ia}$ can be found in Eq. (3.66) thus permitting the construction of the n paths of the solution that describe the symmetric motion of the system over a half period, i.e., $\Omega t = \pi$. (The conditions in the second half are the negative of those in the first half).

Applying the aforementioned general results to the specific case of sinusoidal excitation, it will be found that if $F(t) = F_0 \sin \Omega t$, then

$$I(t_i, t) = -\frac{A}{\eta}e^{-\xi\omega(t-t_i)}[(r\cos\theta_i + \xi\sin\theta_i)\sin\omega\eta(t - t_i) + \eta\sin\theta_i\cos\omega\eta(t - t_i)]$$

$$+ A\sin(\Omega t + \beta) \quad (3.74)$$

$$\dot{I}(t_i, t) = \frac{A\omega}{\eta}e^{-\xi\omega(t-t_i)}[(\xi r\cos\theta_i + \sin\theta_i)\sin\omega\eta(t - t_i) - \eta r\cos\theta_i\cos\omega\eta(t - t_i)]$$

$$+ A\Omega\cos(\Omega t + \beta) \quad (3.75)$$

in which $A \equiv (F_0/k)\Big/\sqrt{(1 - r^2)^2 + (2\xi r)^2}$; $\psi \equiv \tan^{-1}[2\xi r/(1 - r^2)]$; $\beta \equiv \Omega T_p - \psi$; $\theta_i \equiv \beta + \alpha_i$.

By employing approaches similar to the one used herein, Warburton [14], and later Masri and Caughey [11] derived the exact solution for the symmetric 2 impacts per cycle motion of a single-particle impact damper attached to a sinusoidally excited primary system. Substituting Eqs. (3.74) and (3.75) into Eq. (3.73) and applying the resulting equation to the case $n = 1$ yields Eq. (3.74) of Ref. [11]. If, in addition, the primary system is undamped (i.e., $\xi = 0$) Eq. (3.61) of Ref. [14] is obtained.

For the case of a one-unit (i.e., $n = 1$) impact damper attached to a primary mass that is subjected to periodic excitation, Eqs. (3.67) and (3.68) simplify to

$$x_i = F_{5_i} + F_{6_i} + \frac{I_{i+1}}{D_{2_i}} \quad i = 1, 2 \quad (3.76)$$

$$F_{9_i} + F_{10_i}I_i + F_{11_i}I_{i+1} + \dot{I}_i = 0 \quad i = 1, 2 \quad (3.77)$$

Equation (3.77) is a system of two coupled transcendental equations involving T_p and t_2, that can be solved by iteration. Integrals I_i and $\dot{I}_i$ can be evaluated for a typical case, as shown in Fig. 3.23.

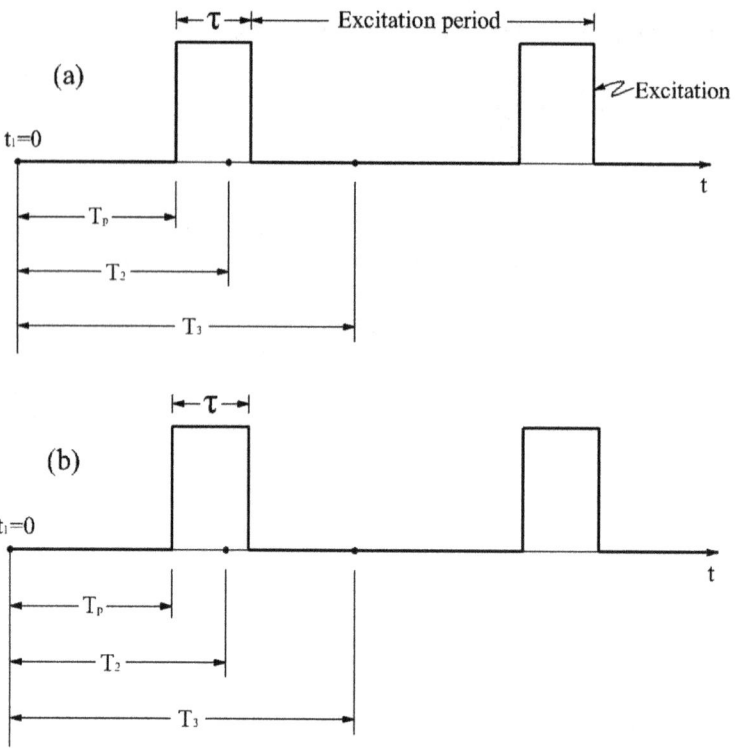

Fig. 3.23 Evaluation of response integrals: **a** $0 < t_2 \le T_p + \tau$, $I_1 = I(T_p, t_2)$, $I_2 = I(t_2, T_p + \tau)$; and **b** $T_p + \tau < t_2 < t_3$, $I_1 = I(T_p, T_p + \tau)$, $I_2 = 0$

If it is known a priori that the motion of the system is symmetric, t_2 is then known (equal to one half the excitation period) and Eq. (3.77) reduces to two uncoupled and identical equations that can be solved explicitly for T_p.

The response integrals corresponding to the periodic pulse excitation shown in Fig. 3.23 are

$$\left. \begin{array}{l} I(t_i, t) = \frac{F_0}{\omega \eta M} \left\{ e^{-\xi \omega (t - t_v)} \left[\xi \sin \omega \eta (t - t_v) + \eta \cos \omega \eta (t - t_v) \right] \right\} \Big|_{t_v = t_i}^{t_f} \\[2mm] \dot{I}(t_i, t) = -\frac{F_0 \omega}{\eta M} \left[e^{-\xi \omega (t - t_v)} \sin \omega \eta (t - t_v) \right] \Big|_{t_i}^{t_f} \end{array} \right\} \qquad (3.78)$$

Referring to Fig. 3.23, t_i and t_f are to be evaluated as

$$\begin{array}{l} 0 < t < T_p; \ t_i = t_f; \ I(t_i, t) = 0; \ T_p < t < T_p + \tau; \\[1mm] t_i = T_p, t_f = t; \ I(t_i, t) = I(T_p, t); \ T_p + \tau < T_p + period; \\[1mm] t_i = T_p, t_f = T_p + \tau; \ I(t_i, t) = I(T_p, T_p + \tau) \end{array} \qquad (3.79)$$

Predictions of the theory developed were confirmed by experimental studies with an electronic analog computer and numerical studies on a digital computer using circuits and techniques similar to those of Ref. [13].

Figure 3.24 shows the construction of a typical solution corresponding to the symmetric motion of two particles in parallel, attached to a sinusoidally excited primary system. Stable solution curves of single and two-unit impact dampers as a function of some of their pertinent variables are shown in Figs. 3.25 and 3.26. In regard to the response of a primary system provided with a single-unit impact damper and subjected to periodic pulse excitation of the type shown in Fig. 3.23, construction of a typical solution is given in Fig. 3.27, and the analog computer output is exhibited in Figs. 3.28 and 3.29.

The traces shown in Fig. 3.28 are (from top to bottom, in order of the first appearance on the left-hand side of photograph) the periodic excitation $f(t)$, $\dot{x}$, x, $\dot{y}$, and y. Note that the value of e can be verified directly from the trace of $y(e = \dot{y}_+/\dot{y}_-)$.

Fig. 3.24 Motion of sinusoidally excited system provided with two-unit impact damper

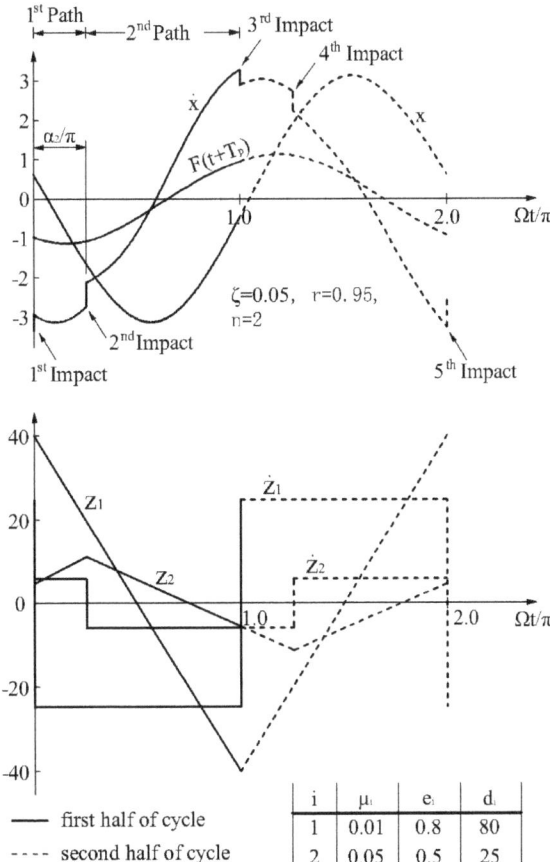

i	μ_i	e_i	d_i
1	0.01	0.8	80
2	0.05	0.5	25

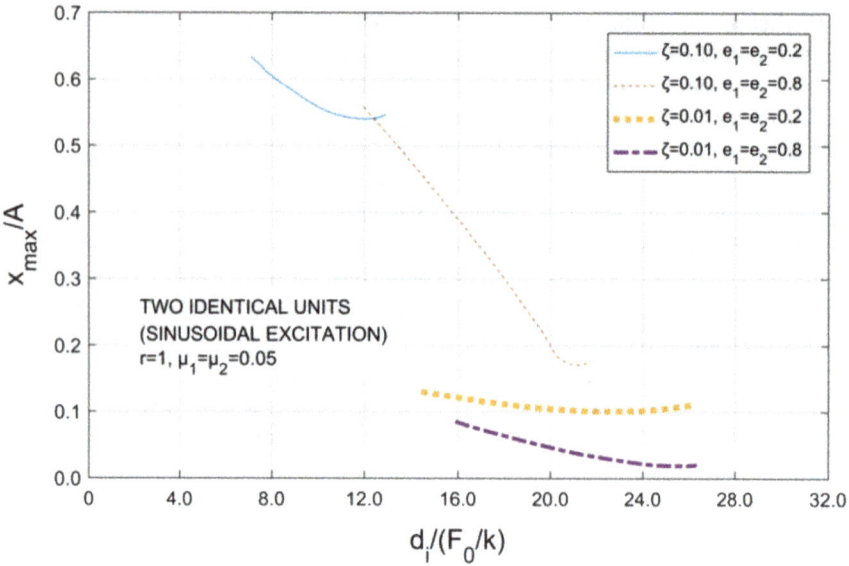

Fig. 3.25 Effects of viscous damping and coefficient of restitution on impact dampers

Figure 3.29 shows the drastic reduction in the displacement and velocity of the primary system that can be attained by a properly designed impact damper. The typical response curves shown in Fig. 3.30 show the good agreement between the predictions of the theory and the experimental results.

Unlike the dynamic vibration neutralizer (Frahm damper), the effectiveness of a properly designed impact damper is insensitive to slight changes in the frequency ratio. However, even this slight dependence on frequency can be reduced further by employing equivalent multiple-unit impact dampers. Such dampers offer added flexibility in the choice of design parameters (even while retaining the same total μ) so that a nearly flat response curve, within a certain band of excitation frequencies, may be achieved. The asymptotic stability of the solutions can be investigated analytically by applying the method used in Ref. [11] to determine the stability of single-unit impact dampers subjected to sinusoidal excitation.

3.1.3 Steady-State Motion of Impact Damper—Tuned Absorber Combination

This part presents the exact solution for the steady-state motion and the asymptotic stability of the highly nonlinear two-degree-of-freedom nonautonomous system shown in Fig. 3.31. In addition to being coupled to the primary mass by means of linear spring k_2 and dashpot c_2, the auxiliary mass m_2 is constrained to oscillate with

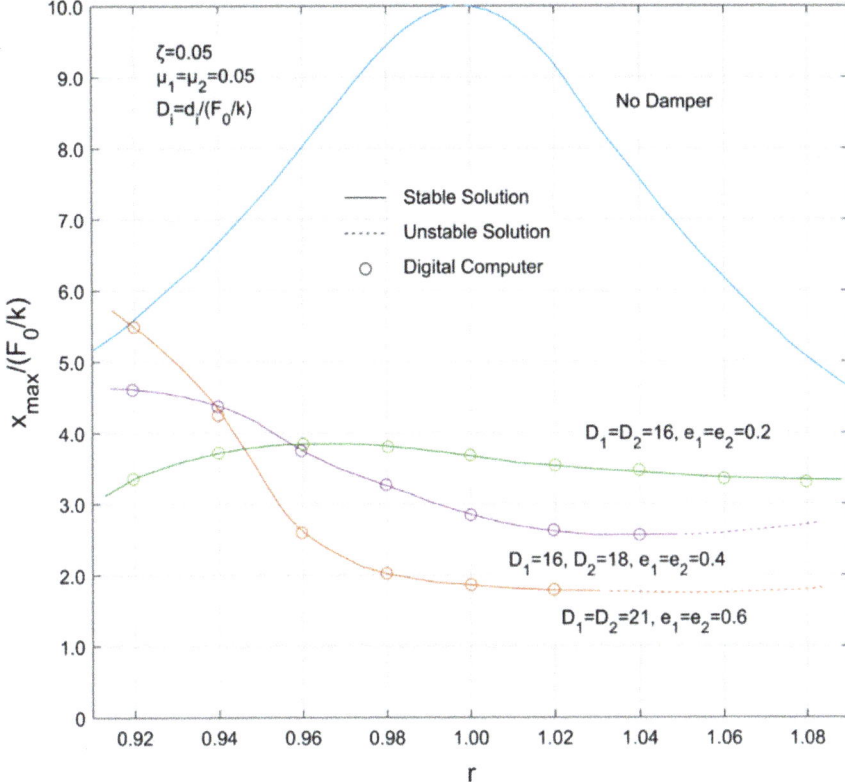

Fig. 3.26 Frequency dependence of sinusoidally excited two-unit impact damper

clearance d with respect to m_1. The amount of mechanical energy dissipated during the collision of m_2 with the rigid stops (assumed to be infinitely stiff) is governed by the coefficient of restitution e, which may assume values $0 \leq e \leq 1$, i.e., from the completely plastic up to the completely elastic impact. Note that the nonlinearity in this problem involves the relative velocity as well as the relative displacement.

Depending upon the choice of parameters, this system can model various types of auxiliary mass dampers. For example, if $d/(F_0/k_1) \gg 1$, $c_2 = 0$ the system reduces to the well-known dynamic vibration neutralizer; if $d/(F_0/k_1) \gg 1$, $k_2 = 0$ and $c_2 \neq 0$, the system assumes the form of the Lanchester damper; and if the clearance d is of the same order as the amplitude of m_1 and $c_2 = k_2 = 0$, the system then represents the impact damper.

It will be shown that, with a proper choice of its parameters, the system under consideration alleviates some of the deficiencies inherent in, and possesses response characteristics superior to, other well-known dampers.

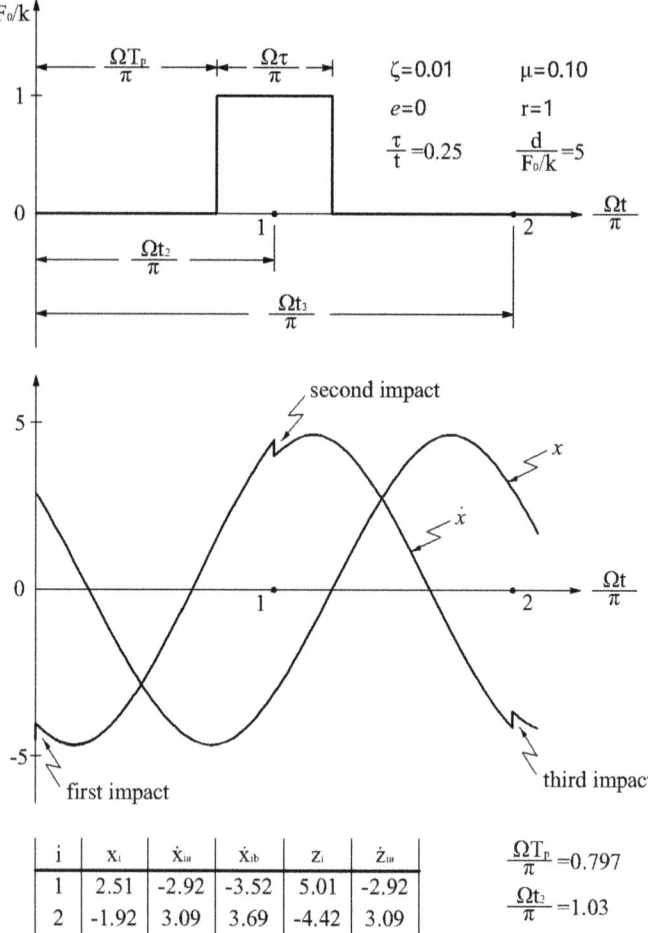

Fig. 3.27 Construction of typical solution

The equations of motion for such systems are

$$a_{11}\ddot{x}_1 + b_{11}\dot{x}_1 + c_{11}x_1 + b_{12}\dot{x}_2 + c_{12}x_2 = Q_1(t)$$
$$b_{12}\dot{x}_1 + c_{12}x_1 + a_{22}\ddot{x}_2 + b_{22}\dot{x}_2 + c_{22}x_2 = 0 \qquad (3.80)$$

where $a_{11} = m_1$; $a_{22} = m_2$; $b_{11} = c_1 + c_2$; $b_{12} = -c_2$; $b_{22} = c_2$; $c_{11} = k_1 + k_2$; $c_{12} = -k_2$; $c_{22} = k_2$; m_1 is mass of primary system, m_2 is mass of auxiliary system (damper); c_1, c_2 are damping constants; k_1, k_2 are spring constants; x_j is displacement of jth mass; $\dot{x}_j$ is velocity of jth mass; and $Q_1(t) = F_0 \cos \Omega t$; F_0 is maximum force of excitation; Ω is excitation frequency.

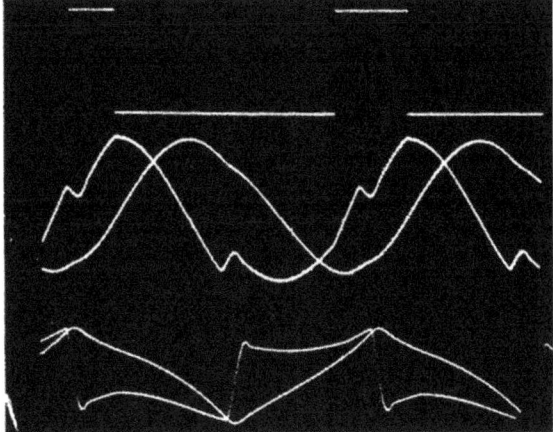

Fig. 3.28 Representative sample of analog computer results: $\xi = 0.01$; $\mu = 0.10$; $e = 0.75$; $d / (F_0/k) = 9.5$; $r = 1$; $\tau / T = 0.25$; $n = 1$. (Vertical sensitivity differs for each trace; horizontal sensitivity 2 ms/div)

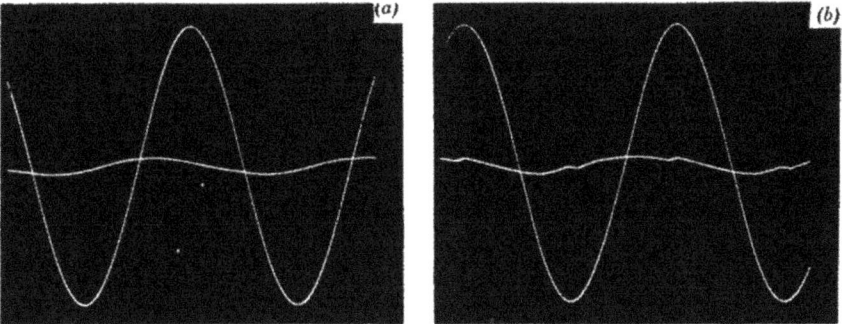

Fig. 3.29 Comparison between steady-state motion of periodically pulse excited primary system with and without the impact damper: **a** Displacement; and **b** Velocity. (Traces have same vertical and horizontal sensitivity. System parameters same as in Fig. 3.28)

Consider the steady-state motion of the system, and shift the origin of the time axis so that at $t = 0$, $|x_2 - x_1| = d/2$, d is clearance within which auxiliary mass can oscillate. This shift will modify the form of the excitation force so that $Q_1(t) = F_0 \cos(\Omega t + \alpha_0) = \text{Re}\{F_0 \exp[i(\Omega t + \alpha_0)]\}$, where $i = \sqrt{-1}$, α_0 is phase angle (initially unknown).

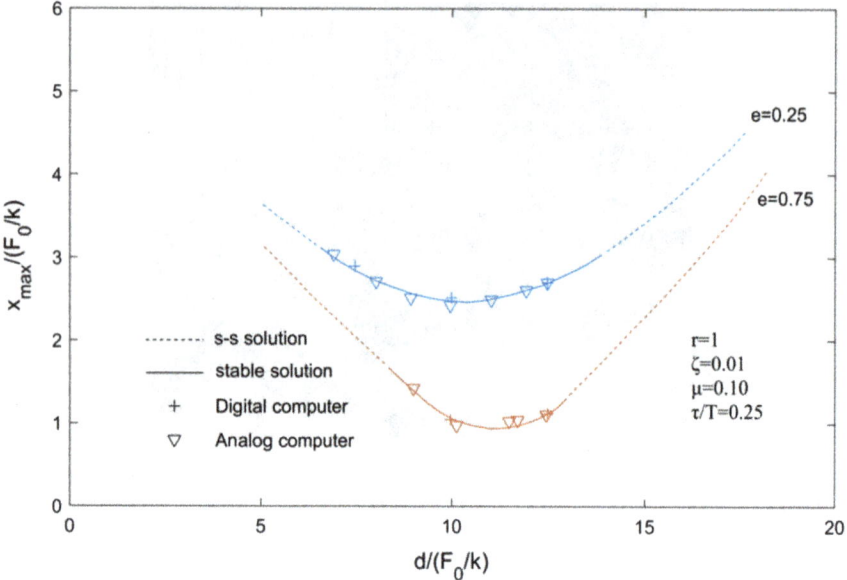

Fig. 3.30 Comparison between theoretical and experimental results for periodically pulse excited single-unit impact damper

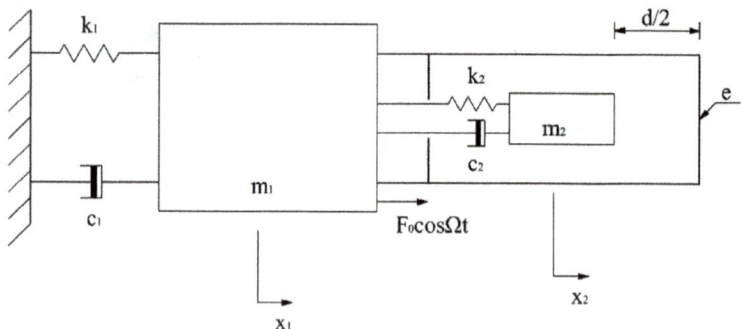

Fig. 3.31 Model of system

Then the complete solution of (3.80) is

$$x_1(t) = \sum_{j=1}^{4} \lambda_{1j} \exp(s_j t) + \mathrm{Re}\{A_1 \exp[i(\Omega t + \alpha_0)]\}$$

$$x_2(t) = \sum_{j=1}^{4} r_j \lambda_{1j} \exp(s_j t) + \mathrm{Re}\{A_2 \exp[i(\Omega t + \alpha_0)]\} \qquad (3.81)$$

where s_1, s_2, s_3 and s_4 are the roots of the characteristic equation, A is displacement amplitude of primary system in absence of damper, A_1 is displacement amplitude of primary mass with damper operating without motion-limiting stops, A_2 is displacement amplitude of auxiliary mass operating without motion-limiting stops,

$$\left(a_{11}s^2 + b_{11}s + c_{11}\right)\left(a_{22}s^2 + b_{22}s + c_{22}\right) - \left(b_{12}s + c_{12}\right)^2 = 0,$$

the $\lambda's$ are related by

$$\frac{\lambda_{1j}}{\lambda_{2j}} = -\frac{b_{12}s_j + c_{12}}{a_{11}s_j^2 + b_{11}s_j + c_{11}} \equiv r_j \quad (j = 1, 2, 3, 4),$$

and

$$\begin{aligned}
\lambda_{1s} &= F_0 \exp(i\alpha_0)\left[-\Omega^2 a_{22} + i\Omega b_{22} + c_{22}\right]/D_s \equiv A_1 \exp(i\alpha_0) \\
\lambda_{2s} &= -F_0 \exp(i\alpha_0)\left[i\Omega b_{12} + c_{12}\right]/D_s \equiv A_2 \exp(i\alpha_0)
\end{aligned},$$

where

$$D_s \equiv \left(-\Omega^2 a_{11} + i\Omega b_{11} + c_{11}\right)\left(-\Omega^2 a_{22} + i\Omega b_{22} + c_{22}\right) - \left(i\Omega b_{12} + c_{12}\right)^2.$$

Using the initial conditions

$$x_1(0) = x_{10}, \dot{x}_1(0)_+ = \dot{x}_{1a}, x_2(0) = x_{20}, \dot{x}_2(0)_+ = \dot{x}_{2a} \tag{3.82}$$

where the $+$ subscript indicates conditions immediately after the prescribed time.

In conjunction with (3.81), the relationships between the $\lambda's$ and the initial conditions can be expressed as $\lambda = [M]^{-1}v$, where

$$[M] = \begin{pmatrix} 1 & 1 & 1 & 1 \\ s_1 & s_2 & s_3 & s_4 \\ r_1 & r_2 & r_3 & r_4 \\ r_1 s_1 & r_2 s_2 & r_3 s_3 & r_4 s_4 \end{pmatrix},$$

$$\lambda = \text{col}\{\lambda_{11}, \lambda_{12}, \lambda_{13}, \lambda_{14}\},$$

$$\begin{aligned}
v = \text{col}\{&x_{10} - \text{Re}[A_1 \exp(i\alpha_0)], \dot{x}_{1a} - \text{Re}[i\Omega A_1 \exp(i\alpha_0)], \\
&x_{20} - \text{Re}[A_2 \exp(i\alpha_0)], \dot{x}_{2a} - \text{Re}[i\Omega A_2 \exp(i\alpha_0)]\}
\end{aligned}$$

Let $\mathbf{E}(t) = \text{col }\{\exp(s_1 t), \exp(s_2 t), \exp(s_3 t), \exp(s_4 t)\}$,

$$[R] = \begin{pmatrix} r_1 & & & 0 \\ & r_2 & & \\ & & r_3 & \\ 0 & & & r_4 \end{pmatrix},$$

$$[S] = \begin{pmatrix} s_1 & & & 0 \\ & s_2 & & \\ & & s_3 & \\ 0 & & & s_4 \end{pmatrix}.$$

Then the displacements and velocities of the system can be expressed as

$$x_1(t) = \mathbf{I}^T \mathbf{E}(t) + \mathrm{Re}\{A_1 \exp[i(\Omega t + \alpha_0)]\}$$
$$\dot{x}_1(t) = \mathbf{I}^T [S]\mathbf{E}(t) + \mathrm{Re}\{i\Omega A_1 \exp[i(\Omega t + \alpha_0)]\}$$
$$x_2(t) = \mathbf{I}^T [R]\mathbf{E}(t) + \mathrm{Re}\{A_2 \exp[i(\Omega t + \alpha_0)]\}$$
$$\dot{x}_2(t) = \mathbf{I}^T [R][S]\mathbf{E}(t) + \mathrm{Re}\{i\Omega A_2 \exp[i(\Omega t + \alpha_0)]\} \tag{3.83}$$

Experiments with this type of damper indicate that, except in rare instances, when steady-state motion is achieved, the damper undergoes two impacts on opposite sides of its stops per cycle of the excitation. Furthermore, in the overwhelming majority of cases, this type of motion simplifies to the case of two symmetric impacts/cycle that occur at equal time intervals.

In steady-state motion with two symmetric impacts/cycle of the forcing function

$$x_1(\pi/\Omega) = -x_{10} \qquad \dot{x}_1(\pi/\Omega)_- = -\dot{x}_{1b}$$
$$x_2(\pi/\Omega) = -x_{20} \qquad \dot{x}_2(\pi/\Omega)_- = -\dot{x}_{2b} \tag{3.84}$$

where the $(-)$ subscript indicates conditions immediately before the prescribed time.

Employing conditions (3.84) in conjunction with system (3.83), it will be found that

$$-x_{10} = g_{11}x_{10} + g_{12}\dot{x}_{1a} + g_{13}x_{20} + g_{14}\dot{x}_{2a} + g_{31}\cos\alpha_0 + g_{32}\sin\alpha_0$$
$$-\dot{x}_{1b} = g_{15}x_{10} + g_{16}\dot{x}_{1a} + g_{17}x_{20} + g_{18}\dot{x}_{2a} + g_{33}\cos\alpha_0 + g_{34}\sin\alpha_0$$
$$-x_{20} = g_{21}x_{10} + g_{22}\dot{x}_{1a} + g_{23}x_{20} + g_{24}\dot{x}_{2a} + g_{35}\cos\alpha_0 + g_{36}\sin\alpha_0$$
$$-\dot{x}_{2b} = g_{25}x_{10} + g_{26}\dot{x}_{1a} + g_{27}x_{20} + g_{28}\dot{x}_{2a} + g_{37}\cos\alpha_0 + g_{38}\sin\alpha_0 \tag{3.85}$$

where the $g's$ are functions of $E(\pi/\Omega)$, $[M]$, $[R]$, and $[S]$.

Idealizing the collision between m_1 and m_2 to be a discontinuous process, then from the momentum equation and the definition of the coefficient of restitution

$$\dot{x}_{1a} = h_1\dot{x}_{1b} + h_2\dot{x}_{2b} \tag{3.86}$$

$$\dot{x}_{2a} = h_3\dot{x}_{1b} + h_4\dot{x}_{2b} \tag{3.87}$$

where the h's are functions of μ and e.

Since an impact occurs when the relative displacement between m_1 and m_2 is $\pm d/2$

$$x_{20} - x_{10} = d/2 \tag{3.88}$$

System (3.85) with Eqs. (3.86) and (3.87) furnish 7 equations among the 7 unknowns $x_{10}, \dot{x}c_{1a}, \dot{x}_{1b}, x_{20}, \dot{x}_{2a}, \dot{x}_{2b}$, and α_0.

By elimination, the previous equations can be put in the form

$$
\begin{aligned}
\dot{x}_{2b} &= -(g_{25}x_{10} + g_{26}\dot{x}_{1a} + g_{27}x_{20} + g_{28}\dot{x}_{2a} + g_{37}\cos\alpha_0 + g_{38}\sin\alpha_0)\\
\dot{x}_{1b} &= -(g_{15}x_{10} + g_{16}\dot{x}_{1a} + g_{17}x_{20} + g_{18}\dot{x}_{2a} + g_{33}\cos\alpha_0 + g_{34}\sin\alpha_0)\\
\dot{x}_{2a} &= g_{57}x_{10} + g_{58}\dot{x}_{1a} + g_{59}x_{20} + g_{60}\cos\alpha_0 + g_{61}\sin\alpha_0\\
\dot{x}_{1a} &= -\frac{1}{g_{73}}(g_{72}x_{10} + g_{74}x_{20} + g_{75}\cos\alpha_0 + g_{76}\sin\alpha_0)\\
x_{20} &= x_{10} + d/2\\
x_{10} &= g_{85}d/2 + g_{86}\cos\alpha_0 + g_{87}\sin\alpha_0\\
g_{88}\sin\alpha_0 &+ g_{89}\cos\alpha_0 = g_{90}d
\end{aligned}
\tag{3.89}
$$

From the last equation in (3.89)

$$\alpha_0 = \tan^{-1}\left[\frac{g_{88}g_{90}d \pm g_{89}\left(g_{88}^2 + g_{89}^2 - g_{90}^2 d^2\right)^{1/2}}{g_{88}g_{90}d \mp g_{88}\left(g_{88}^2 + g_{89}^2 - g_{90}^2 d^2\right)^{1/2}}\right] \tag{3.90}$$

With α_0 determined from (3.90), the rest of the unknowns can be found by back-substitution.

Applying the method used by Masri and Caughey [11] to investigate the stability of the impact damper, it can be shown that the solution will be asymptotically stable if, and only if, all the eigenvalues of a matrix $\boldsymbol{P}$ have modulus less than unity, where $\boldsymbol{P}$ is a constant matrix that relates ξ_i, the ith perturbation vector to ξ_{i+1}.

Discussions are as followed:

(1) Modified dynamic vibration neutralizer

Response curves for a dynamic vibration neutralizer (DVN) without any motion-limiting stops (i.e., $d/(F_0/k_1) = \infty$) are presented in Fig. 3.32 as a function of the dimensionless frequency ratio Ω/ω_1 for a typical case where $\xi_1 = 0.01$, $\xi_2 = 0.01$, $\mu = 0.10$, $\omega_2/\omega_1 = 1$. Curves $|A_1|/(F_0/k_1)$ and $|A_2 - A_1|/(F_0/k_1)$ correspond to the displacement amplitude of the primary mass and the amplitude of the relative displacement between the two masses, respectively, and curve $A/(F_0/k_1)$ is the amplification ratio of the primary mass m_1 in the absence of the damper. The response

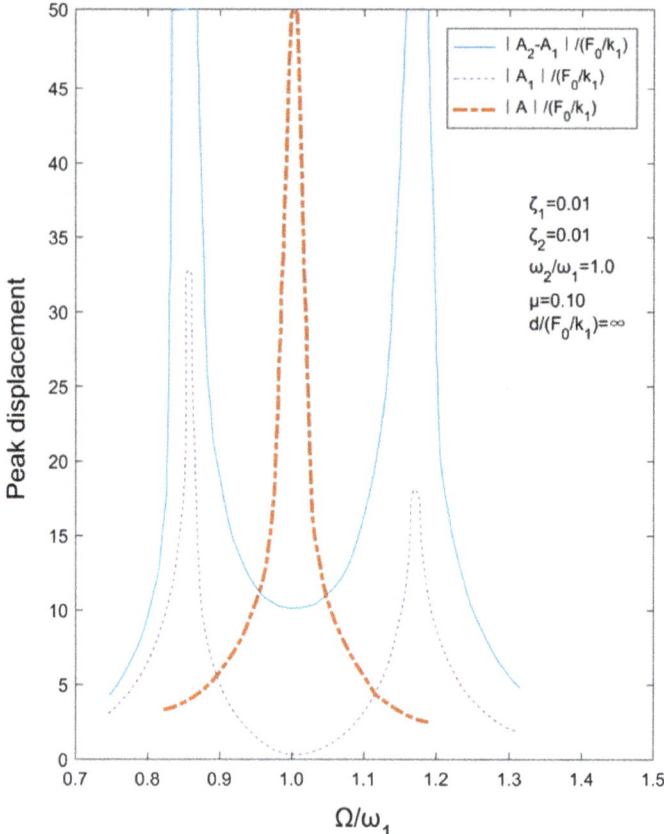

Fig. 3.32 Peak displacement of primary mass and relative displacement of auxiliary mass in a dynamic vibration neutralizer

curves in Fig. 3.32 agree with available analytical results concerning the behavior of conventional DVN [15].

When the DVN under consideration is modified by the addition of motion-limiting stops that are characterized by a clearance ratio $d/(F_0/k_1) = 15$ and a coefficient of restitution $e = 0$, two steady-state solutions exist, each with two symmetric impacts/cycle corresponding to the two solutions for α_0 in Eq. (3.90). The curves in Fig. 3.33 labeled α_0^+ and α_0^- correspond to the upper and lower sets of signs, respectively, in Eq. (3.90). The stable regions of the solutions exhibited in Fig. 3.33 were determined by finding the regions in which all the eigenvalues of matrix P have modulus less than unity.

The ordinates in Fig. 3.33 represent the ratio of x_{max}, the peak displacement amplitude of the primary mass when provided with the modified DVN, divided by the static deflection F_0/k_1. The corresponding response curve $|A_1|/(F_0/k_1)$ obtained with a conventional DVN (i.e., $d/(F_0/k_1) = \infty$) is also shown for comparison. The

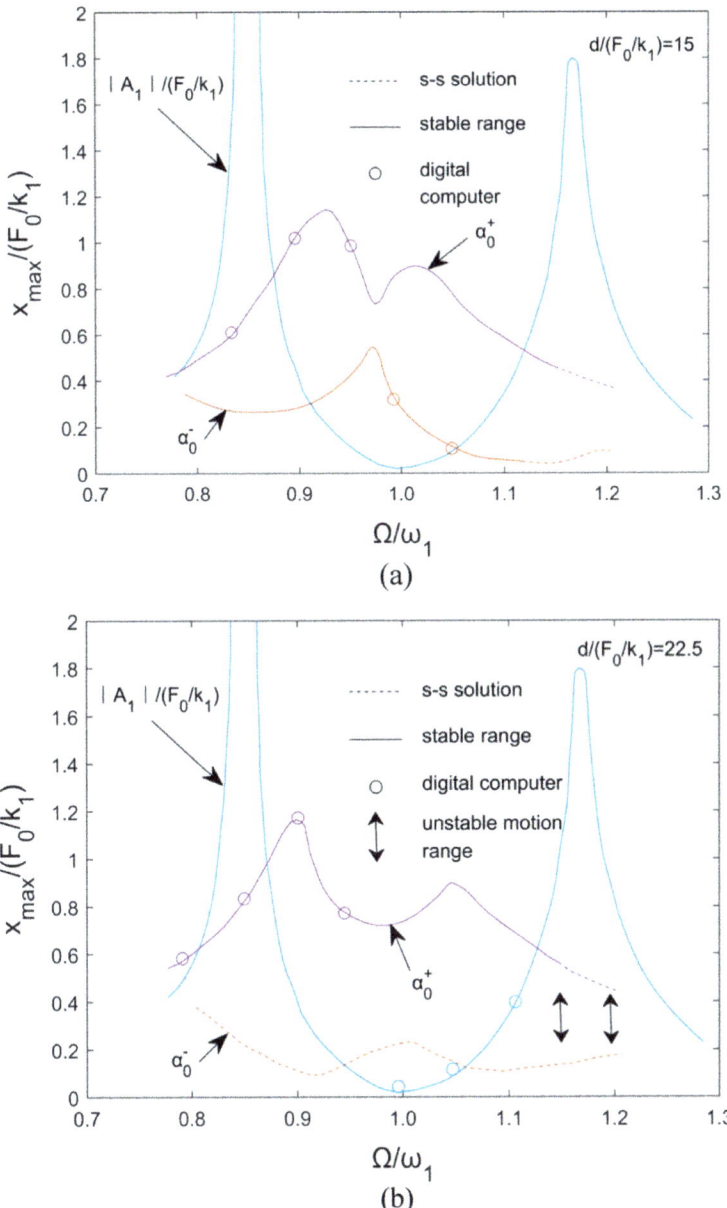

Fig. 3.33 Steady-state motion of a dynamic vibration neutralizer with motion-limiting stops, $\xi_1 = \xi_2 = 0.01$, $\mu = 0.1$, $e = 0$, $\omega_2/\omega_1 = 1.0$: **a** $d/((F_0/k_1)) = 15$; and **b** $d/((F_0/k_1)) = 22.5$

value of x_{max} is determined from Eq. (3.83) by finding the maximum value that $x_1(t)$ will assume during the time interval $0 \leq \Omega t \leq \pi$.

Examination of the solution curves in Fig. 3.33 shows that at a given excitation frequency Ω/ω_1, more than one mode of steady-state motion is possible, and for some frequencies more than one mode of stable steady-state motion is possible.

In fact, for cases where the clearance d is greater than the peak-to-peak relative displacement (i.e., $2|A_2 - A_1|$) required for operation of the system as a conventional DVN, three stable solutions exist. For example, if $d/(F_0/k_1) = 22.5$, it is seen from Fig. 3.32 that over the frequency ranges $\Omega/\omega_1 < 0.80$, $0.95 < \Omega/\omega_1 < 1.05$ and $\Omega/\omega_1 > 1.27$ the peak relative displacement of m_2 is such that $|A_2 - A_1| < d/2$; hence, the presence of the stops will not interfere with the performance of the system as a conventional DVN. At frequencies that would result in $|A_2 - A_1| > d/2$, the stops will become involved in the operation of the system, thus limiting the relative motion of the auxiliary mass to the preset level and, at the same time, considerably attenuating the displacement amplitude of the primary mass as compared with what it would have been, had the stops not been incorporated in the system.

The existence of multiple steady-state solutions in this highly nonlinear system is in accordance with the observations made by Arnold [16] who noted the 'existence of up to three modes of oscillation for a single value of disturbing frequency' in the steady-state motion of systems provided with nonlinear dynamic vibration neutralizers.

Keeping in mind that the stability analysis presented here applies to the asymptotic stability of the solutions, when more than one mode of motion is deemed stable, with two symmetric impacts/cycle, then the type of motion that will actually occur, depends on the initial conditions. Furthermore, since our discussion is mainly concerned with symmetric-type motion if, for a particular set of parameters no steady-state solution with two symmetric impacts/cycle is stable, this does not preclude the possibility that another mode of motion (e.g., one with two unsymmetric impacts/cycle) may be stable.

For example, in the case shown in Fig. 3.33a both solutions are stable over the range $0.75 \leq \Omega/\omega_1 \leq 1.15$ while in the case of Fig. 3.33b where $d/(F_0/k_1)$ has been increased to 22.5, the solution α_0^- is completely unstable in the indicated range while solution α_0^+ is stable in the range $0.8 \leq \Omega/\omega_1 \leq 1.15$.

Experimental studies by means of an analog computer and numerical solution of the governing differential equations of motion by means of a digital computer using circuits and methods similar to those of reference [11] confirmed the predictions of the theory and yielded the data shown in Fig. 3.33.

With $d/(F_0/k_1) = 22.5$, the experimental results matched the prediction of the theory and the stability analysis that the solution α_0^+ was the only one leading to two symmetric impacts/cycle, except in the neighborhood of resonance $\Omega/\omega_1 = 1$ where $|A_2 - A_1| < d/2$ in which case the motion of the system resorted to that of a conventional DVN since the available clearance did not interfere with the relative displacement requirements of this type of motion. The data points in Fig. 3.33b beyond the stable range correspond to unstable motion (in the sense of two symmetric impacts/cycle).

Time histories of three representative modes of motion are presented in Fig. 3.34 for a typical set of parameters. Figure 3.34a and b illustrate unsymmetric and symmetric modes of motion, respectively, while Fig. 3.34c corresponds to a case where the stops are not engaged in action, thus leading to a simple harmonic motion of the system. The value of the coefficient of restitution for the case shown in Fig. 3.34 can be verified directly from the traces corresponding to y since $e \equiv -\dot{y}_+/\dot{y}_-$.

The auxiliary mass in the case of Fig. 3.33 is tuned so that $\omega_2/\omega_1 = 1$. The response of the system in Fig. 3.33a when ω_2/ω_1 is changed, is given in Fig. 3.35 for two ratios $\omega_2/\omega_1 = 0.50$ and 0.01. Comparison among Figs. 3.33a, 3.35a and b indicates that a $\omega_2/\omega_1 \approx 0.5$ will lead to a more efficient damping system regarding stability and the amplification ratio over the suppression band.

The improvement in the modified nonlinear system can be attributed to the fact that when the stops are engaged in the motion a new vibration damping mechanism is introduced. Instead of relying solely on the force exerted by m_2 to counteract the exciting force (which is what occurs in a DVN), the addition of the stops introduces the mechanism of momentum transfer and energy dissipation during impact, which enhances the performance of the system.

(2) *Modified Lanchester damper*

As in the case of the DVN, the operation of a system employing a Lanchester damper can be improved by incorporating into it a limit on the relative motion between the two masses. Rather than depending purely on energy dissipation during the relative motion, the presence of the stops will introduce impact damping with its own characteristic features.

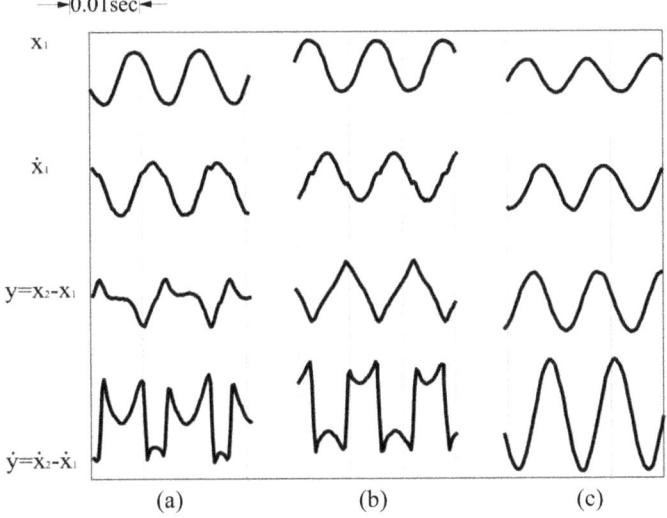

Fig. 3.34 Analog computer results showing steady-state motion of a dynamic vibration neutralizer with motion limiting stop: **a** Unsymmetric; **b** Symmetric; and **c** No impacts

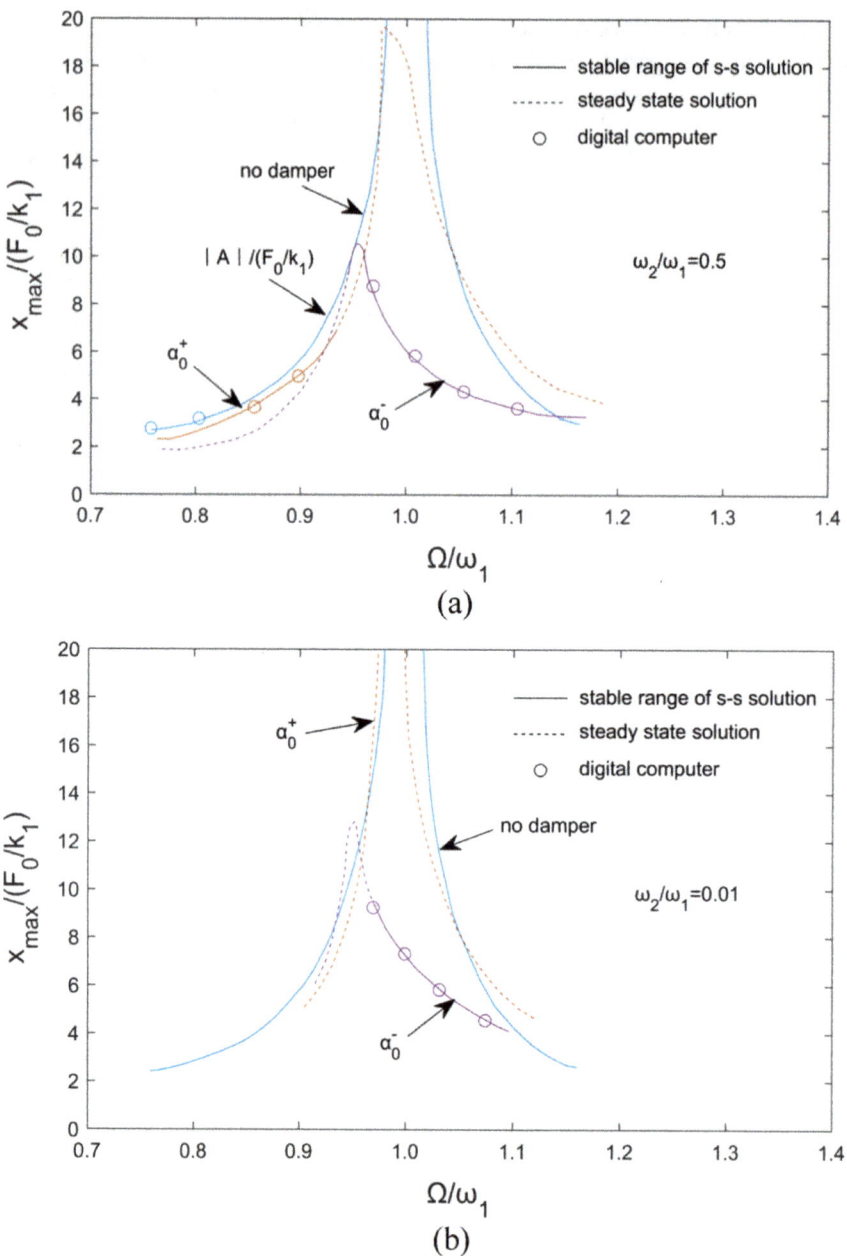

Fig. 3.35 Steady-state response of system, $\xi_1 = \xi_2 = 0.01$, $\mu = 0.1$, $e = 0$, $d/((F_0/k_1)) = 15$: **a** $\omega_2/\omega_1 = 0.50$; and **b** $\omega_2/\omega_1 = 0.01$

Figure 3.36 shows the response of a system provided with a Lanchester damper with, and without, stops. It is clear that better performance within the suppression band can be achieved if the relative motion is limited and if ξ_2 is decreased in order to allow a greater proportion of the mechanical energy to be utilized for momentum transfer, as evidenced by the response curves shown in Fig. 3.36b. Unlike the DVN, the absence of a coupling spring between the two masses results in single-valued stable solution curve with two impacts/cycle, when this mode of motion is stable.

(3)   Modified impact damper

Elimination of the coupling spring k_2 and the dashpot c_2 will put the system under discussion in the form of the impact damper. Application of the present theory to a particular case that is discussed in reference [11] where $\xi_1 = 0.10, \xi_2 = 0, \mu = 0.10$, $e = 0.8, \omega_2/\omega_1 = 0.01, \Omega/\omega_1 = 1$ leads to solution curves that completely match those of reference [11] in regard to the steady-state solution as well as the stability boundaries.

If the combination $\xi_2 = 0, e = 0.8$ in the foregoing case is replaced by the combination $\xi_2 = 0.01, e = 1$, the response of the primary system will be virtually the same in both cases. In other words, the highly nonlinear form of damping represented by the coefficient of restitution (plastic deformation during the collision process) can be replaced by an equivalent linear type of energy dissipating mechanism (c_2) without significantly altering the behavior of the primary system. Bearing in mind that the analytical treatment of problems with nonlinear damping is in general more involved than the treatment of similar systems with nonlinear springs, it is usually advantageous to be able to replace the nonlinear damping mechanism by an equivalent viscous dashpot [17].

(4)   Effects of various system parameters

The influence of the independent parameters on the response of the system was determined, and it can be summarized as follows:

Solution phase angle α_0. In general, if for a given set of parameters the radical in Eq. (3.90) is not negative, there are two possible solutions α_0^+ and α_0^- corresponding to two distinct steady-state solutions. The two solutions will coalesce at $d = 0$ and at $d = d*$ that will render the radical in Eq. (3.90) equal to zero. The nonexistence of solutions beyond $d = d*$ is due to the fact that for these conditions the auxiliary mass does not have enough energy to enable it to travel from one restraining stop to the other twice per cycle in steady-state motion.

Clearance ratio $d/(F_0/k_1)$. For any given set of parameters there is an optimum d that will lead to the maximum attenuation of the response of the primary system. The sensitivity of the system response to d is increased as the energy dissipation of the system is decreased. If $d/|A_1|$ is too small, multiple impacts/cycle will occur and if $d/|A|$ is too large, the auxiliary mass will not acquire momentum between impacts to sustain two collisions/cycle on opposite ends of its stops. In general, the stability boundaries enclose the optimum d.

Coefficient of restitution e. Since e measures the elasticity of the collisions between m_2 and the stops, the more elastic the impact is the more effective that

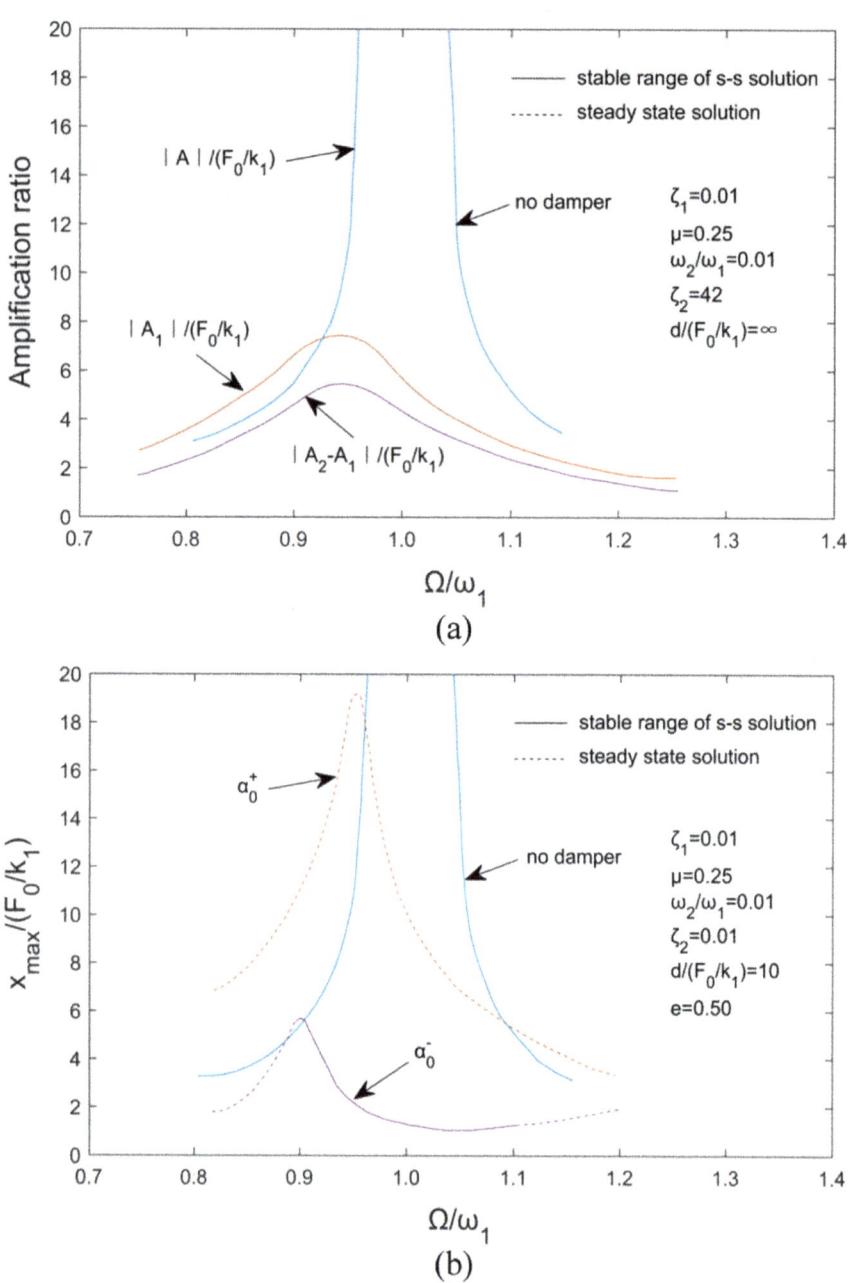

Fig. 3.36 Steady-slate response of a Lanchester damper: **a** Conventional damper; and **b** Modified damper

the damper will be, provided the parameters of the system, are fixed. However, a low e is more suitable if the parameters (particularly Ω/ω_1) change. Using a low e with a system in which the excitation frequency varies substantially, improves the stability of the system and 'flattens' the frequency response curve [18].

Auxiliary mass frequency ratio ω_2/ω_1. With a system operating at a fixed excitation frequency Ω/ω_1, the optimum response will be achieved if ω_2/ω_1 is set equal to Ω/ω_1. If the system is to operate primarily about $\Omega/\omega_1 = 1$ setting $d \geq 2|A_2 - A_1|_{\Omega/\omega_1=1}$ will permit the system to behave as a DVN at $\Omega/\omega_1 = 1$ in addition to preventing the coupling spring from being damaged due to the extreme excursions of m_2 at frequencies Ω/ω_1 slightly below or above 1. For optimum operation over a wide frequency range, a low value of ω_2/ω_1 is more suitable. In that case, spring k_2 in conjunction with the infinitely rigid stops will behave as a nonlinear spring that is alternatively 'soft' and 'hard' as compared to k_1. It was pointed out in Ref. [19] that such an arrangement of the coupling spring in a nonlinear dynamic vibration absorber will improve the response.

Auxiliary mass critical damping ratio ξ_2. In cases where $d/(F_0/k_1)$ is such that impacts will occur, an increase in ξ_2 will have approximately the same effect as a reduction in e. However, when the stops are not engaged, ξ_2 will improve the performance of the DVN with respect to the excitation frequency. The optimum ξ_2 for these cases can be determined analytically [15].

Mass ratio μ. The system under discussion is effective even with mass ratios on the order of 5%. An increased mass ratio simplifies the optimization procedure, especially in regards to 'flatness' of the frequency response curve over a wide frequency band. In lightly damped primary systems, approximately the same results can be obtained with two different, $\mu's$ if the following relationship is maintained: $\mu_1 d_1 = \mu_2 d_2$, i.e., if the mass ratio is doubled, the clearance ratio should be halved. The effectiveness of the damper per unit μ decreases as μ increases [18].

Excitation frequency ratio Ω/ω_1. In an undamped DVN two natural frequencies exist at which the response of the system is not bounded or very large, depending on whether ξ_1 is zero or not. In a properly tuned Lanchester damper, one peak exists. The system under discussion usually has one peak below resonance at $\Omega/\omega_1 \approx (1 + \mu)^{-1/2}$ and the sharpness of this peak is primarily dependent on d and e. These effects of the highly nonlinear coupling spring agree with the observations of Pipes [20]. With a proper choice of parameters this peak can be considerably flattened thus permitting a nearly flat response curve around $\Omega/\omega_1 \approx 1$.

Primary mass critical damping ratio ξ_1. The addition of damping to the primary mass will tend to extend the stable ranges of operation of the system. As ξ_1 increases, the system can accommodate larger values of e since an increase in ξ_1, or decrease in e indicates an increase in mechanical energy dissipation. The effectiveness of the damper for a given μ decreases as ξ_1 increases [18].

(5) *Nonsymmetric motion*

Although steady-state unsymmetric motion does exist in some cases, in the overwhelming majority of cases where periodic motion occurs, it corresponds to the symmetric variety under consideration here.

Values of d near the lower bound of the stability boundaries usually lead to steady-state motion with two unsymmetric (unequally spaced in time) impacts/cycle which, in the context of the stability analysis under discussion, are classified as unstable. Experimental data points representing unsymmetric motion are usually bounded by the two symmetric solutions.

The existence of subharmonics [21] was noted in some of the experimental studies with systems having strong spring coupling between the two masses and small clearance ratios.

The analytical approach presented here can be easily extended to cases where the motion is repeated for values of $\Omega t \neq \pi$. For example, if the motion has two nonsymmetric impacts/cycle, then the period is $\Omega t = 2\pi$ which will result in twice as many unknowns and equations similar to those of Eq. (3.89). These equations will in turn lead to two-coupled transcendental equations whose solution will yield α_0, and the time spacing of the impacts.

(6) *Unstable motion*

Operation of the system with parameters leading to unstable solutions usually resulted in irregular motion whose amplitude did not exceed the corresponding motion that the system would have had in the absence of the damper. For the example, the termination of the stable range in Fig. 3.33 will lead to irregular-type motion (sometimes with beats) which has relatively small amplification ratio; see RHS in Fig. 3.33b.

3.1.4 Steady-State Solution and Stability Analysis of Particle Dampers

This part presents an analytical investigation of the steady-state motion and asymptotic stability of a two-particle, single-container impact damper attached to a sinusoidally excited primary system.

A model of the system under discussion is shown in Fig. 3.37. It is assumed that the two particles m_1 and m_2 are identical and that each particle has a single degree of freedom. Furthermore, it will be assumed that, if there is a collision between, let us say, m_1 and the RHS of the container at time $t = 0$, then the next impact will occur at $\Omega t = \alpha_0$ between the two particles. At $\Omega t = \pi$, there will be a collision between m_2 and the LHS of the container; and at $\Omega t = \pi + \alpha_0$, the two particles will collide again. The phase plane representation of the corresponding periodic motion is shown in Fig. 3.38. Experiments with a mechanical model showed that this type of symmetric 4 impacts/cycle motion predominates in steady-state response of the system to sinusoidal excitation.

Following a method due to Warburton [14] and assuming the disturbing force to $F_0 \sin(\Omega t + \alpha)$, where α is unknown phase angle, the equation of motion of the primary mass M on the analytic trajectory AB (Fig. 3.38) becomes.

$$M\,\ddot{x} + c\,\dot{x} + k\,x = F_0 \sin(\Omega t + \alpha) \qquad (3.91)$$

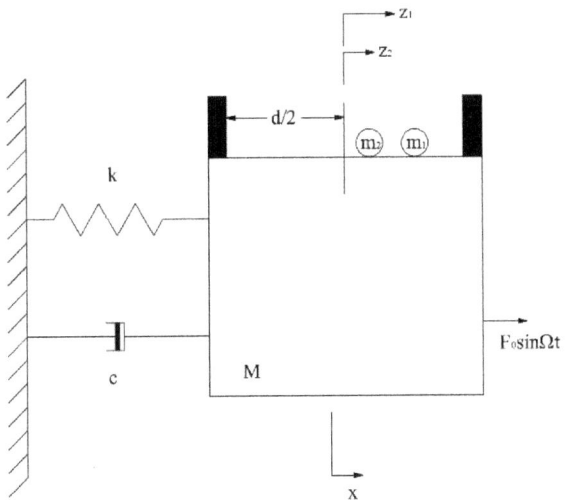

Fig. 3.37 Model of system

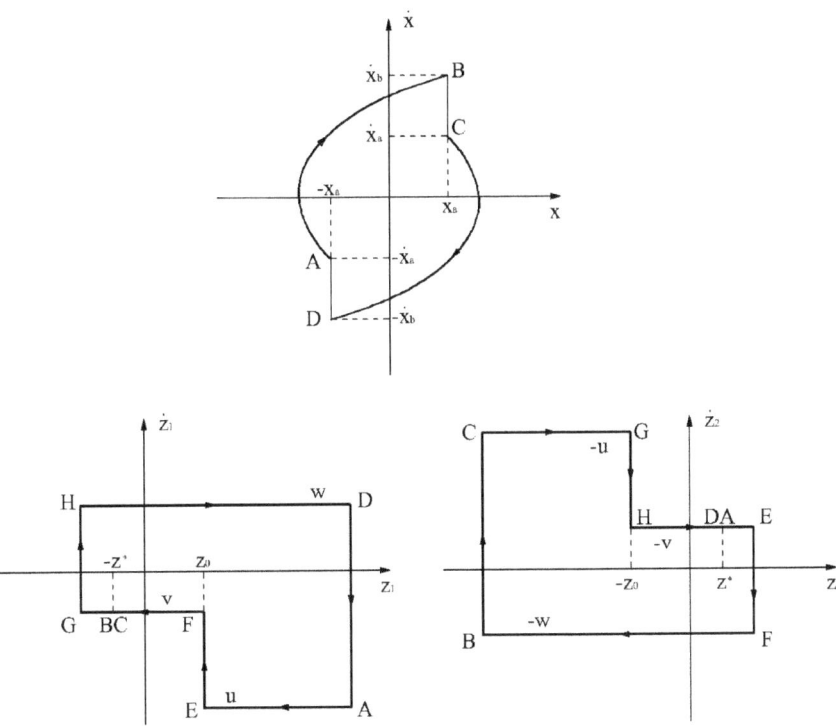

Fig. 3.38 Phase plane representation of periodic 4 impacts/cycle motion

and its complete solution is

$$x = e^{-\delta\omega t}[B_1 \sin \eta\omega t + B_2 \cos \eta\omega t] + A \sin(\Omega t + \tau) \quad (0 \le \Omega \le \pi) \quad (3.92)$$

where δ, ω, η, A, and τ are as given by Masri and Caughey [11], and B_1 and B_2 are unknown constants.

Evaluating Eq. (3.92) and its derivative at $\Omega t = 0$ and π, and by using the definition of the coefficient of restitution and the momentum equation, we obtain six relationships involving the six unknowns x_a, $\dot{x}_b$, $\dot{x}_a$, B_1, B_2 and τ. The value of B_1, B_2, and τ in the present case will be given by Eqs. (20), (21), and (23) Masri and Caughey [11] if their constants σ_1 and σ_2 are replaced by the new constants C_{19} and C_{20}, respectively, where

$$C_{19} \equiv (\alpha_0 + \pi \ C_9/2)[\Omega(k_7 C_{10} + k_8)]^{-1} \quad C_{20} \equiv (\alpha_0 + \pi \ C_9/2)[\Omega(k_9 C_{10} + k_{10})]^{-1}$$
$$C_9 \equiv (1 - e_3)/(3 + e_3) \qquad\qquad C_{10} \equiv -(1 + 3e_3)/(3 + e_3)$$
$$k_7 \equiv (e - \mu)/(1 + e) \qquad\qquad k_8 \equiv (1 + \mu)/(1 + e)$$
$$k_9 \equiv e(1 + \mu)/(1 + e) \qquad\qquad k_{10} \equiv (1 - \mu e)/(1 + e)$$
$$\alpha_0 \equiv \pi \ e_3/(1 + e_3) \qquad\qquad \mu \equiv m/M = m_1/M = m_2/M$$

where e is coefficient of restitution between container and m_i, e_3 is coefficient of restitution between m_1 and m_2.

Let the solution curve be perturbed right after an impact, e.g., at point A (see Fig. 3.38). The perturbations at point C are continuously related to the perturbations at point A by a relation of the form

$$\boldsymbol{\xi}_c = P_1 \ \boldsymbol{\xi}_{(0)} + \mathbf{R}_c(\boldsymbol{\xi}_{(0)})$$

where $\boldsymbol{\xi}_c$ is perturbation vector at C, P_1 is constant matrix, $\boldsymbol{\xi}_{(0)}$ is initial perturbation vector, and $\mathbf{R}_c$ contains all terms of $\boldsymbol{\xi}_{(0)}$ higher than the first power. Similarly, the perturbations at the end of a complete cycle, i.e., at A, will be related to those at C by

$$\boldsymbol{\xi}_{(1)} = P_2 \ \boldsymbol{\xi}_c + \mathbf{R}_A(\boldsymbol{\xi}_c) = P\boldsymbol{\xi}_{(0)} + \mathbf{R}^*(\boldsymbol{\xi}_{(0)}),$$

where P_2 is a constant matrix, $P = P_2 P_1$, and $\mathbf{R}^* = P_2\mathbf{R}_c + \mathbf{R}_A$.

By following the perturbed solution from one impact to the next one, we obtain the continuous transformation

$$\boldsymbol{\xi}_{(n)} = P^n \ \boldsymbol{\xi}_{(0)} + \sum_{i=0}^{n-1} P^{n-1-i}\mathbf{R}^*(\boldsymbol{\xi}_{(i)}).$$

It can be shown [11] that the solution will be asymptotically stable if and only if all the eigenvalues of P have modulus less than unity. The perturbed values of the variables starting with the initial perturbation $\Omega t = \Delta t_0$ are summarized for five consecutive impacts in Table 3.1. Note that

$$\xi_{(0)} = \text{col.}\{\eta_1, \eta_2, \eta_4, \eta_5, \eta_6, \eta_9\},$$
$$\xi_c = \text{col.}\{\eta_1'', \eta_2'', \eta_3'', \eta_4'', \eta_5'', \eta_6'', \eta_9''\}.$$

With some effort, it will be found that

$$P_1 = \begin{bmatrix} c_{33} & c_{34} & c_{35} & c_{36} & c_{37} & c_{38} \\ c_{50} & c_{51} & c_{52} & c_{53} & c_{54} & c_{55} \\ c_{62} & c_{63} & c_{64} & c_{65} & c_{66} & c_{67} \\ 0 & 0 & k_5 & 0 & -k_6 & 0 \\ c_{33} & c_{34} & c_{35} & c_{36} & c_{37} & c_{38} \\ c_{56} & c_{57} & c_{58} & c_{59} & c_{60} & c_{61} \\ c_{39} & c_{40} & c_{41} & c_{42} & c_{43} & 1 + c_{44} \end{bmatrix},$$

where

$$c_{33} \equiv -c_1 - c_3 c_{39} \qquad c_{34} \equiv -c_2 - c_3 c_{40}$$
$$c_{35} \equiv -c_3 c_{41} \qquad c_{36} \equiv -c_3 c_{42}$$
$$c_{37} \equiv -c_3 c_{43} \qquad c_{38} \equiv -c_4 - c_3 c_{44}$$
$$c_{50} \equiv -k_1(\rho_1 + c_{39}\rho_3) \qquad c_{51} \equiv -k_1(\rho_2 + c_{40}\rho_3)$$
$$c_{52} \equiv -k_1 c_{41}\rho_3 - k_2 k_6 \qquad c_{53} \equiv -k_1 c_{42}\rho_3$$
$$c_{54} \equiv -k_1 c_{43}\rho_3 + k_2 k_5 \qquad c_{55} \equiv -k_1(c_{44}\rho_3 + \rho_4)$$
$$c_{56} \equiv -k_3(\rho_1 + c_{39}\rho_3) \qquad c_{57} \equiv -k_3(\rho_2 + c_{40}\rho_3)$$
$$c_{58} \equiv -k_3 c_{41}\rho_3 - k_4 k_6 \qquad c_{59} \equiv -k_3 c_{42}\rho_3$$
$$c_{60} \equiv -k_3 c_{43}\rho_3 + k_4 k_5 \qquad c_{61} \equiv -k_3(\rho_4 + c_{44}\rho_3)$$
$$c_{62} \equiv -c_{39}v/\Omega \qquad c_{63} \equiv -c_{40}v/\Omega$$
$$c_{64} \equiv [vc_{41} - (\pi - \alpha_0)k_5]/\Omega \quad c_{65} \equiv (2c_{22} - c_{42})/\Omega - 1$$
$$c_{66} \equiv [(2c_{21} - c_{43})v + \alpha_0 + (\pi - \alpha_0)k_6]/\Omega \qquad c_{67} \equiv -c_{44}v/\Omega$$
$$u = \Omega[z_0 - (x_a + d/2)]/\alpha_0 \quad v = -2\Omega z_0/\pi$$

and

$$w = \Omega[z_0 + x_a + d/2]/(\pi - \alpha_0) \quad z_0 = (x_a + d/2)\left(1 - e_3^2\right)/\left(1 + 6e_3 + e_3^2\right)$$

$$c_0 \equiv (w + \Omega c_3)^{-1}; \ c_{21} \equiv -\alpha_0/(u + v); \ c_{22} \equiv \Omega/(u + v)$$

$$c_{39} \equiv -\Omega c_1 c_0; \ c_{40} \equiv -\Omega c_2 c_0; \ c_{41} \equiv [(w - v)c_{21} + k_6(\pi - \alpha_0)]c_0$$

Table 3.1 Summary of perturbed conditions

Ωt	x	$\dot{x}$	z_1	$\dot{z}_1$	z_2	$\dot{z}_2$	τ
Δt_{0+}	$x_a + \eta_1$	$\dot{x}_a + \eta_2$	$x_a + \frac{d}{2}$	$\mu + \eta_4$	$z^* + \eta_5$	$-(v + \eta_6)$	$\tau_0 + \eta_9$
$(\alpha_0 + \eta_7)_+$			$z_0 + \eta'_3$	$v + \eta'_4$	$z_0 + \eta'_5$	$-(w + \eta'_6)$	
$(\pi + \Delta t'_0)_+$	$-(x_a + \eta''_2)$	$-(\dot{x}_a + \eta''_2)$	$-(z^* + \eta''_3)$	$v + \eta''_4$	$-\left[x_a + \frac{d}{2} + \eta''_5\right]$	$-(u + \eta''_6)$	$\tau_0 + \eta''_9$
$(\pi + \alpha_0 + \eta'_7)_+$			$-(z_0 + \eta'''_3)$	$w + \eta'''_4$	$-(z_0 + \eta'''_5)$	$-(v + \eta'''_6)$	
$(2\pi + \Delta t''_0)_+$	$x_a + \eta^*_1$	$\dot{x}_a + \eta^*_2$	$x_a + \frac{d}{2} + \eta^*_3$	$u + \eta^*_4$	$z^* + \eta^*_5$	$-(v + \eta^*_6)$	$\tau_0 + \eta^*_9$

$$c_{42} \equiv [\Omega + (w - v)c_{22}]c_0; \; c_{43} \equiv [(w - v)c_{21} - \alpha_0 - k_5(\pi - \alpha_0)]c_0$$

$$c_{44} \equiv -\Omega c_4 c_0; \; k_5 \equiv (1 - e_3)/2; \; k_6 \equiv (1 + e_3)/2$$

and where $c_1, c_2, c_3, c_4, k_1, k_2, k_3, k_4, \rho_1, \rho_2, \rho_3$ and ρ_4 are as defined in a previous paper [11].

Following a similar procedure, it will be found that

$$P_2 = \begin{bmatrix}
c_{33} & c_{34} & c_{36} & c_{37} & d_{18} & c_{35} & c_{38} \\
c_{50} & c_{51} & c_{53} & c_{54} & d_{25} & c_{52} & c_{55} \\
c_{56} & c_{57} & c_{59} & c_{60} & d_{32} & c_{58} & c_{61} \\
c_{62} & c_{63} & c_{65} & c_{66} & d_{39} & c_{64} & c_{67} \\
0 & 0 & 0 & -k_6 & 0 & k_5 & 0 \\
c_{39} & c_{40} & c_{42} & c_{43} & d_{11} & c_{41} & 1 + c_{44}
\end{bmatrix},$$

where $d_{11} \equiv \Omega c_0(v - w)/(u + v); \; d_{18} \equiv -c_3 d_{11}; \; d_{25} \equiv -k_1 \rho_3 d_{11}; \; d_{32} \equiv -k_3 \rho_3 d_{11}; \; d_{39} \equiv -(2c_{22} + d_{11}) v/\Omega.$

If the same total mass is used in a single-particle as in a two particle impact damper, the former will be about twice as effective in reducing the magnification factor. In general, the effectiveness of impact vibration dampers improves as the particle mass ratio and coefficient of restitution increase and as the viscous damping in the primary system decreases.

Results of the steady-state solution and its stability analysis were verified through step-by-step construction of solutions on a digital computer.

3.2 Simplified Analytical Method

Because multiple particle damper shows strongly nonlinear performance when colliding, it is quite difficult to accurately analyze the dynamical system with multiple particle damper. In view of the fact that theoretical analysis of the single-particle damper is well developed and the motion characteristic of the damper, a simplified analytical solution in which the particles in a PTMD are equivalent to a single particle based on certain principles is used to simulate the vibration reduction effects of PTMDs [22]. In the method, the collision between the particles is neglected, and the collision between particles and the container wall is emphasized. Because the collision force between the particle and the container wall is the key control force affecting the vibration responses of the main structure. Considering the characteristic of the particle damper vibration simulation, the simplified analytical method has certain precision and can be used to be referred during engineering practical design. This section mainly takes the single-degree-of-freedom (SDOF) structure as an example. Meanwhile, the section introduces its implementation. More examples will be introduced in Chaps. 6 and 7.

3.2.1 Equivalent Principles

1. Equivalent principles

Based on the following equivalent principles, the particle group in the PTMD is replaced by a single particle:

(1) The void volume of the PTMD (V_{epd}) equals that of the single-particle damper (V_{eid}).
(2) The total mass of the particles in the PTMD (m) equals that of the single particle (m_p).
(3) The particles in both the PTMD and single-particle damper are spheres, and their densities are kept constant as ρ.
(4) The container of the PTMD is a cuboid with a height that equals the diameter of the particles (D_p), whereas the container of the single-particle damper is a cylinder and its bottom diameter equals the diameter of a single particle (D).

The diagrams of the PTMD and equivalent single-particle damper according to the above principles are shown in Fig. 3.39. The parameters and symbols of the PTMD and equivalent single-particle damper are shown in Table 3.2.

The simplified method is as follows:

The volume of the particles in the PTMD V_{spd} is given by Eq. (3.93):

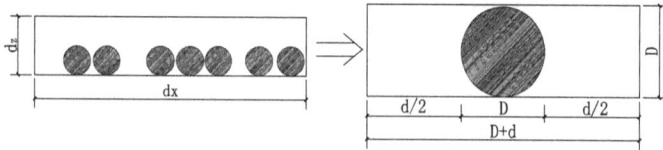

Fig. 3.39 Diagrams of the PTMD and equivalent single-particle damper

	Physical meaning	Symbol
PTMD	Cavity size (Length, width and height)	(dx, dy, dz)
	The mass of single-particle	m_{1p}
	The mass of all particles	m
	The diameter of single-particle	D_p
Single-particle damper	The mass of particle	m_3
	The diameter of particle	D
	Free moving distance of particle	d

Table 3.2 The parameters and symbols of the PTMD and the equivalent single-particle tuned damper

$$V_{spd} = 1/6N\pi D_p^3 \qquad\qquad (3.93)$$

where N is the number of particles, $N = m/m_{1p}$, and m_{1p} is the mass of a single particle in the PTMD.

The void volume of the PTMD is given by Eq. (3.94):

$$V_{epd} = V_{pd} - V_{spd} = (1/\rho_p - 1)V_{spd} \qquad\qquad (3.94)$$

where V_{pd} is the volume of the PTMD and $\rho_p = V_{spd}/V_{pd}$ is the packing density of the PTMD.

The void volume of the single-particle damper V_{eid} is:

$$V_{eid} = \pi/12D^3 + \pi/4D^2d \qquad\qquad (3.95)$$

where d is the clearance of the simplified single-particle damper.

Based on the equivalent principle (1), the clearance of the simplified single-particle damper can be obtained from Eq. (3.96):

$$\left(\frac{1}{\rho_p} - 1\right)\frac{m}{\rho} = \frac{m_3}{2\rho} + \frac{\pi}{4}\left(6\frac{m_3}{\pi\rho}\right)^{\frac{2}{3}}d \qquad\qquad (3.96)$$

The study of Hales shows that the volume ratio of balls with the same radium cannot exceed 0.74 when densely stuffed, i.e. $\rho_p \leq 0.74$. According to the equivalent principle (2), $m = m_3$. The expression of the length of particles freely moving in the equivalent single-particle damper is established, namely, the Eq. (3.96).

2. *Particle-container wall collision simulation*

In the simplified analysis method, the most critical step is the collision simulation of the particle-container wall, which is segmented linearly, as is shown in Fig. 3.40.

y is the relative displacement of the particle and the cavity. $G(y)$ and $H(y, \dot{y})$ are nonlinear equations of the elastic force and damping force when colliding. When

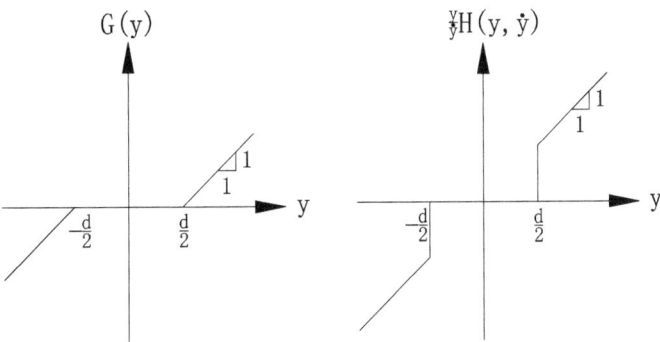

Fig. 3.40 $G(y)$ and $H(y, \dot{y})$

particle moves within the length d of the container, no collision occurs, $G(y)$ and $H(y, \dot{y})$ are zero. When the particles hit the wall of the container $(\pm d/2)$, the collision force is generated, that is, $G(y)$ and $H(y, \dot{y})$ are generated.

3. Equation of motion

Taking the shaking table test of the SDOF structure with the particle damper as an example. In the test, a particle damper is suspended by a rope at the top of the structure. The diagram calculation mode of the system is shown in Fig. 3.41.

The governing equations for the entire system of the additional particle damper are expressed in matrix as Eq. (3.97):

$$[M]\{\ddot{X}(t)\} + [C]\{\dot{X}(t)\} + [K]\{X(t)\} = \{F_p(t)\} \qquad (3.97)$$

where $[M]$ is mass matrix; $[C]$ is damping matrix; $[K]$ is stiffness matrix. $\{\ddot{X}(t)\}$ is acceleration vector; $\{\dot{X}(t)\}$ is velocity vector; $\{X(t)\}$ is displacement vector; $\{Fp(t)\}$ is load vector.

Assumed that the displacement of the shaking table is $x_g(t)$; the displacement of the main structure is $x_1(t)$; the mass of the main structure is m_1; the displacement of the cavity is $x_2(t)$); the mass of the cavity is m_2; the displacement of the particle is $x_3(t)$, and the mass of the particle is m_3. The system control equation is expanded as shown in Eq. (3.98):

$$\begin{cases} m_1(\ddot{x}_1 + \ddot{x}_g) + c_1\dot{x}_1 + k_1x_1 - c_2(\dot{x}_2 - \dot{x}_1) - k_2(x_2 - x_1) = 0 \\ m_2\ddot{x}_2 + c_2(\dot{x}_2 - \dot{x}_1) + k_2(x_2 - x_1) - c_3H(y, \dot{y}) - k_3G(y) = 0 \qquad (3.98) \\ m_3\ddot{x}_3 + c_3H(y, \dot{y}) + k_3G(y) = 0 \end{cases}$$

Fig. 3.41 The diagram of calculation model

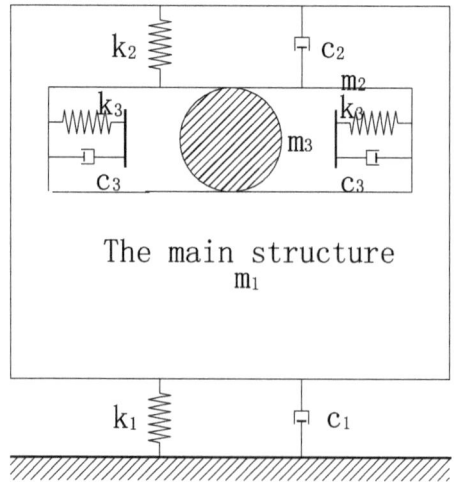

where $y = x_3 - x_2$ represents the relative displacement of the particles and the cavity. The expression of $G(y)$ and $H(y, \dot{y})$ is shown in Fig. 3.40, which is the nonlinear equation characterizing the model.

3.2.2 Parameter Selection

The mass of the main structure can be obtained by measuring the mass of the model. The circular frequency and damping ratio can be obtained by the white noise sweep and the half power method. In the same way, the mass of the damper cavity can be obtained. Considering the tuning, the circular frequency of the damper is the same as the self-vibrating circular frequency of the main structure. The damping ratio of the cavity is difficult to measure, which is determined by trial calculation in numerical simulation.

According to the study of Masri [23] when the spring stiffness k_3 is much larger than the stiffness of the main structure, the interaction between the particles and the vessel wall can be simulated suitably. So the particle circular frequency $\omega_3 = 20\omega_1$ is generally taken. The damping ratio ξ_3 of the particle and the recovery coefficient e of the particle are obtained according to the physical properties of the particle material.

3.2.3 Program Compiling

Taking the shaking table test of the (SDOF) structure with the particle damper as an example to illustrate the implementation and application of the equivalent simplified analytical method.

The main structure is a single-layer steel frame. The total mass of the structure is 7.6 kg, and the natural frequency is 1.37 Hz.

The particle-filled container is suspended vertically from the top of the structure by four ropes. Taking into account the dynamic performance of the damper, the damper is simplified to a single-pendulum. Considering the condition that the damper and the main structure frequency are the same, the suspension length of the rope is calculated according to the rope length calculation equation of the single-pendulum. The box has the same cross-sectional dimensions, 60 mm × 60 mm, and length is 80 mm. The box contains 26 steel balls whose diameters are 10 mm and the particle density ρ is 7644 kg/m^3. The El Centro is chosen as the seismic wave with the sampling period of 0.02 s.

According to the results from program, the calculated value and experimental value of the acceleration and displacement response of the top layer of the single-layer steel frame model with the additional particle damper under the El-Centro seismic excitation is shown in Fig. 3.42. Overall, the calculated value and experimental value fit well with each other.

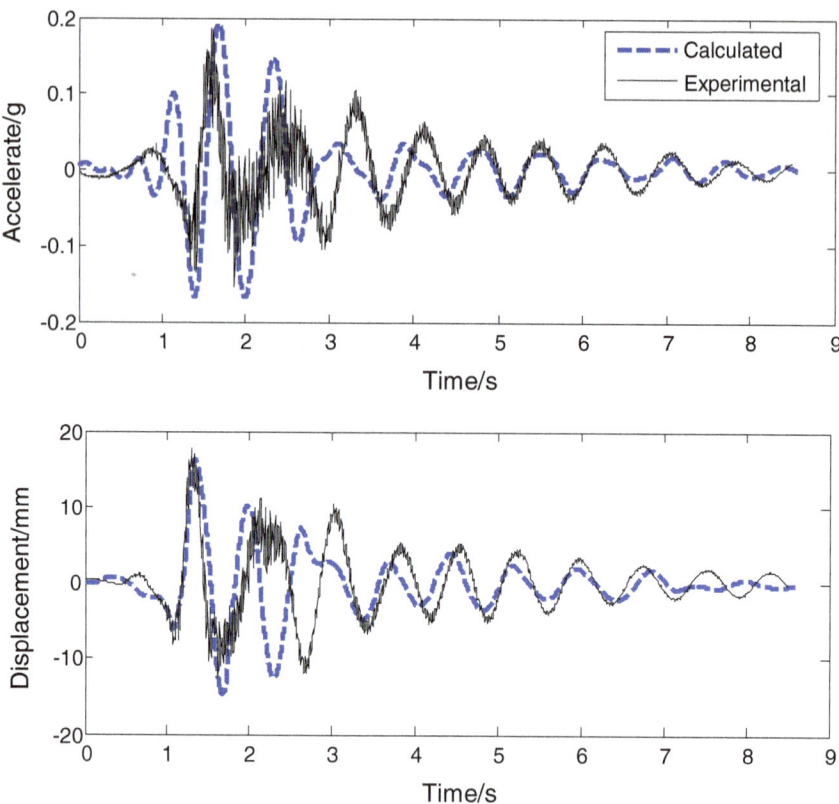

Fig. 3.42 The acceleration and displacement response of the top layer of the single-layer steel frame model with the additional particle damper under the El-Centro seismic excitation

In addition, in order to verify the feasibility and accuracy of the simplified analytical method, by transforming the particle forms, the peak and root-mean-square (R.M.S.) acceleration responses of the structural model with different particle forms under different excitations are obtained. The calculated value is compared with the experimental value as shown in Table 3.3. It could be found that the peak acceleration fits well, and the R.M.S. acceleration error can also be limited within an acceptable range.

3.3 Spherical Discrete Element Simulation

The discrete element method (DEM) is a discontinuous numerical calculation method proposed by Cundall P.A., an American scholar, in 1979 [24]. The DEM was originally used to analyze the motion of rock slopes. With the deepening of research

Table 3.3 The comparison between the calculated value and experimental value of the peak and R.M.S. acceleration responses of the structural model with particle tuned damper

Seismic wave	Particle	The peak acceleration			The R.M.S. acceleration		
		Calculated value (m/s^2)	Experimental value (m/s^2)	Error (%)	Calculated value (m/s^2)	Experimental value (m/s^2)	Error (%)
El-Centro	4 × 20 mm steel ball	2.6024	2.5645	1.46	0.5696	0.5378	5.58
	5 × 20 mm steel ball	2.6049	2.6949	−3.46	0.5582	0.5538	0.79
	19 × 10 mm steel ball	1.9674	1.9299	1.90	0.5771	0.5494	4.81
	26 × 10 mm steel ball	1.8833	1.8259	3.05	0.5763	0.5607	2.71
Kobe	4 × 20 mm steel ball	3.5215	3.5518	−0.86	0.5056	0.5377	−6.34
	5 × 20 mm steel ball	3.7400	3.7354	0.12	0.5345	0.5799	−8.49
	19 × 10 mm steel ball	3.8869	3.8562	0.79	0.5245	0.5751	−9.64
	26 × 10 mm steel ball	3.6410	3.7484	−2.95	0.4983	0.5435	−9.07

and the development of computer technology, in addition to the study of stability of slopes, mining and roadways, and the analysis of the microstructure of granular, the DEM has been extended to be used for not only the dynamic processes of earthquakes and explosions but also the physical processes such as infiltration and heat transfer [25, 26]. In the past decade, the DEM has been gradually introduced into the field of structural engineering [27, 28]. As far as the DEM itself is concerned, it can meticulously simulate the interaction between discrete units, and is a powerful tool for simulation calculation. For different subjects, different discrete unit models must be established. Common discrete unit models are block units (two-dimensional or three-dimensional), disk units (two-dimensional), elliptical units (two-dimensional), and spherical units (three-dimensional). For different unit models, the principle and calculation process of DEM are the same. However, the specific calculation method and data structure are different.

3.3.1 *Fundamental Principles of Discrete Element Method*

The basic principle of the DEM is to divide the research object into discrete blocks or sphere units. During the process of deformation and motion, the unit could be in contact with its neighboring units or separated. The units only need to satisfy the constitutive relation, the equilibrium relationship and the boundary conditions. There is no constraint relationship between the units. Therefore, the DEM is especially suitable for solving large deformation and discontinuous structure problems.

Different from the meaning of the constitutive relation in the finite element method, the constitutive relation of the DEM is used to determine the relationship between the relative displacement and the interaction force between the units, that is, the force-displacement relationship. According to the DEM, the units can interact with each other through contact points, contact faces or connecting springs. The value of the interaction force is determined by the contact constitutive equation or the constitutive relationship of the connecting springs. The motion of some units is determined by the imbalance and unbalanced moments of the unit based on Newton's law of motion [29].

1. *Basic assumption*

   (1) The contact force of each particle and the displacement of the particle combination can be obtained by performing a series of calculations on the trajectories of the respective particles. These motions are the result of dynamic propagation of the self-weight, external load or disturbance source on the boundary in the medium. The propagation velocity is the function of the physical properties of the discrete medium. In the description of the motion numerical properties of the particle the velocity and acceleration are assumed to be constant in a one-time step in the display iterative calculation. Meanwhile, this time step can be chosen to be so small that during each time step, the disturbance cannot propagate from either particle to its neighboring particles. So, at any time, the resultant force of any particles can be uniquely determined by the interaction of the particles in contact with it. In addition, the display algorithm does not need to calculate a structural stiffness matrix to avoid complex matrix operations. Therefore, when considering large displacements and nonlinearities, the DEM is very stable compared to the finite element method.

   (2) When the particles interact, it is assumed that there is a overlapping amount at the contact point. This superposition characteristic reveals the deformation of a single particle. The amount of the overlapping is directly related to the

Fig. 3.43 a Normal contact force model between particle and wall; and **b** Normal contact force model between particle and particle

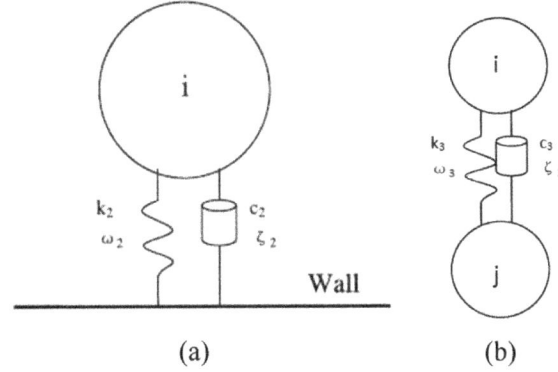

(a)　　　　　　　(b)

contact force. However, these overlappings are much smaller relative to the particle size. The deformation of the particles themselves is much smaller than the translation and rotation of the particles. So the particles are treated as rigid bodies. This not only simplifies the calculation, but also does not cause excessive errors. The motion properties of the particles are represented by the gravity centers, and the contact forces between the particles follow the laws of action and reaction [30].

(3) There is no tensile force after particle separating.

2. The force-displacement relationship

There are scholars from various countries have proposed a variety of contact force models of discrete units to quantitatively determine normal and tangential forces. However, this is still a hot topic, especially for the determination of tangential forces [31–33]. In this paper, the simplest force-displacement model is used: the normal contact linear force model and the tangential Coulomb friction model.

Figure 3.43a is a normal linear contact force model of the particle and container wall. k_2 is the spring stiffness. The variable $\omega_2 = \sqrt{k_2/m}$ is the natural frequency, which can be used to simulate a container wall. The ratio of $\omega_2/\omega_n \geq 20$ [34] is appropriate to represent the container wall (ω_n is the fundamental frequency of the primary system). The parameter $\zeta_2 = c_2/(2m\omega_2)$ is the damping ratio, which can be used to simulate inelastic impacts, ranging from the completely plastic up to the elastic one, so that the value of any desired coefficient of restitution e can be adjusted by selecting the proper value for ζ_2. Similarly, k_3, ω_3, c_3 and ζ_3 are the stiffness, natural frequency of the spring, damping coefficient and damping ratio of the damper, respectively, in the inter-particle contact model along the normal direction. Hence, the normal contact force is expressed by

$$F_{ij}^n = \begin{cases} k_2\delta_n + 2\zeta_2\sqrt{mk_2}\dot{\delta}_n, & \delta_n = r_i - \Delta_i \quad (particle - wall) \\ k_3\delta_n + 2\zeta_3\sqrt{\frac{m_im_j}{m_i+m_j}}k_3\dot{\delta}_n, & \delta_n = r_i + r_j - |\mathbf{p}_j - \mathbf{p}_i| \quad (particle - particle) \end{cases}$$

$$(3.99)$$

where δ_n and $\dot{\delta}_n$ are the relative displacement and relative velocity of particle i relative to particle j in the normal direction, respectively, and Δ_i is the distance from the center of particle i to the wall.

Considering Coulomb's friction, the tangential contact force is expressed by

$$F_{ij}^t = -\mu_s F_{ij}^n \dot{\delta}_t / |\dot{\delta}_t| \qquad (3.100)$$

where μ_s is the coefficient of friction between any two particles or between a particle and the wall of the container, and $\dot{\delta}_t$ is the velocity of particle i relative to particle j or the wall, in the tangential direction.

First, selecting a particle unit i. Then, according to the mutual contact relationship (relative displacement) between the units and the contact constitutive relationship (force-displacement relationship), all contact forces acting on the particle unit can be obtained. With other force received by other units (such as gravity), the resultant force and resultant moment could be calculated. According to Newton's second law of motion, the equation of motion of particle i can be obtained.

$$m_i \ddot{\boldsymbol{p}}_i = m_i \boldsymbol{g} + \sum_{j=1}^{k_i} (\boldsymbol{F}_{ij}^n + \boldsymbol{F}_{ij}^t), \quad \boldsymbol{I}_i \ddot{\varphi}_i = \sum_{j=1}^{k_i} \boldsymbol{T}_{ij} \qquad (3.101)$$

where m_i is the mass of particle i, $\boldsymbol{I}_i$ is the moment of inertia of particle i and g is the acceleration vector due to gravity; $\boldsymbol{p}_i$ is the position vector of the center of gravity of particle i, φ_i is the angular displacement vector, $\boldsymbol{F}_{ij}^n$ is the normal contact force between particle i and particle j (if particle i is in contact with container wall, then j denotes that wall) and $\boldsymbol{F}_{ij}^t$ is the tangential contact force. The contact forces act at the contact point between particle i and particle j rather than the particle center, and they will generate a torque, $\boldsymbol{T}_{ij}$, causing particle j to rotate. For a spherical particle of radius r_i, $\boldsymbol{T}_{ij}$ is given by $\boldsymbol{T}_{ij} = r_i \boldsymbol{n}_{ij} \times \boldsymbol{F}_{ij}^t$, where $\boldsymbol{n}_{ij}$ is the unit vector from the center of particle i to the center of particle j. $\times$ denotes the cross product. These inter-particle forces are summed over the ki particles in contact with particle i.

3.3.2 Spherical Discrete Element Modeling

1. *Force calculation between units*

As shown in Fig. 3.44, the global coordinate system is set to OXYZ and the local coordinate system is set to OXYZ. The line taking the point i-j is the X-axis. In the plane perpendicular to the X-axis, a line parallel to the x-y plane is taken as the Y-axis, and the Z-axis is determined by the right-handed screw rule. So, the X-axis is along the normal of the two particles, the Y-axis and the Z-axis are on the tangent plane. The transformation matrix of the two sets of coordinate systems is:

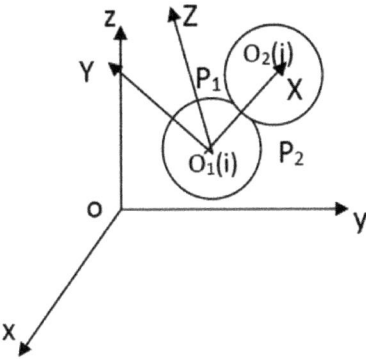

Fig. 3.44 Diagram of coordinate system and relative position of particles

$$
\left\{ \begin{array}{c} X \\ Y \\ Z \end{array} \right\} = \begin{pmatrix} l_{Xx} & l_{Xy} & l_{Xz} \\ l_{Yx} & l_{Yy} & l_{Yz} \\ l_{Zx} & l_{Zy} & l_{Zz} \end{pmatrix} \left\{ \begin{array}{c} x \\ y \\ z \end{array} \right\} = [Te] \left\{ \begin{array}{c} x \\ y \\ z \end{array} \right\} \tag{3.102}
$$

where l_{Xx} is the direction cosine of the X-axis and the x-axis and $[Te]$ is the coordinate transformation matrix.

There are two balls i and j in the space. The line connects the ball i to P1, and the ball j to P2. When the distance between the two balls is smaller than the sum of the radius of the two balls, it is seemed that the two spheres are in contact collision. The speed at point P1 is synthesized by the translational speed of the ball i and the rotational speed around the center of the ball. It is assumed that translation speed of the ball i be $(V_{olx}, V_{oly}, V_{olz})$, the rotation speed be $(\dot{\theta}_{o1x}, \dot{\theta}_{o1y}, \dot{\theta}_{o1z})$, the translation speed of the ball j is $(V_{o2x}, V_{o2y}, V_{o2z})$, and the rotation speed is $(\dot{\theta}_{o2x}, \dot{\theta}_{o2y}, \dot{\theta}_{o2z})$. According to kinematics of a rigid body, for the ball i, the speed of the P1 point is

$$
\mathbf{V}_{p1} = \mathbf{V}_{o1} + \boldsymbol{\omega}_{o1} \times \bar{r} \tag{3.103}
$$

where $\mathbf{V}_{o1}$ is the translation speed of O_1 and $\boldsymbol{\omega}_{o1}$ is the rotation speed of the ball, $\bar{r} = \overline{O_1 P_1}$. The component form of speed $\mathbf{V}_{p1}$ is

$$
\begin{cases} V_{p_1x} = V_{o_1x} + (\dot{\theta}_{o_1y} r_1 \cos \gamma - \dot{\theta}_{o_1z} r_1 \cos \beta) \\ V_{p_1y} = V_{o_1y} - (\dot{\theta}_{o_1x} r_1 \cos \gamma - \dot{\theta}_{o_1z} r_1 \cos \alpha) \\ V_{p_1z} = V_{o_1z} + (\dot{\theta}_{o_1x} r_1 \cos \beta - \dot{\theta}_{o_1y} r_1 \cos \alpha) \end{cases} \tag{3.104}
$$

where r_1 is the radius of ball i, *and* $e = (e1, e2, e3) = (\cos \alpha, \cos \beta, \cos \gamma)$ is the direction cosine of $O_1 P_1$, which is also the contact normal vector. Similarly, the speed of point P_2 $(\mathbf{V}_{p2})$ is

$$\begin{cases} V_{p_2x} = V_{o_2x} - (\dot{\theta}_{o_2y}r_2 \cos \gamma - \dot{\theta}_{o_2z}r_2 \cos \beta) \\ V_{p_2y} = V_{o_2y} + (\dot{\theta}_{o_2x}r_2 \cos \gamma - \dot{\theta}_{o_2z}r_2 \cos \alpha) \\ V_{p_2z} = V_{o_2z} - (\dot{\theta}_{o_2x}r_2 \cos \beta - \dot{\theta}_{o_2y}r_2 \cos \alpha) \end{cases} \tag{3.105}$$

where r_2 is the radius of j and $(-\cos \alpha, -\cos \beta, -\cos \gamma)$ is the direction cosine of O_2P_2.

The speed $(\Delta V_x, \Delta V_y, \Delta V_z)$ of the contact point is obtained.

$$\begin{cases} \Delta V_x = V_{p_1x} - V_{p_2x} \\ \Delta V_y = V_{p_1y} - V_{p_2y} \\ \Delta V_z = V_{p_1z} - V_{p_2z} \end{cases} \tag{3.106}$$

Transform the speed to the local coordinate system:

$$\begin{Bmatrix} \Delta V_X \\ \Delta V_Y \\ \Delta V_Z \end{Bmatrix} = [Te] \begin{Bmatrix} \Delta V_x \\ \Delta V_y \\ \Delta V_z \end{Bmatrix} \tag{3.107}$$

Then, the normal velocity and the tangential velocity are obtained, and the contact force between the normal direction and the tangential direction can be obtained according to the contact model. The method to calculate the force when the particles are in contact with the container wall is completely consistent with the method to calculate the interaction force between the spherical particle units.

2. *Determination of the calculation step dt*

Since DEM is an iterative method, the selection of iteration time step is key, which directly influence the stability of the calculation process. In general, under the premise of ensuring stability and calculation accuracy, it is desirable to take a longer time step as much as possible, which can reduce the amount of calculation process. In terms of DEM, the basic equation of motion for a particle unit is:

$$m\ddot{x}(t) + c\dot{x}(t) + kx(t) = F(t) \tag{3.108}$$

where m is the mass of the particle unit, x is the displacement, t is the time, c is the viscous damping coefficient, k is the stiffness coefficient, and $F(t)$ is the external force to which the element is subjected. To stabilize the solution, the following equation must be satisfied:

$$dt \leq 2\sqrt{\frac{m}{k}}(\sqrt{1 + \varsigma^2} - \varsigma) \tag{3.109}$$

where $\varsigma = c/(2\sqrt{mk})$ is the damping ratio of the system.

3. *Determination of normal damping coefficient*

The normal damping coefficient can be obtained from the recovery coefficient of the collision between the ball and the wall. The collision model is shown in Fig. 3.43a. The dynamic equation after the ball is in contact with the vessel wall is:

$$m\ddot{x}(t) + c\dot{x}(t) + kx(t) = 0 \qquad (3.110)$$

The initial position of the particle is $x_0 = 0$, the incident velocity before the contact is $\dot{x}_0^- = \dot{x}_0$. The displacement response and velocity response are as follows:

$$x(t) = \exp(-\zeta_2\omega_n t)\frac{\dot{x}_0}{\omega_d}\sin(\omega_d t) \qquad (3.111)$$

$$\dot{x}(t) = \exp(-\zeta_2\omega_n t)(\dot{x}_0\cos(\omega_d t) - \frac{\zeta_2\dot{x}_0}{\sqrt{1-\zeta_2^2}}\sin(\omega_d t)) \qquad (3.112)$$

where $\omega_d = \omega_n\sqrt{1-\zeta_2^2}$.

Assumed that at the end of the collision, $t = t_p$, the following should be satisfied:

$$x(t) = 0 \qquad (3.113)$$

The time of collision:

$$t_p = \frac{\pi}{\omega_d} \qquad (3.114)$$

According to the definition of particle recovery coefficient:

$$e = \left|\frac{\dot{x}_0^+}{\dot{x}_0^-}\right| = \left|\frac{\exp(-\zeta_2\omega_n t_p)(\dot{x}_0\cos(\omega_d t_p) - \frac{\zeta_2\dot{x}_0}{\sqrt{1-\zeta_2^2}}\sin(\omega_d t_p))}{\dot{x}_0^-}\right| = \exp(\frac{-\zeta_2\pi}{\sqrt{1-\zeta_2^2}}) \qquad (3.115)$$

Then, the relationship between the normal damping coefficient and the particle recovery coefficient can be obtained. By adjusting the value of ξ_2 can simulate the elastic state of various material particles, as shown in Fig. 3.45.

4. *Contact detection algorithm*

Since the DEM assumes that within one time step, the disturbance of one particle will not transmitted to other particles than the particle it is in contact with, the iterative time step needs to be short. Meanwhile, the contact relationship between each particle should be judged at each time step. So it is a research method that needs large amount of calculation. In DEM, the contact detection algorithm is especially critical.

Fig. 3.45 Relationship
between the normal damping
coefficient and the particle
recovery coefficient

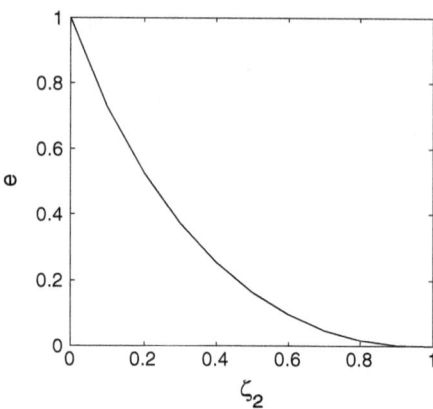

The most intuitive and simple algorithm of contact detection is traversal algorithm, and the complexity of the detection time T_d is N^2 (N is the number of particles). The most commonly used algorithm [35–39] is generally binary search, whose complexity is

$$T_d \propto N \ln(N) \tag{3.116}$$

The paper choose (*No Binary Search, NBS*) algorithm proposed by Munjiza [40]. The collision detection algorithm, and its complexity is:

$$T_d \propto N \tag{3.117}$$

The NBS is suitable for the case where the particles radius differ little. There is no limit to the degree of density of the particles. The NBS is briefly introduced by collision detection of two-dimensional circular particles.

The equal-sized particle units with the labels {0, 1, 2, …, $N - 1$} are located within a rectangular range, which is divided into square grids with the side length of $2r$ (r is the particle radius). The NBS is based on this spatial meshing. The reason that the side length is $2r$ is to ensure that each particle can only be located within one grid.

Each square grid is represented by two-dimensional coordinates (ix,iy) ($ix = 0, 1, 2, …, ncelx$-1; $iy = 0, 1, 2, …, ncely$-1), where $ncelx$ and $ncely$ are the total number of grids along the X and Y directions, respectively.

$$ncelx = \frac{x_{max} - x_{min}}{2r} \tag{3.118}$$

$$ncely = \frac{y_{max} - y_{min}}{2r} \tag{3.119}$$

$$ix = Int\left(\frac{x - x_{min}}{2r}\right) \qquad (3.120)$$

$$iy = Int\left(\frac{y - y_{min}}{2r}\right) \qquad (3.121)$$

where x_{min}, x_{max}, y_{min}, y_{max} are the four boundaries of the rectangular area.

In this way, all particle sets $E_p = \{0, 1, 2, \ldots, N - 1\}$ can be mapped to the square meshes set C:

$$C = \left\{ \begin{array}{lll} (0, 0), & (0, 1), & (0, ncely - 1) \\ (1, 0), & (1, 1), & (1, ncely - 1) \\ (ncelx - 1, 0), & (ncelx - 1, 1), \ldots & (ncelx - 1, ncely - 1) \end{array} \right\} \qquad (3.122)$$

In order to save memory, a linked list structure is used to store data. First, the particles are cycled once, and the particles are mapped to the linked list according to the y-coordinates of the particles. The linked list is formed by two arrays, one is the heady array, which stores the last particle label in the row, so the size of the heady array is ncely; the other is the *nexty array*, for any particle, the array stores the labels of particles in the same row of the particle. So the size of the *nexty* array is N. Both arrays end up with −1. In addition, if the mesh has no particles, it is also marked with −1. For example, if the particles 0, 4, 5, 6, and 7 are all on the second line, then *heady*[41] = 7, *nexty*[42] = 6, *nexty*[43] = 5, *nexty*[44] = 4, *nexty*[45] = 0, *next*[0] = −1. If there is no particle in line 0, then *heady*[0] = −1, and Y_{iy} is marked as new.

Second, all particle are cycled to detect new Y_{iy} and mark it old. Each particle located in the chain maps to the (X_{ix}, Y_{iy}) chain according to its x coordinate and is labeled new. According to the same method, a *headx* and *nextx* array is created. Then, all the particles are mapped one-to-one on the linked list.

(1) *Contact detection*

For a square grid, the particle contact can be determined by detecting the grid adjacent to it. For example, if a particle is mapped in the grid, then contact of particles only in the grid (ix, iy), $(ix − 1, iy)$, $(ix − 1, iy − 1)$, $(ix, iy − 1)$ and $(ix + 1, iy − 1)$ should be detected.

Therefore, in order to ensure that (X_{ix}, Y_{iy}) only corresponds to the particles of adjacent rows, a *headsx* array is created, which is a two-dimensional array with a size of 2ncelx. For example, *headsx*[2] [ncelx], where the array *headsx*[0] corresponds to a single linked list (X_{ix}, Y_{iy}), and the array headsx[1] corresponds to a single linked list (X_{ix}, Y_{iy-1}).

(2) *Execution process*

According to the above, the execution process of the NBS is as follows: [41].

1. Cycle for all particles

{

Calculate the coordinates of each particle to get *ix* and *iy*

Put the current particle into the linked list Y_{iy}

Mark the linked list Y_{iy} as new

}

2. Cycle for all particles

{

If the particle belongs to the new list Y_{iy}

   {

      Mark the linked list Y_{iy} as old

      3. Cycle for the particles in linked list Y_{iy}

        {

           Put the current particle into the linked list $(X_{ix},\ Y_{iy})$

           Mark the linked list $(X_{ix},\ Y_{iy})$ as old

        }

      4. Cycle for the particles in linked list Y_{iy-1}

        {

           Put the current particle into the linked list $(X_{ix},\ Y_{iy-1})$

        }

      5.Cycle for the particles in linked list Y_{iy}

        {

           If the particle belongs to the new list *$(X_{ix},\ Y_{iy})$*

           {

              Mark the $(X_{ix},\ Y_{iy})$ linked list as old

              Detect collision of the particles belong to linked list $(X_{ix},\ Y_{iy}), (X_{ix}, Y_{iy}),$

                    $(X_{ix-1}, Y_{iy}), (X_{ix-1}, Y_{iy-1}), (X_{ix}, Y_{iy-1})$ and (X_{ix+1}, Y_{iy-1})

           }

        }

      6. Cycle for the particles in linked list Y_{iy}

        {

           Remove linked list$(X_{ix},\ Y_{iy})$ by setting *headsx*[0][281] = -1

        }

      7. Cycle for the particles in linked list Y_{iy-1}

        {

           Remove linked list $(X_{ix},\ Y_{iy-1})$ by setting *headsx*[1][281] = -1

        }

   }

}

8. Cycle for all particles

{

Remove linked list Y_{iy} by setting *heady*[iy] = -1

}

3.3.3 Program Compiling

1. *Generation of particle combinations*

The first step of DEM is to simulate generate a certain number of particle combinations that follow the certain distribution law (space geometry, gradation and shape, etc.) and certain boundary conditions. In this paper, spherical particles of the same radius are considered. Assumed that the spatial geometric position is randomly distributed. After a certain number of particles are generated in the container, they fall freely and piled up only under the action of gravity, which is an initial state of calculation. The block diagram is shown in Fig. 3.46.

After the Spherical bulk unit model is established, the DEM can be used to simulate the mechanical properties of the particle damper. First, calculate the relative position not only between the particles but also between the particles and the wall. If $\delta_n > 0$, the contact force acting on the particle can be computed from Eqs. (3.99) and (3.100); while if $\delta_n \leq 0$, no contact force is generated. Second, sum all the contact forces exerting on a particle, including inter-particle forces and particle-wall forces, if they exist. Then, compute the motion of the particle by Eq. (3.101); the same procedure is repeated for all the particles. Finally, sum all the contact forces between particles and the container wall, and the combined force of the contact forces on the wall is obtained, which is the resultant force acting on the main body system for each time step. Then solve the dynamic equation of the main system and get the response of the system. The block diagram is shown in Fig. 3.47.

3.4 Verification of Spherical Discrete Element Simulation

Section 3.2 establishes the spherical DEM that simulates a particle damper. This section verifies its correctness to lay the foundation for the analysis in the following chapters. Two methods are generally used to verify the program. One is the theoretical verification of the limit case. In other words, the calculation result of the program should be logical common sense. The other is that the model is compared with the real model, usually using experiments.

3.4.1 Ideal Test Verification

Several special cases were examined to check the asymptotic behavior of the model, and a summary of these validation tests is presented below.

1. *Test 1: Normal elastic force, vertical*

To test for the particle–wall impact case, the test simulates a free-falling particle under gravity hitting the floor of the container. Tangential forces and damping are

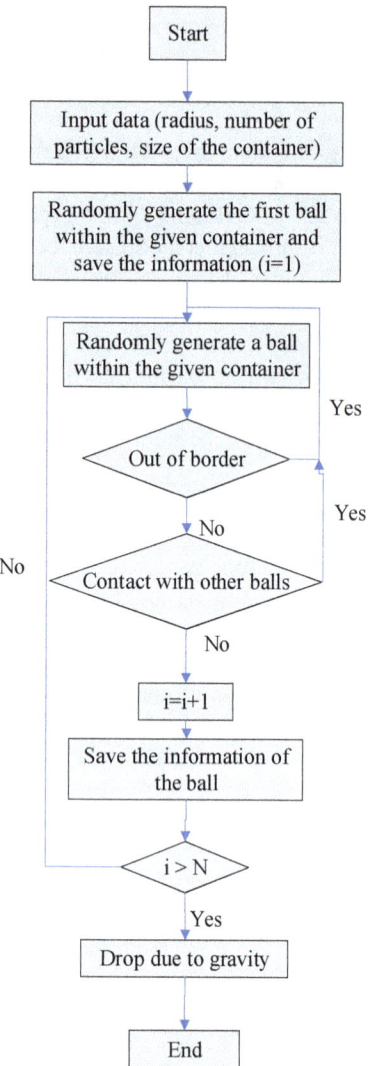

Fig. 3.46 Block diagram of generation of particle combinations by computer simulation

set to zero. To test for the particle–particle force implementation, an identical test is conducted, but the particle impacts with a stationary particle instead of the floor of the container. The particle is dropped from the same height in both cases, as shown in Fig. 3.48a.

Since the particle is dropped from the same height, the results of both cases are identical. Figure 3.48b shows how in both cases the particle rebounds to its original height and the normal elastic force reaches a peak during contact. Note that no movement exists in the x-y directions and no rotation occurs.

Fig. 3.47 Block diagram of the DEM based procedure for particle damper system simulation

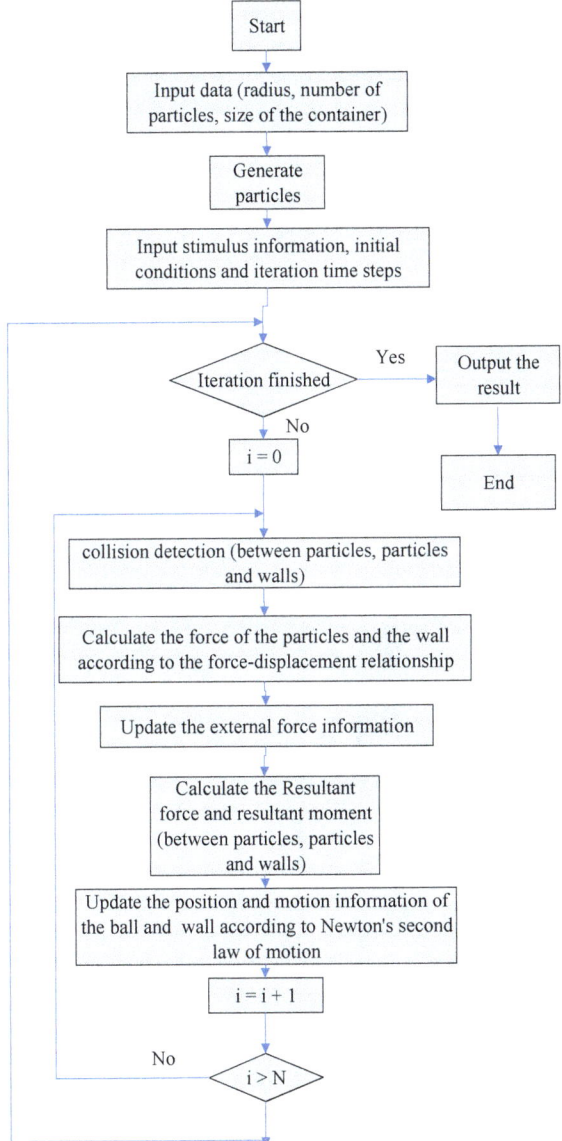

2. *Test 2: Normal elastic force, horizon*

This test is identical to Test 1. One particle moving with initial velocity either in the x-direction, colliding back and forth with the container wall, as is shown in the Fig. 3.49a. Gravitational, tangential, and damping forces are set to zero.

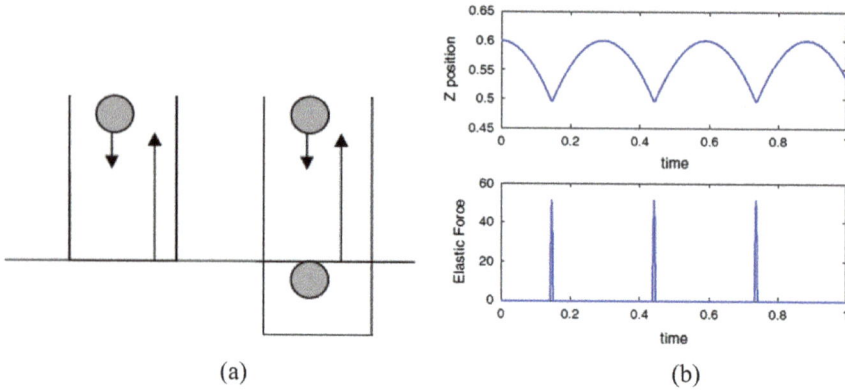

Fig. 3.48 Test 1: Normal elastic force, vertical: **a** Schematic diagram and **b** Z position of the particle and elastic force

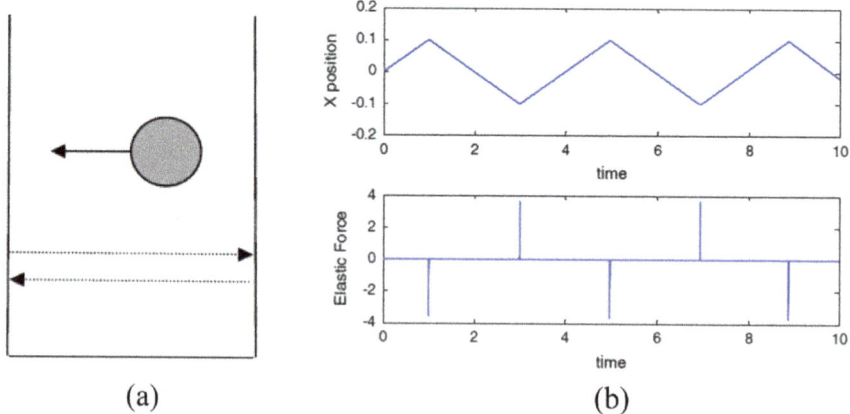

Fig. 3.49 Test 2: Normal elastic force, horizontal: **a** Schematic diagram and **b** X position of the particle and elastic force

Figure 3.49b shows that the particle rebounds horizontally between the two walls of the container with no energy loss and the normal elastic force reaches a peak during contact. There is no movement in the y-z directions and no rotation occurs.

3. *Test 3: Normal damping force*

This test is identical to Test 1, but only accounting for the normal damping force. The normal critical damping ratio used is 0.3. The schematic diagram can be referred to Fig. 3.48a.

Figure 3.50a shows how in both cases when the particle rebounds, it fails to reach the original height and its height decays due to damping, and finally it reaches a static equilibrium. Figure 3.50c shows the velocity immediately after the impact is

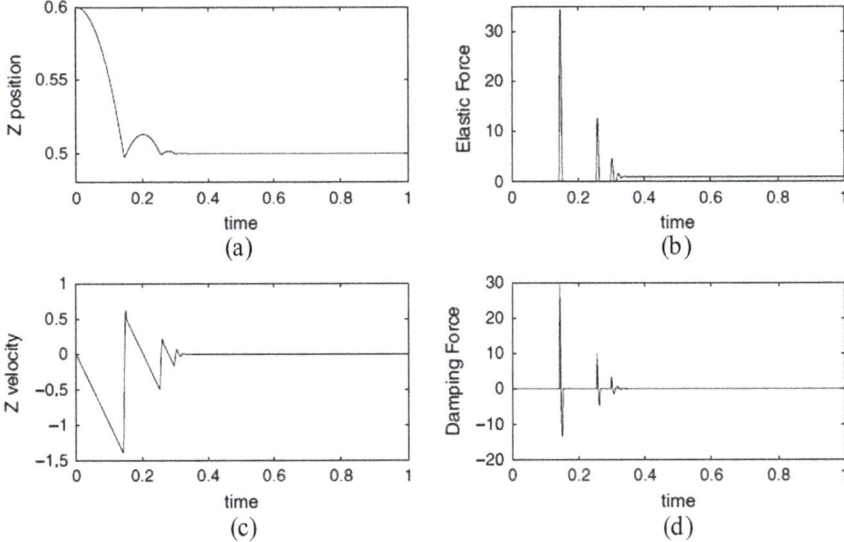

Fig. 3.50 Test 3: **a** Z position of the particle; **b** Normal elastic force; **c** Z velocity of the particle; and **d** Normal damping force

smaller than that before the collision, until it becomes zero. Also, the normal elastic force and damping force at consecutive contacts are successively less, as shown in Fig. 3.50b, d. There is no movement in the x-y direction and no rotation.

4. *Test 4: Normal damping force with stacking*

This test is identical to Test 3 but with two particles dropped from different heights with zero initial velocity at the same time, as shown in Fig. 3.51a.

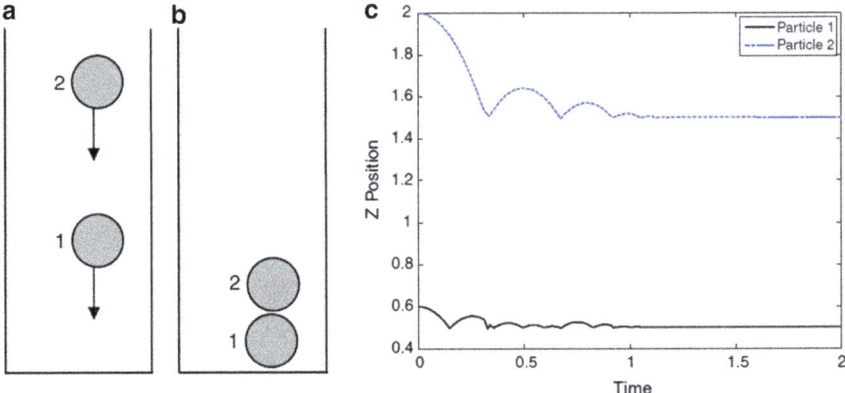

Fig. 3.51 Test 4: Normal damping force with stacking: **a** Schematic diagram; **b** Stack diagram; and **c** Z position of particle 1 and particle 2

Figure 3.51b shows the final condition, in which the higher particle stacks on the lower particle in a column, and the lower particle rests on the ground. The separation between the lower particle and the ground is its radius, while that between two particles equals to the summation of their radii. This can also be found in Fig. 3.51c, which shows their position time histories. It also validates the normal interaction model between the particles.

5. *Test 5: Particle impacts two particles (walls) simultaneously*

This test is designed to test the situation in which a particle collides with two initially stationary particles or walls at the same time. The particle 1 is placed in the center of a square container. It is used as the impacting particle, which has the same initial velocities in the x- and y- directions. Particle 2 and particle 3 are positioned besides particle 1 in the x- and the y- directions, as shown in Fig. 3.14a.

At the beginning, particle 1 impacts particle 2 and particle 3 simultaneously, then it moves back in the southwest direction until it reaches the south wall and the west wall at the same time. After collision with these two walls, the particle moves back along the northeast direction, which is the original track before the collision with walls. Hence, the track of particle 1 is a straight line in the diagonal direction, which is shown in Fig. 3.52b.

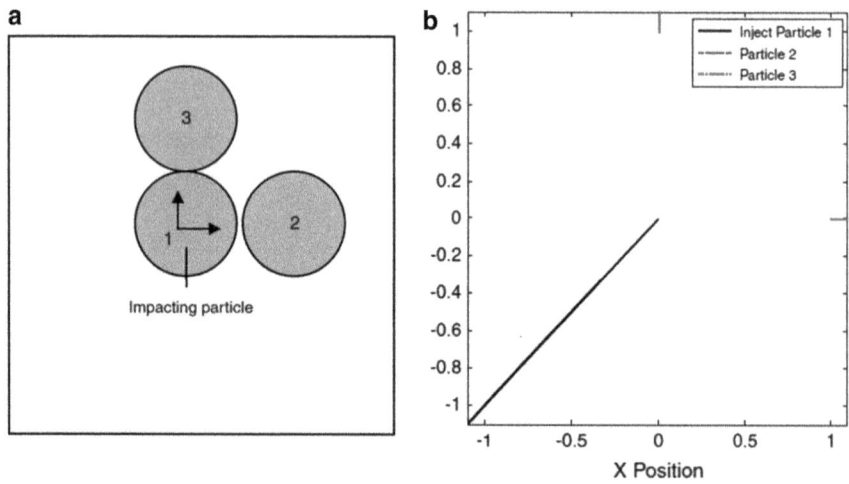

Fig. 3.52 Test 5: Particle impacts with two particles (walls) simultaneously: **a** Schematic diagram and **b** X-Y positions of particles

3.4.2 Shaking Table Test Verification

1. *Verification test of one-unit-multi-particle damper by shaking table test*

Saeki [42] made a shaking table test of one-unit-multi-particle damper in 2002. Hundreds of spherical particles are placed in a rectangular container. The device is attached to a SDOF primary system and apply horizontal harmonic excitation to the base. The experimental calculation model is shown in Fig. 3.53.

Figure 3.54 shows the dimensionless curve of the experimental results and the simulation results. The abscissa is the frequency ratio, where f_n is the natural frequency of the primary system. The ordinate is the ratio of R.M.S. of the displacement response of the primary system to harmonic excitation amplitude. The calculation

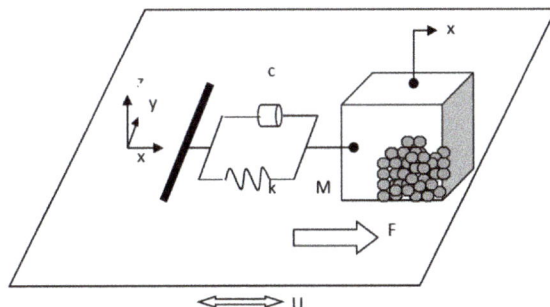

Fig. 3.53 The diagram of shake table test experiment of one-unit-multi-particle damper

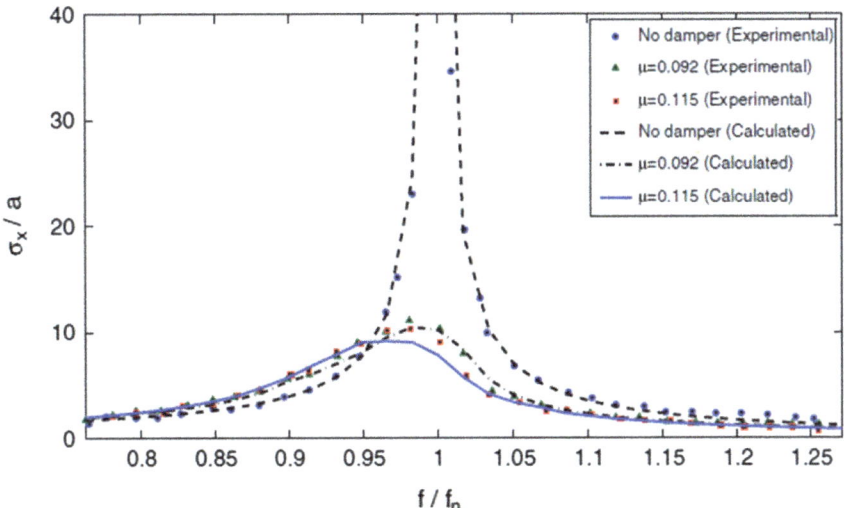

Fig. 3.54 Comparison between experimental results and calculated results for a harmonically excited primary system provided with a one-unit-multi-particle damper

Table 3.4 Values of system parameters in Figs. 3.54 and 3.56

Parameter	Figure 3.54	Figure 3.56
Unit number	1	5
Total particle number	200 ($\mu = 0.092$); 250 ($\mu = 0.115$)	192×5 ($\mu = 0.098$)
Diameter of the particle (m)	0.006	0.006
Density of the particle (kg/m^3)	1190	1190
Volumetric filling ratio	0.27 ($\mu = 0.092$); 0.34 ($\mu = 0.115$)	0.26 ($\mu = 0.098$)
Coefficient of friction	0.52	0.52
Critical damping ratio of the primary system	0.0027	0.0065
Critical damping ratio of the damper	0.1	0.1
Stiffness of the spring between particle and particle (N/m)	1.0×10^5	1.0×10^5
Stiffness of the spring between particle and wall (N/m)	1.3×10^5	1.3×10^5
Amplitude of excitation (m)	0.0005	0.0005

parameters are shown in Table 3.4. As can be seen from the figure, the calculation results are in good agreement with the test results.

2. *Verification test of multi-unit-multi-particle by shaking table test*

In 2005, Saeki [43] conducted a shaking table test of a multi-unit-multi-particle damper. Thousands of spherical particles are evenly placed in five identically symmetric cylindrical containers. The device is attached to a SDOF primary system. Harmonic excitation is also applied to the substrate. The experimental calculation model is shown in Fig. 3.55. Figure 3.56 shows the curves of the test results and the simulation results, which are in good agreement. The calculation parameters are shown in Table 3.4.

Fig. 3.55 The diagram of shake table test experiment of multi-unit-multi-particle damper

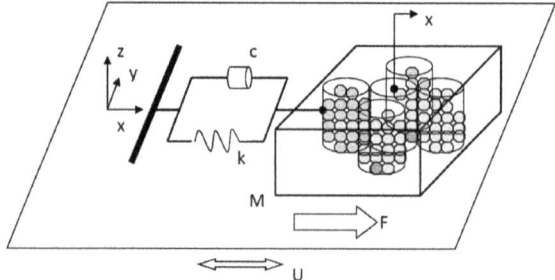

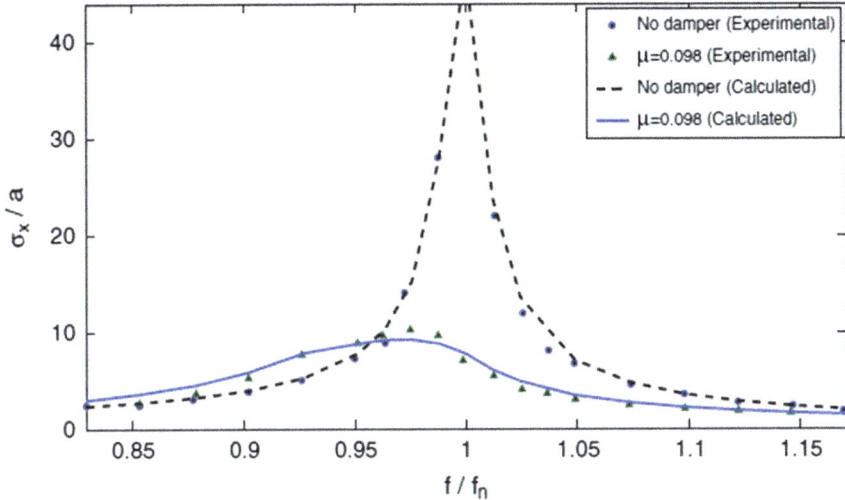

Fig. 3.56 Comparison between experimental results and calculated results for a harmonically excited primary system provided with a multi-unit-multi-particle damper

Chapter 6 of this book will detail a shaking table test of a three-layer steel frame the model with an additional rectangular multi-unit-multi-particle damper, and further verify the numerical model.

References

1. Araki, Y., I. Yokomichi, and Y. Jinnouchi. 1986. Impact damper with granular materials : 4th report, frequency response in a horizontal system. *Transactions of the Japan Society of Mechanical Engineers C* 29 (258): 4334–4338.
2. Papalou, A., and S.F. Masri. 1996. Performance of particle dampers under random excitation. *Journal of Vibration and Acoustics, ASME* 118 (4): 614–621.
3. Friend, R.D., and V.K. Kinra. 2000. Particle Impact damping. *Journal of Sound and Vibration* 233 (1): 93–118.
4. Lu, Z., and X.L. Lv. 2013. Numerical simulation of vibration control effects of particle dampers. *Journal of Tongji University (Natural Science)* 41 (8): 1140–1144.
5. Yan, W.M., et al. 2014. Experimental and theoretical research on the simplified mechanical model of a tuned partical damper, its parameter determination method and earthquake-induced vibration control of bridge. *Engineering Mechanics* 31 (6): 79–84.
6. Hu, S., Q.B. Huang, and Z.S. Xu. 2008. Regression analysis of particle damping. *China Mechanical Engineering* 19 (23): 2834–2837.
7. Jiang, H., and Q. Chen. 2007. Application of restoring force surface method to particle damping research. *Journal of Vibration, Measurement & Diagnosis* 27 (3): 228–231.
8. Bathe, K., E.L. Wilson, and F.E. Peterson. 1973. SAP IV, A structural analysis program for static and dynamic response of linear systems. In *EERC Report No. 73-11*. Berkeley: University of California.
9. Masri, S.F. 1972. Theory of the dynamic vibration neutralizer with motion-limiting stops. *Journal of Applied Mechanics, ASME* 39 (2): 563–568.

10. Meirovitch, L. 1967. *Analytical methods in vibrations*. New York: The Macmillan Co.
11. Masri, S.F., and T.K. Caughey. 1966. On the stability of the impact damper. *Journal of Applied Mechanics* 33 (3): 586–592.
12. Kuroiwa, J.H. 1967. *Vibration test of a multistory building*. California Institute of Technology.
13. Masri, S.F. 1965. *Analytical and experimental studies of impact dampers*. Pasadena: California Institute of Technology.
14. Warburton, G.B. 1957. On the theory of the acceleration damper. *Journal of Applied Mechanics* 79: 322–324.
15. Thomson, W.T. 1965. *Vibration theory and applications*, vol. 1, 184 and 197. Englewood Cliffs, N. J.: Prentice Hall.
16. Arnold, F.R. 1955. Steady-state behaviour of systems provided with nonlinear dynamic vibration absorbers. *Journal of Applied Mechanics* 22: 487–492.
17. Iwan, W.D. 1969. Application of an equivalent nonlinear system approach to dissipative dynamical systems. *Journal of Applied Mechanics* 36 (3): 412.
18. Masri, S.F. 1970. Numerical response characteristics of systems provided with impact dampers. In *Structural mechanics laboratory report no. USC-CE 101*. University of Southern California.
19. Roberson, R.E. 1952. Synthesis of a nonlinear dynamic vibration absorber. *Journal of The Franklin Institute-Engineering and Applied Mathematics* 254 (3): 205–220.
20. Pipes, L.A. 1953. Analysis of a nonlinear dynamic vibration absorber. *Journal of Applied Mechanics* 79: 515–518.
21. Caughey, T.K. 1954. The existence and stability of ultraharmonics and subharmonics in forced non-linear oscillations. *Journal of Applied Mechanics Transactions of the ASME* 21 (4): 327–335.
22. Hales, T.C. 1992. The sphere packing problem. *Journal of Computational and Applied Mathematics* 44 (1): 41–76.
23. Masri, S.F., and A.M. Ibrahim. 1973. Response of the impact damper to stationary random excitation. *The Journal of the Acoustical Society of America* 53 (1): 200–211.
24. Cundall, P.A., and O. Strack. 1979. A distinct element model for granular assemblies. *Geotechnique* 29: 47–65.
25. Wei, Q. 1991. *Numerical principle and procedure of the basic principle of the discrete element method [M]*. Beijing: Science Press.
26. Wang, Y.J., and J.B. Xing. 1991. *Discrete element method and its application in geotechnical mechanics [M]*. Shenyang: Northeastern University Press.
27. Wang, Q., and X.L. Lv. 2003. Discrete element method and its application status in construction engineering [A]. In *Modern Civil Engineering Theory and Practice*, 656–661. Wuhan: Hohai University Press.
28. Zhang, F.W., and X.L. Lv. 2009. Numerical simulation and analysis of different collapse modes of frame structures. *Journal of Building Structures* 30 (5): 119–125.
29. Wang, Q. 2005. Nonlinear analysis and earthquake collapse response analysis of reinforced concrete frame structure based on discrete element method. In *Structural engineering*. Shanghai: Tongji University.
30. Mao, K.M. 1999. Theoretical study and application of mechanical mechanism of non-obstructive particle damping. In *Mechanical engineering*. Xi'an: Xi'an Jiaotong University.
31. Du, Y.C., and S.L. Wang. 2009. Energy dissipation in normal elastoplastic impact between two spheres. *Journal of Applied Mechanics* 76 (6): 061010–061017.
32. Elperin, T., and E. Golshtein. 1997. Comparison of different models for tangential forces using the particle dynamics method. *Physica A: Statistical and Theoretical Physics* 242 (3–4): 332–340.
33. Di Renzo, A., and F.P. Di Maio. 2004. Comparison of contact-force models for the simulation of collisions in DEM-based granular flow codes. *Chemical Engineering Science* 59 (3): 525–541.
34. Masri, S.F. 1973. Steady-state response of a multidegree system with an impact damper. *Journal of Applied Mechanics-Transactions of the Asme* 40 (1): 127–132.
35. Oldenburg, M., and L. Nilsson. 1994. The position code algorithm for contact searching. *International Journal for Numerical Methods in Engineering* 37: 359–386.

36. Bonet, J., and J. Peraire. 1991. An alternating digital tree (ADT) algorithm for 3D geometric searching and intersection problems. *International Journal for Numerical Methods in Engineering* 31: 1–17.
37. Connor, R.O., J. Gill, and J.R. Williams. 1993. A linear complexity contact detection algorithm for multi-body simulation. In *Proceedings of 2nd U.S. conference on discrete element methods.* MIT, MA.
38. Preece, D.S., and S.L. Burchell. 1993. Variation of spherical element packing angle and its influence on computer simulations of blasting induced rock motion. In *Proceedings of 2nd U.S. conference on discrete element methods.* MIT, MA.
39. Mirtich, B. 1988. *Impulse-based dynamic simulation of rigid body systems.* Berkeley: University of California.
40. Munijiza, A., and K.R.F. Andrews. 1998. NBS contact detection algorithm for bodies of similar size. *International Journal for Numerical Methods in Engineering* 43: 131–149.
41. Shan, J., et al. 2016. Seismic data driven identification of linear models for building structures using performance and stabilizing objectives. *Computer-Aided Civil and Infrastructure Engineering* 31 (11): 846–870.
42. Saeki, M. 2002. Impact damping with granular materials in a horizontally vibrating system. *Journal of Sound and Vibration* 251 (1): 153–161.
43. Saeki, M. 2005. Analytical study of multi-particle damping. *Journal of Sound and Vibration* 281 (3–5): 1133–1144.

Chapter 4
Performance Analysis of Particle Dampers Attached to Single-Degree-of-Freedom (SDOF) Structures

Based on the numerical model of the particle damper established in the previous chapter, starting from this chapter, the performance of different structures with additional particle dampers (including their variants) under different excitations will be discussed according to the order from shallow to deep, from simple to complex, and from single-degree of freedom to multi-degree of freedom. This chapter first derives the analytical solution of the single-degree-of-freedom system with a single particle impact damper under simple excitation, then introduces the vertical dynamic characteristics of the particle damper, and finally systematically investigates the influence of the parameters of the particle damper under horizontal harmonic excitation.

4.1 Analytical Solution to a Particle Damper Attached to a SDOF Structure

4.1.1 Computational Model

According to the discussion in Chap. 2, single particle dampers are divided into single-unit-single-particle damper and multi-unit-single-particle damper in terms of the number of units (computational model is shown in Fig. 4.1). This type of dampers only have particle-wall collision without collision between particles. It is relatively simple to analyze, so its analytical solution under simple excitation can be obtained.

In fact, single-unit-single-particle damper can be regarded as a special form of multi-unit-single-particle damper. The governing equation of multi-unit single-particle damper is discussed below.

© China Machine Press and Springer Nature Singapore Pte Ltd. 2020
Z. Lu et al., *Particle Damping Technology Based Structural Control*,
Springer Tracts in Civil Engineering,
https://doi.org/10.1007/978-981-15-3499-7_4

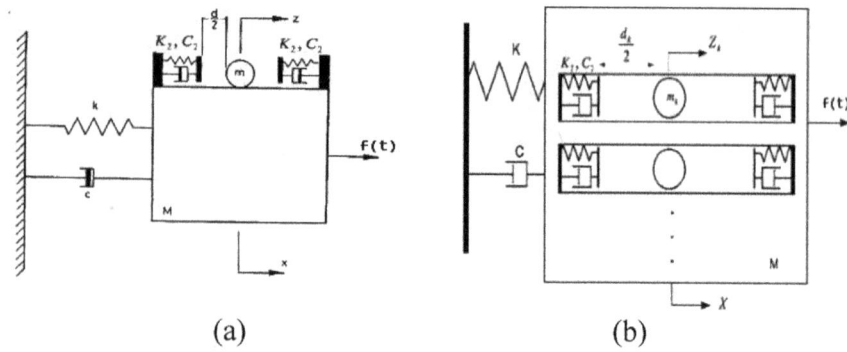

Fig. 4.1 Computational model of a particle damper attached to a SDOF structure: **a** single-unit-single-particle damper; **b** multi-unit-single particle damper

$$\ddot{x} = -2\zeta\omega_n\dot{x} - \omega_n^2 x + \frac{f(t)}{M} + \sum_{k=1}^{N}(\mu_k[\omega_2^2 G(z_k) + 2\zeta_2\omega_2 H(z_k, \dot{z}_k) + \mu_s g\,\mathrm{sgn}(\dot{z}_k)])$$

$$\ddot{z}_k = -\ddot{x} - [\omega_2^2 G(z_k) + 2\zeta_2\omega_2 H(z_k, \dot{z}_k) + \mu_s g\,\mathrm{sgn}(\dot{z}_k)], \quad k = 1, 2, \ldots, N \qquad (4.1)$$

where d is the size of the single-particle damper, that is, the gap clearance of particles. $x, \dot{x}, \ddot{x}$ are displacement, velocity and acceleration of the main system respectively. $z_k, \dot{z}_k, \ddot{z}_k$ are relative displacement, relative velocity and acceleration of the kth particle and the main system, respectively, $f(t)$ is the external excitation, M is the quality of the main system, μ_k is the mass ratio of the first k particle to the main system, μ_s is the friction coefficient, g is the gravity acceleration, sgn is the symbolic function, $G(z_k), H(z_k, \dot{z}_k)$ are the functions of the non-linear spring force and the non-linear damping force of the single particle damper respectively, as shown in Fig. 4.2.

4.1.2   Analytical Solution Method

As the non-linear spring force and the non-linear damping force function of single-particle damper are divided into two stages, Eq. (4.1) is also divided into two stages to solve. Assuming that the external excitation is sinusoidal and the effect of friction is neglected, the equation of motion between two adjacent collisions is as follows:

$$M\ddot{x} + c\dot{x} + kx = F_0 \sin \Omega t$$
$$\dot{y}_k = 0, k = 1, 2, \ldots, N \qquad (4.2)$$

where y is the displacement of the particle, so the relative displacement $z = y - x$. Suppose that at the moment $t = t_i$, the ith collision completes, when

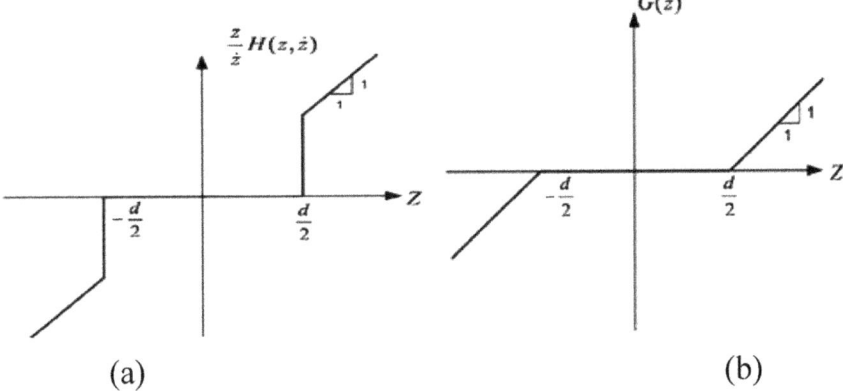

Fig. 4.2 a Function of the non-linear spring force $G(z)$ and **b** function of non-linear damping force $H(z, \dot{z})$

$$x(t_i) = x_i, \quad \dot{x}(t_{i+}) = \dot{x}_{ia}, \quad y_k(t_i) = y_{ki}, \quad \dot{y}_k(t_{i+}) = \dot{y}_k, \quad k = 1, 2, \ldots, N$$

Thus, the motion of the system from t_{i+} to $t_{(i+1)-}$ in time interval can be obtained:

$$
\begin{aligned}
x(t) &= e^{-(\zeta/r)(\Omega t - \alpha_i)}[a_i \sin(\eta/r)(\Omega t - \alpha_i) + b_i \cos(\eta/r)(\Omega t - \alpha_i)] + A \sin(\Omega t - \psi) \\
\dot{x}(t) &= \omega e^{-(\zeta/r)(\Omega t - \alpha_i)}[-(\zeta a_i + \eta b_i) \sin(\eta/r)(\Omega t - \alpha_i) \\
&\quad + (\eta a_i - \zeta b_i) \cos(\eta/r)(\Omega t - \alpha_i)] + A\Omega \cos(\Omega t - \psi) \\
y_k(t) &= \dot{y}_{ki}(t - t_i) + y_{ki} \\
\dot{y}_k(t) &= \dot{y}_{ki}, \quad k = 1, 2, \ldots, N; \quad t_{i+} \le t \le t_{(i+1)-}
\end{aligned}
$$

$$(4.3)$$

where,

$$\zeta = c/2\sqrt{kM}, \quad r = \Omega/\omega, \quad \omega = \sqrt{k/M}, \quad \alpha_i = \Omega t_i,$$

$$A = \frac{F_0/k}{\sqrt{(1 - r^2)^2 + (2\zeta r)^2}}, \quad \psi = \tan^{-1}[2\zeta r/(1 - r^2)],$$

$$b_i = x_i - A \sin(\alpha_i - \psi), \quad \eta = \sqrt{1 - \zeta^2},$$

$$a_i = (1/\eta)[(1/\omega)\dot{x}_{ia} - Ar \cos(\alpha_i - \psi) + \zeta b_i]$$

When a particle (assuming the jth particle) touches the wall of the container, the first $(i + 1)$th collision occurs, and the system enters the second stage.

$$|z_j| = |y_j - x| = d_j/2 \tag{4.4}$$

So at the moment $t = t_{i+1}$,

$$x(t_{(i+1)+}) = x(t_{(i+1)-}),$$

$$y_k(t_{(i+1)+}) = y_k(t_{(i+1)-}), \quad k = 1, 2, \ldots, N$$
$$\dot{y}_k(t_{(i+1)+}) = \dot{y}_k(t_{(i+1)-}), \quad k = 1, 2, \ldots, N; \quad k \neq j \qquad (4.5)$$

According to the law of conservation of momentum and the definition of the coefficient of restitution, the formula can be obtained

$$\dot{x}_+ = k_{1j}\dot{x}_- + k_{2j}\dot{y}_{j-}$$
$$\dot{y}_{j+} = k_{3j}\dot{x}_- + k_{4j}\dot{y}_{j-} \qquad (4.6)$$

where,

$$\mu_j = m_j/M, \quad k_{1j} = (1 - \mu_j e_j)/(1 + \mu_j), \quad k_{2j} = \mu_j(1 + e_j)/(1 + \mu_j),$$
$$k_{3j} = (1 + e_j)/(1 + \mu_j), \quad k_{4j} = (\mu_j - e_j)/(1 + \mu_j)$$
$$e_j = \text{Restitution Coefficient of Particle J}$$

Equations (4.5) and (4.6) can be used as new initial conditions for Eq. (4.2) between the time intervals $t_{(i+1)+}$ to $t_{(i+2)-}$. Conduct the above two processes repeatedly and sequentially, the motion pattern of the particle damper system in the whole time course can be obtained.

Masri's study [1] points out that when the system reaches steady-state vibration, if each particle collides with the container twice in each cycle, the system is stable, and the effect of vibration reduction is optimal at this time. In addition, compared with single-unit-single-particle damper, multi-unit-single-particle damper can greatly reduce the collision force between particles and the container, thus reducing the plastic deformation of the container wall and noise, as shown in Fig. 4.3.

4.2   A Particle Damper Attached to a SDOF Structure Under Free Vibration

The biggest difference between vertical particle damper and horizontal particle damper is gravity, which has its own motion characteristics. In this section, the motion characteristics of particles in this kind of damper are investigated in terms of its free vibration form. Friend [2] fixed a box containing metal particles at the end of a beam in 2000 and made corresponding experiments. The natural frequency of the main system is 17.8 Hz, the mass is 0.0376 kg, the critical damping ratio is 0.012, the friction coefficient is 0.55, the diameter of each particle is 1.2 mm, and the coefficient of restitution is 0.75. The total mass of 512 particles is 0.004 kg.

Figure 4.4 shows the free vibration displacement of the primary system for an initial displacement (A_0). The displacement (Z) is normalized by dividing by the static deflection (Z_{st}). It clearly shows a dramatic attenuation of the transient vibration due to the presence of particles, and the equivalent damping ratio increases from 0.012

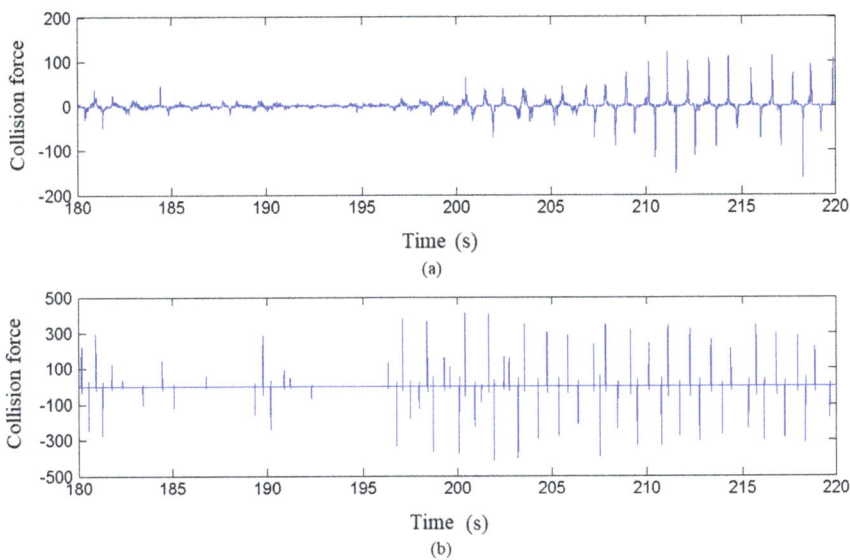

Fig. 4.3   Collision force contrast diagram. **a** Multi-unit-single-particle damper; **b** single-unit-single-particle damper

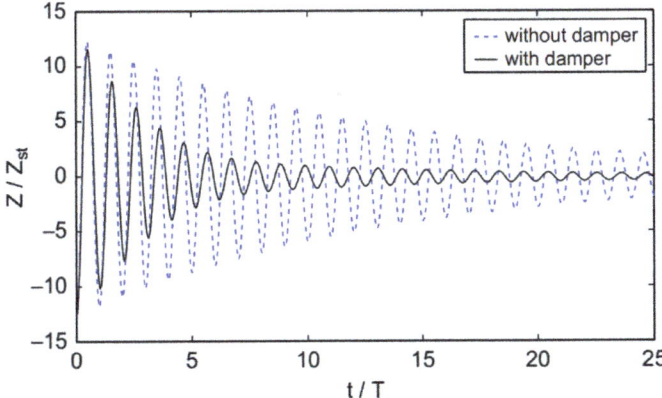

Fig. 4.4   Free vibration displacement time history curve of main system with vertical particle damper. The system parameters are: $\mu = 0.1$, $\zeta = 0.012$, $\mu_s = 0.55$, $dx/Z_{st} = 9$, $dy/Z_{st} = 9$, $dz/Z_{st} = 32$, $d/Z_{st} = 1.5$, $e = 0.75$, $A_0/Z_{st} = 13$

to 0.048. Moreover, the natural frequency of the primary system becomes smaller because of the additional particle mass.

Figure 4.5 gives a series of snapshots to show the motion of the structural mass with the particles. These snapshots illustrate the approximate damping regimes during the course of free vibration [3]. In the first regime approximately between 0.0 and

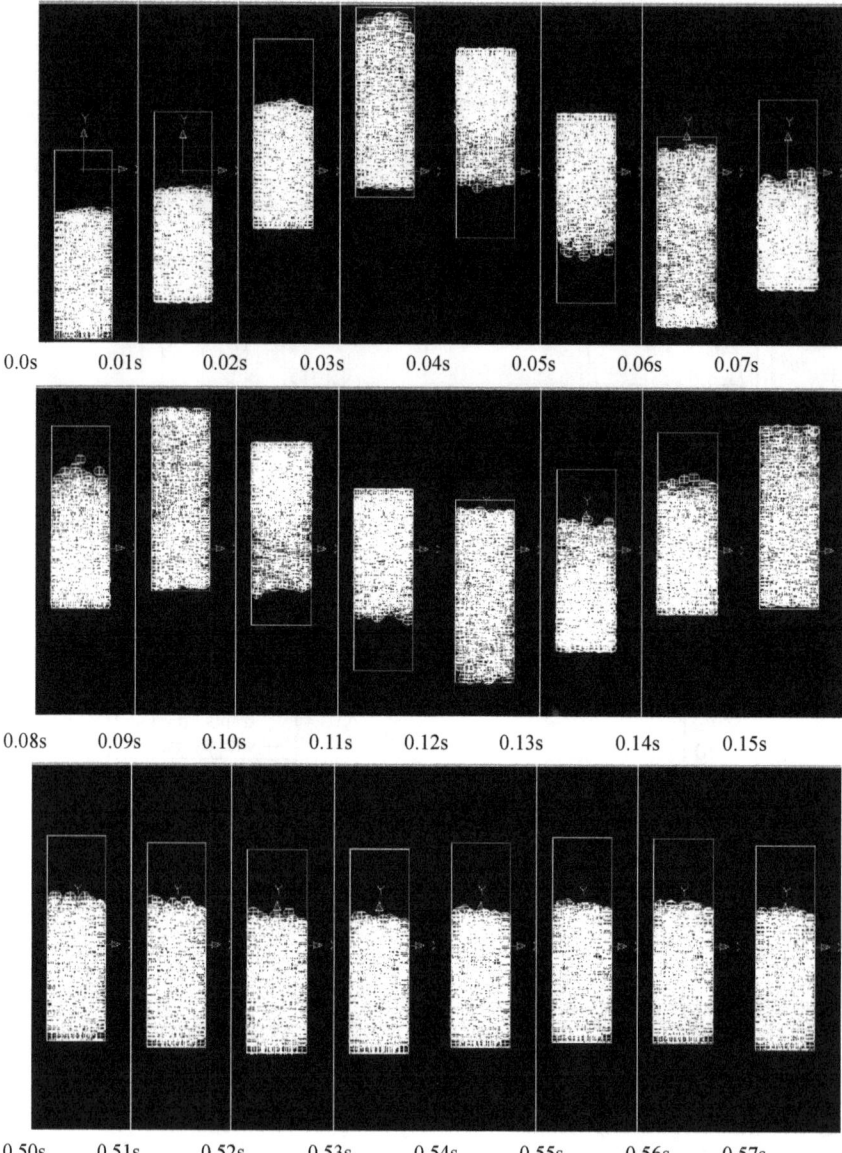

Fig. 4.5 A snapshot of the motion of the particle damper in free vibration. The system parameters are the same as those in Fig. 4.4

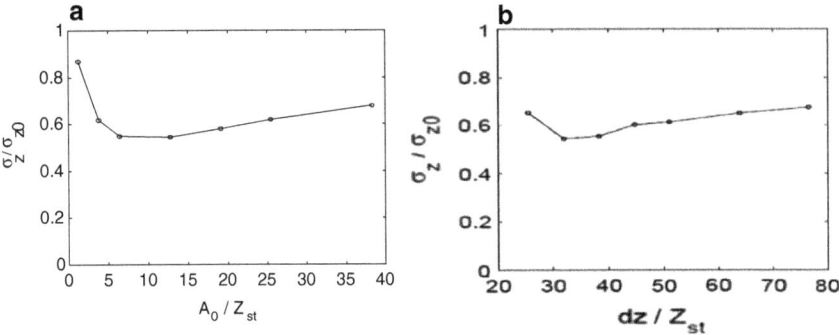

Fig. 4.6 The R.M.S. response of the free vibration displacement of the main system, the system parameters are the same as those in Fig. 4.4. **a** The influence of amplitude; **b** the influence of container size

0.05 s, the damper container walls impact the particles and transfer momentum to the particles. During the second regime, significant damping effect is achieved, not only due to a large number of collisions between particles and the container walls, but also to the mechanical energy dissipation through impact and friction. After that, in the third regime, the particles accumulate at the bottom of the container under gravity, which only produces little damping effect.

In addition to the effect of particle damping shown in Fig. 4.4, the effect of vibration control can also be examined by the ratio of R.M.S. response of the displacement (σ_z/σ_{z0}) of the main system with and without additional dampers. This is a common method in stochastic processes and another response of the effect of particle damping. Figure 4.6a, b show that the particle damping depends on the vibration amplitude and the gap clearance, respectively. The optimum design strategies can be found in both cases. Consequently, it can be seen that a properly designed vertical particle damper could provide additional significant damping to the primary system, and appreciably attenuate its response. Besides these two factors, many other parameters also influence the performance of particle dampers, and will be discussed in the following chapters.

4.3 A Particle Damper Attached to a SDOF Structure Under Simple Harmonic Vibration

This section mainly discusses the system response of a single-degree-of-freedom system with an additional particle damper under horizontal harmonic excitation, and study the influence of various system parameters, including the size of the container, the number of particles (N), the size (d) and the material of the particles, the mass ratio of particles to the main system (μ) and the external excitation frequency, and so forth. The schematic diagram of the calculation model is shown in Fig. 4.1. In the numerical simulation, the natural frequency of the main structure is 11.4 Hz and the mass is 0.573 kg. The calculation time exceeds 250 times of the natural vibration period of the main system to eliminate the influence of transient vibration. The initial position of particles is normal random distribution. The effect of vibration reduction is measured by the ratio of the R.M.S. displacement response (σ_x/σ_{x0}) of the main system with and without additional dampers.

There are three types of collision between particles and container wall according to the velocity direction of the two before collision.

(1) The absolute velocities of both particle and the primary system are opposite to each other at the instant immediately before contact, which is the face-to-face impact.
(2) Although the particle and the primary system's absolute velocities have the same direction, the relative velocity of the particle is opposite prior to contact, which is the case for the primary system catching up with the particle.
(3) The particle and system's absolute velocities and the particle's relative velocity have the same direction just before contact, which is the case for the particle catching up with the primary system.

Collision types (1) and (2) can reduce the response of the main structure, because the collision force is opposite to the direction of motion of the main structure and will prevent its movement. This type of collision is 'Beneficial Impact'. The corresponding momentum exchange is defined as 'Beneficial Momentum Exchange'. On the other hand, the collision type (3) will increase the response of the main structure. It should be accelerated. This type of collision is 'Adverse Impact', and the corresponding momentum exchange is defined as 'Adverse Momentum Exchange'. The concept of 'Effective Momentum Exchange (EME)' ('Effective Momentum Exchange' = 'Useful Momentum Exchange' – 'Harmful Momentum Exchange') is proposed herein to describe the combined effects of the two. From the discussion below, it can be seen that this quantity plays an important role in representing the physical nature of dampers.

4.3.1 Effects of Particles' Number, Size and Material

Other parameters, such as external excitation level and mass ratio, remain unchanged. The mass ratio of particles to the main system is defined as:

$$\mu = m/M = N\rho\pi d^3/(6M) \tag{4.7}$$

where ρ and d are the density and diameter of particle, respectively. Consequently, for a given M of the main system, simultaneously changing two parameters of N, ρ and d can result in the same μ.

(1) *Effect of particle size and number*

This test keeps ρ as a constant and changes N and d, which is for the case of choosing to use a few large steel particles or many small steel particles in design process.

Figure 4.7 shows a sample results. Three curves overlap with each other in Fig. 4.7a, this is because dy has no effect on the performance of the damper when the

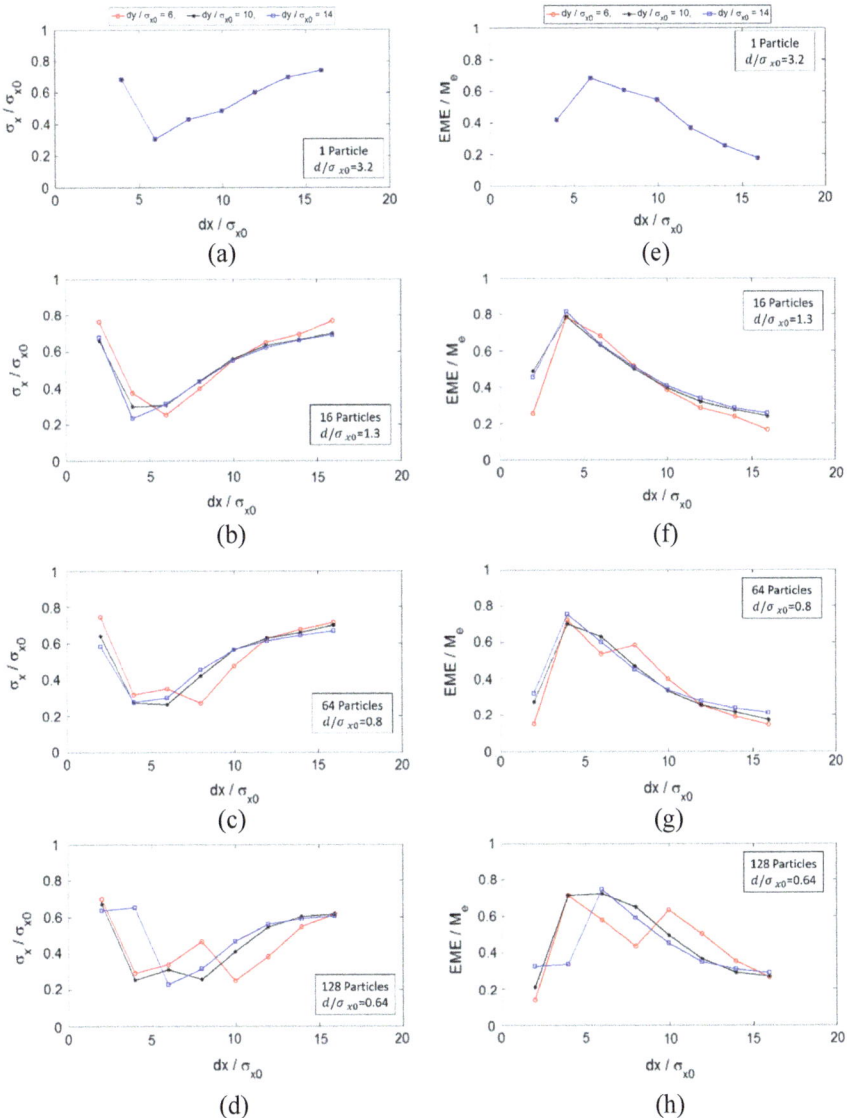

Fig. 4.7 a–d R.M.S. response of the main system displacement and **e–h** effective momentum exchange of the system with $\mu = 0.027$, $e = 0.75$, $\zeta = 0.004$, $\mu_s = 0.05$ (Effect of particle size and number)

particle is single. It can be seen in Fig. 4.7a–d, when dx is small, the R.M.S. response of the main structure is very large, because the particles are piled together at this time, which minimizes the motion of the lower layers and created a vigorous motion only in the top-most layers. The result is a smaller exchange of effective momentum and a corresponding decrease in the effectiveness of the damper. For medium size dx, the response of the system is more sensitive to the particle size. As the number of particles increases and the particle size decreases, the curve is smooth, which means the sensitivity of vibration attenuation to changes in 'gap clearance' decreases, that is, the robustness of the system is great. For a large dx, the response amplitude is high for all particles used. This is due to the energy of many particles is uselessly consumed in the collision between particles and the collision between particles and the wall of the container along the x-axis. Moreover, it takes a long time for particles to move from one wall to the opposite wall after collision, so relatively fewer impacts occur. Another important observation is that particle damper with more particles is slightly more effective in reducing the R.M.S. level of the response, as opposed to a single particle damper. However, there is not much difference in the best R.M.S. response level between the results of a 16- and a 128-particle damper, with the other parameters being the same. That is to say, a further increase of the number of particles above a certain number would not result in further response reduction. This phenomenon was also observed by Friend and Kinra [2] in their experiments.

If the corresponding effective momentum exchange (EME) is calculated, the above-mentioned phenomenon can be seen more clearly. Figure 4.7e–h show the normalized version of EME, which is divided by the momentum exchange of the excitation force. In each case, a higher EME results in a better reduction of R.M.S. response amplitude. The highest points of EME in the 16- and the 128- particle cases are almost the same, so their maximum reductions are almost the same.

(2) *Effect of particle material and size*

This test keeps N as a constant and changes ρ and d, which is for the case of choosing to use large plastic particles or small steel particles in design process.

The abscissa parameter in Fig. 4.8a is normalized container length dx/σ_{x0}, it seems the optimum length gets smaller as particle size becomes smaller. However, as long as the particle size is small enough, the reduction will not change dramatically any more. That is because in the case for $d/\sigma_{x0} = 3.2$, one particle occupies a lot of space. When dx is the smallest, for instance, the ratio of particle diameter to container length (d/dx) is 0.8, which is 0.2 when dx is at its maximum. If nominal 'gap clearance', which is $(dx - d)/\sigma_{x0}$, which is plotted as abscissa parameter, different particle sizes make little difference in the reduction of the response of the primary system, as shown in Fig. 4.8b. This indicates that for the same mass ratio, particle impact damping is insensitive to the type and size of particles. Friend and Kinra [2] also presented the same results in their experiments.

(3) *Effect of particle material and number*

This test keeps d as a constant and changes ρ and N, which is for the case of choosing to use many plastic particles or a few steel particles in design process.

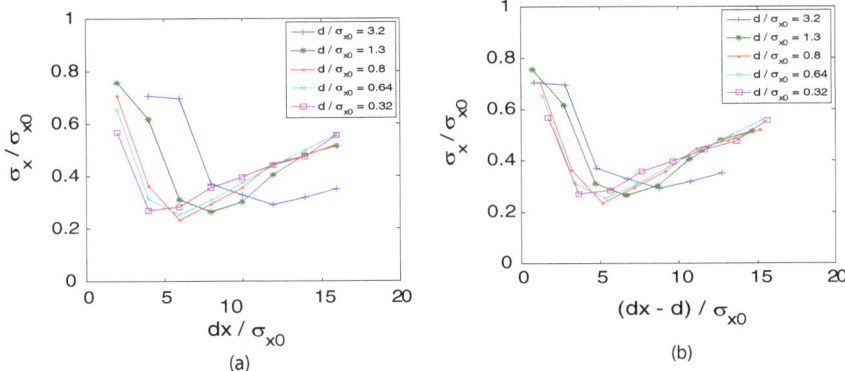

Fig. 4.8 R.M.S. response levels for the primary system with $\mu = 0.027$, $e = 0.75$, $\zeta = 0.004$, $\mu_s = 0.05$, $N = 16$, versus **a** container length and **b** nominal 'gap clearance' (effect of particle material and size)

Figure 4.9a shows that for small sizes of the container, responses of the particle damper with a large number of particles are higher than that with small number of particles. The reason is that in the former case, particles will gather in many layers and lower layers are motionless. However, for large sizes of the container, they are reversed. In such cases, it takes a long time for a single particle traveling from one wall to the opposite wall, which makes fewer impacts. Increasing particle number can increase the possibility of the particle–wall collision. Hence, both optimum range of the container dimension and the efficiency of the particle damper are enhanced.

Figure 4.9b shows the corresponding volumetric filling ratios, which is the fraction of the total volume of all particles and the volume of the container. As can be seen,

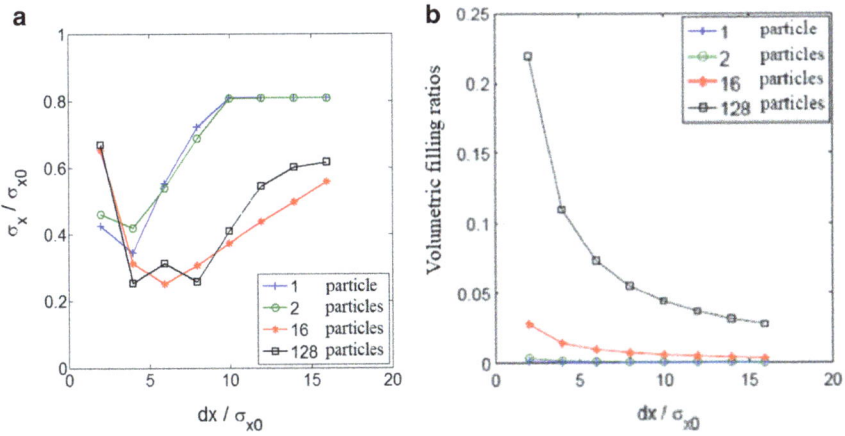

Fig. 4.9 **a** R.M.S. response levels and **b** volumetric filling ratios for the primary system with $\mu = 0.027$, $e = 0.75$, $\zeta = 0.004$, $\mu_s = 0.05$, $d/\sigma_{x0} = 0.64$ (effect of particle material and number)

the ratios for all cases are generally small, that is because a relatively large dz is used in the simulation to get rid of the influence of particles hitting the roof. One can notice that for one-particle and two-particle cases, the volumetric filling ratios are quite small, and this is the reason why they perform better than 128-particle case. In fact, particles are gathered in three layers in 128-particle case for the smallest dx.

The above results clearly indicate that the number of particles plays a very important role in the behavior of a particle damper, given the same mass ratio. Applying more particles, even though the best attenuation of the R.M.S. response level cannot be improved, the optimum range of clearance can be broadened. On the other hand, the particle type and size have minor effects on the primary system performance.

4.3.2 Effects of Container's Size

The length of the container along the excitation direction (dx) has a great influence on the performance of the damper. For different containers' sizes, it is certain that the R.M.S. response level for the particle damped system exhibits a minimum value for a certain clearance ratio, as shown in Fig. 4.10a. The corresponding EME is shown in Fig. 4.10b. In Fig. 4.10c, 'useful impacts' and 'harmful impacts' are compared in non-dimensional manner by dividing them by total impacts, including both particle–wall impacts and particle–particle impacts. One can see that both 'useful impacts' and 'harmful impacts' are only a small fraction of the total impacts. However, if these two are compared with each other, one can find that small dx causes a large number of collisions, but accompanied by many harmful impacts, too. While in large dx cases, although the harmful impacts decrease, or even do not exist, a small number of impacts take place, since the particles do not acquire enough momentum and have longer travel time. Both situations lead to low EME.

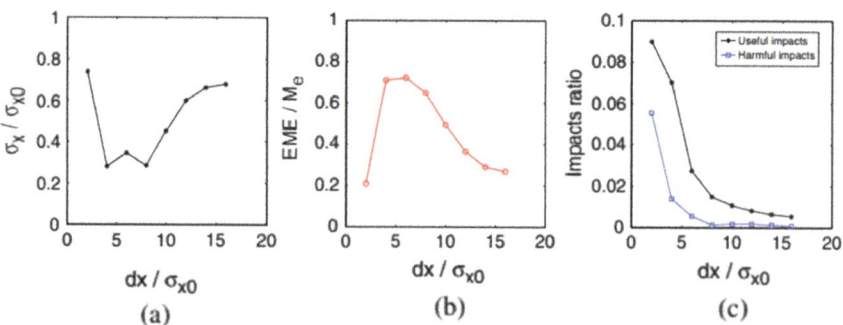

Fig. 4.10 a R.M.S. response levels; **b** EME; **c** impacts ratios for the primary system with $\mu = 0.027$, $e = 0.75$, $\zeta = 0.004$, $\mu_s = 0.05$, $d/\sigma_{x0} = 0.64$, $N = 128$ (effect of container dimensions)

4.3.3 Effects of Particles' Mass Ratio

As shown in Eq. (4.7), three basic methods can be used to investigate the effect of mass ratio:

Keeping ρ and d as constants and changing N, which is the case for having the same material and size of the particles, but using more identical particles.
Keeping ρ and N as constants and changing d, which is the case for having the same material and number of the particles, but using larger size.
Keeping d and N as constants and changing ρ, which is the case for having the same number and size of the particles, but using heavier ones.

Figure 4.11a shows that increasing the mass ratio of particles can reduce the response of the primary system, but the reduction is not directly proportional to the increase in mass ratio. Figure 4.11b illustrates the variation of the minimum R.M.S. response with the mass ratio when the system parameters are optimal. It shows that the effectiveness per unit mass ratio will decrease in a nonlinear manner as the mass ratio increases. Moreover, the optimum values of reduction are almost the same, given a certain mass ratio, no matter which basic method is used to increase μ.

Another interesting observation is that indefinitely increasing the mass of the particles may not reduce the response any further, especially for large sizes of the container. This phenomenon can be explained by considering the conservation of momentum between the particles and the system. As a specific particle's mass increases, its absolute velocity immediately decreases after the impact, which, in turn, reduces its relative velocity, and it takes a longer time to travel towards the other wall of the container. As the mass is increased beyond a certain value, its relative velocity immediately does not allow it to overcome the frictional force while in motion after the impact, and it comes to rest relative to the system prior to reaching the other wall. If a similar situation arises prior to getting to the other container's boundary, the particle reverses its direction once again. It is possible for the particle to reciprocate between the container's boundaries, while the system goes through

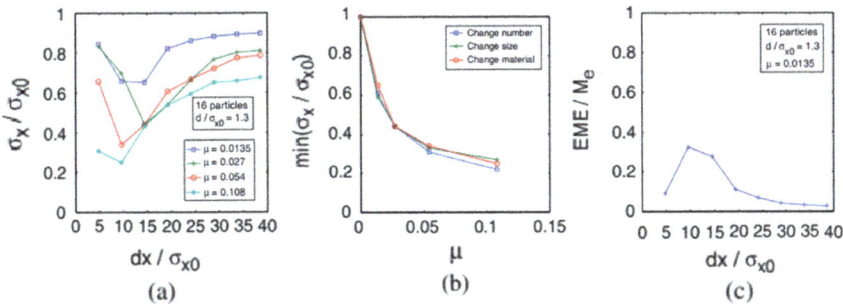

Fig. 4.11 a R.M.S. response levels; **b** Minimum R.M.S. response levels versus m; **c** EME for the primary system with $e = 0.75$, $\zeta = 0.01$, $\mu_s = 0.05$ (effect of mass ratio)

several cycles of motion before making the next impact. This phenomenon is also addressed by Butt in his experiments [4]. From Fig. 4.11c, one can find the EME decreases to a certain low level in large container dimensions.

4.3.4 Effects of External Excitation's Frequency

In the validation Sect. 3.2, Figs. 3.7 and 3.9 show the influence of the excitation frequency. As can be seen, when the external excitation frequency is close to or greater than the natural frequency of the main structure, the particle damper can suppress the vibration response of the main structure over a wider frequency range. However, when the frequency of external excitation is much less than the natural frequency of the main structure, the damper will produce a certain magnification of the response. The reduced peak R.M.S. amplitude of the primary system with particle damper occurs at a lower frequency compared with that without a particle damper, due to the added mass of particles.

4.4 A Particle Damper Attached to a SDOF Structure Under Random Vibration

In the preceding section, the performance analysis of the additional particle damper for a single-degree-of-freedom system is carried out, with emphasis on the case of unidirectional free vibration and simple harmonic vibration. The stiffness of the system in other directions is set to infinite. On this basis, the dynamic performance of the system under bidirectional random excitation will be emphatically analyzed in this chapter. Due to the collision of internal particles, the motion of the system in two directions will be coupled, which will be discussed in detail in this chapter.

4.4.1 Performance Analysis Under Random Vibration with Different Characteristics

The schematic diagram of the computational model of the additional particle damper for the two-degree-of-freedom system is shown in Fig. 4.12. The dynamic governing equation is:

$$M\ddot{x} + kx + c\dot{x} = F_x + f_x$$
$$M\ddot{y} + ky + c\dot{y} = F_y + f_y \qquad (4.8)$$

Or

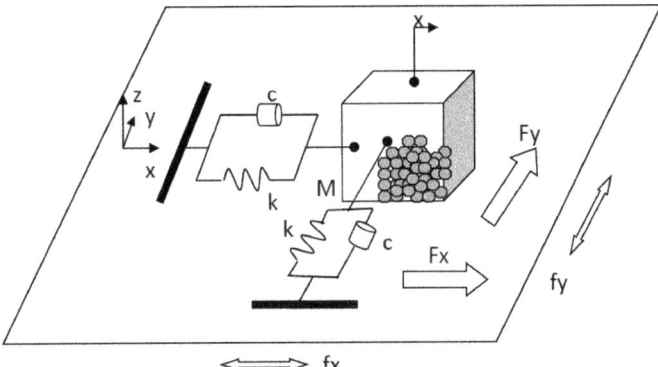

Fig. 4.12 Schematic diagram of calculation model of additional particle damper for two-degree-of-freedom system

$$\ddot{x} = -\omega_n^2 x - 2\zeta\omega_n\dot{x} + (F_x + f_x)/M$$
$$\ddot{y} = -\omega_n^2 y - 2\zeta\omega_n\dot{y} + (F_y + f_y)/M \qquad (4.9)$$

where F_x and F_y are the collision forces of all particles on the container, f_x and f_y are the external excitation.

In the numerical simulation experiment, the parameters of the main system are the same as Sect. 4.3, and the stiffness in x direction and y direction is the same. The random excitation adopts the wideband white noise with Gauss distribution, and the frequency bandwidth is 0–50 Hz.

4.4.2 Stationary Random Excitation in the X Direction Only

This test uses a stationary random excitation in the x direction. Figure 4.13 shows a typical root-mean square (R.M.S.) response time history for different container sizes. After an operating time span of 1000 times the primary system's natural period, the system reaches a stationary response, thus this time duration is used for the following simulations. Choosing a suitable container size, the particle damper can achieve optimum response reductions. As expected, the main response is observed along the direction of the excitation (x direction), and negligible movement exists along the orthogonal direction (y direction). The reason for this behavior is that particles are excited mainly in the x direction; although oblique collisions may cause particles to move in the y direction, they move randomly in the y direction and counteract each other's effects (Some particles move in the direction of $+y$ and some move in the direction of $-y$). The effective contacts of particles and walls in the y direction are consequently at a very low level, and so is the effective momentum exchange.

As to the contact forces, F_x and F_y, they exhibit different behaviors. Figure 4.14

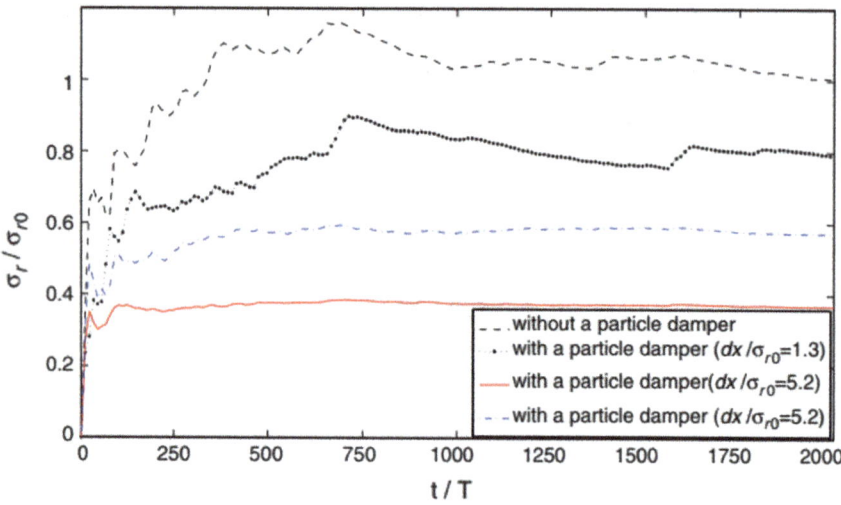

Fig. 4.13 R.M.S. response time history of the displacement magnitude ($\sqrt{x^2 + y^2}$) for the primary system with $\mu = 0.108$, $e = 0.75$, $\zeta = 0.004$, $\mu_s = 0.5$, $dy/\sigma_{r0} = 3.9$, $d/\sigma_{r0} = 0.8$, $N = 16$, under a stationary random excitation in the x direction

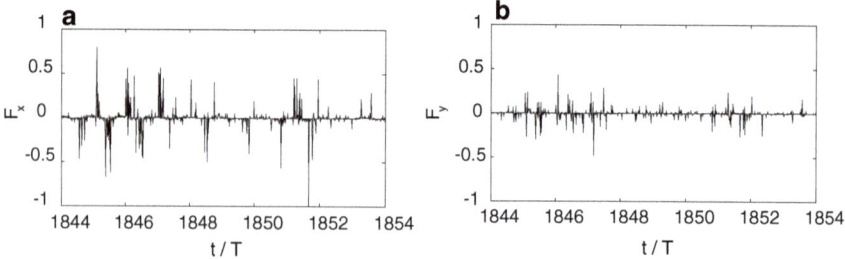

Fig. 4.14 Time history of the contact forces between the primary system and particle container. **a** Force F_x in the x direction; **b** force F_y in the y direction. System parameters are: $\mu = 0.108$, $e = 0.75$, $\zeta = 0.004$, $\mu_s = 0.5$, $dx/\sigma_{r0} = 5.2$, $dy/\sigma_{r0} = 3.9$, $d/\sigma_{r0} = 0.8$, $N = 16$, under a stationary random excitation in the x direction

shows a sample time history for both contact forces, which is made dimensionless by dividing by the maximum force during this sample duration. In Fig. 4.14a, the negative forces are on the left-hand side (LHS) edge of the container, while the positive forces are on the right-hand side (RHS) edge; similarly, in Fig. 4.14b, the negative ones are on the front edge and the positive ones are on the back. One can see that in Fig. 4.14a, b that F_x is larger than F_y, and this occurs at a dominant frequency. Figure 4.15a shows that this F_x frequency is the same as that of the primary system. F_y is smaller and randomly excited, as is shown more clearly in Fig. 4.14b, and that the energy of F_y is generally low in all frequency bandwidths.

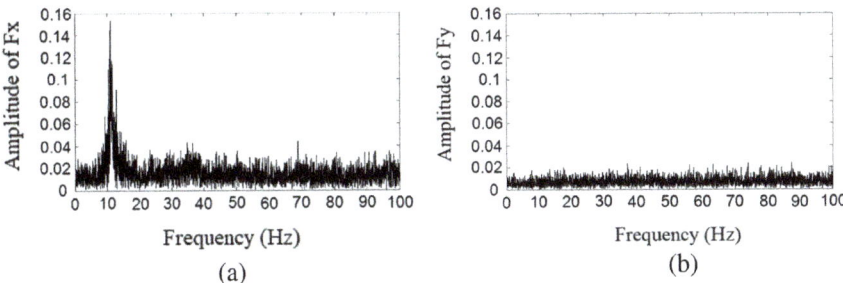

Fig. 4.15 Fast Fourier transforms of the contact forces. **a** Force F_x in the x direction; **b** force F_y in the y direction. System parameters are same with those in Fig. 4.14

4.4.3   Correlated Stationary Random Excitations in the X and Y Directions

This test uses the same stationary random excitation in both the x and y directions, which is equivalent to the situation in which the excitation is applied to the primary system in a diagonal direction. Figure 4.16a shows that the primary system without a particle damper moves back and forth along the diagonal direction. Due to the disturbances induced by the collisions (including inter-particle and particle-wall collisions), the trajectory of the primary system with a particle damper departs from the diagonal direction by a slight amount, but still remains mainly along the diagonal

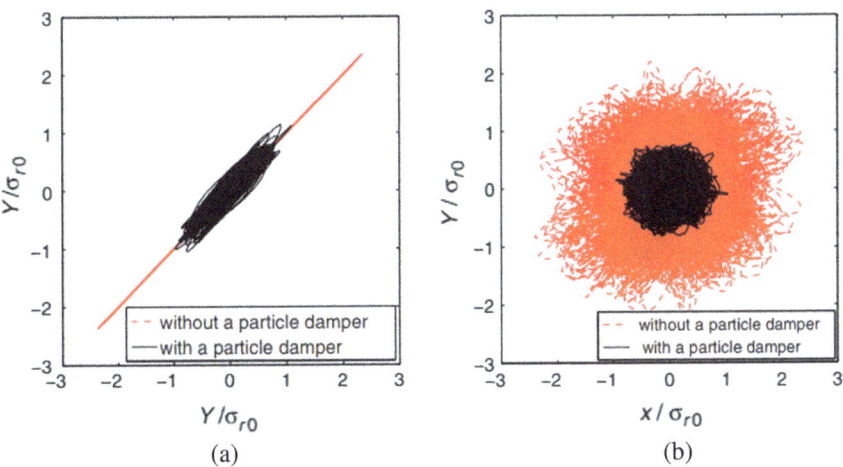

Fig. 4.16 Trajectory of the primary system with $\mu = 0.108$, $e = 0.75$, $\zeta = 0.004$, $\mu_s = 0.5$, $dx/\sigma_{r0} = 5.1$, $dy/\sigma_{r0} = 3.8$, $d/\sigma_{r0} = 0.8$, $N = 16$. Under two simultaneous correlated random excitations in the x and y directions, **b** under two simultaneous uncorrelated random excitations in the x and y directions

line. This is mostly because of the correlated excitation. However, the displacement is reduced a lot in this case, which indicates that particle dampers can be operated in a more robust and efficient way compared to a tuned mass damper (TMD), which can only reduce the response along the installation axis of the TMD.

4.4.4 Uncorrelated Stationary Random Excitations in the X and Y Directions

This test uses two uncorrelated stationary random excitations based on the same probability distribution in both the x and y directions. Compared with previous section, Fig. 4.16b shows that the trajectories of the primary system is different due to the different excitation manner. Both trajectories are centered at the equilibrium position and move in a circular manner. The orbit radius in the particle damper case is much smaller than the corresponding case without a particle damper, which means that the damper has a good vibration reduction effect under this case.

Figure 4.17 shows a sample R.M.S. response of the primary system for different container sizes, which also displays a certain optimum operation. It can be seen that an operating time of about 1000 times the primary system's natural period is sufficient for the primary system to reach a stationary vibration. And for different container sizes, there really exists an optimum value of vibration reduction effect.

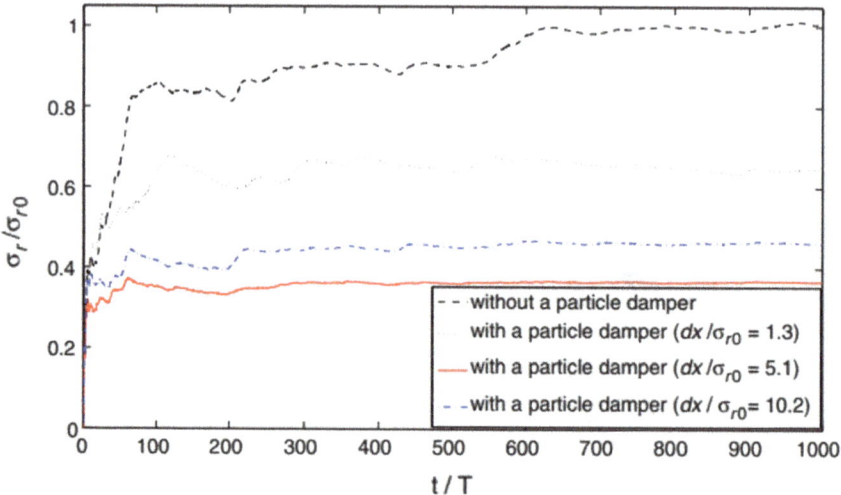

Fig. 4.17 R.M.S. response time history of the displacement magnitude ($\sqrt{x^2 + y^2}$) for the primary system with $\mu = 0.108$, $e = 0.75$, $\zeta = 0.004$, $\mu_s = 0.5$, $dy/\sigma_{r0} = 3.8$, $d/\sigma_{r0} = 0.8$, $N = 16$, under two simultaneous uncorrelated random excitations in the x and y directions

4.4.5   *Discussion*

The nature of the movements of particles are different in the region of the optimum operating conditions compared with other operating conditions. Basically, in the range of high efficiency vibration reduction, plug flow motions of the particles, rather than random movements, can be observed in the optimum cases in each of the three types of excitation which have been discussed before. This is similar qualitative behavior to what occurs in a single-particle impact damper when operating with two-impacts-per-cycle [1]. In plug flow mode, the particles tend to move together, hence cross-correlation functions are used here to indicate the different behavior of the particle dampers.

Figure 4.18a shows the normalized velocity cross correlation in the x direction of two randomly picked particles under a stationary random excitation in the optimum operating range. By dividing the value of the autocorrelation function of the uncontrolled x direction velocity, the cross-correlation $(\overline{R}_{\dot{x}_i \dot{x}_j})$ function can be dimensionless, as follows:

$$\overline{R}_{\dot{x}_i\dot{x}_j}(\tau) = \frac{E[\dot{x}_i(t)\dot{x}_j(t-\tau)]}{E[\dot{x}_{np}(t)\dot{x}_{np}(t)]} = \frac{\frac{1}{T}\int_0^T \dot{x}_i(t)\dot{x}_j(t-\tau)dt}{\frac{1}{T}\int_0^T \dot{x}_{np}(t)\dot{x}_{np}(t)dt} \qquad (4.10)$$

where τ is the time lag, T is the operating time, $\dot{x}_{np}$ is the velocity in the x direction of the primary system without a particle damper, $\dot{x}_i$ is the velocity in the x direction of a randomly picked particle i, and E is the mean-value operator. Figure 4.18a, b show that $\overline{R}_{\dot{y}_i \dot{y}_j}$ is much smaller than $\overline{R}_{\dot{x}_i \dot{x}_j}$. Because particles are moving in a plug flow mode in the x direction, while moving randomly in the y direction.

Corresponding to Fig. 4.17, Fig. 4.19 gives a general impression of how the normalized velocity cross-correlations of two randomly picked particles behave under different operating conditions. The normalized cross correlation here is defined as

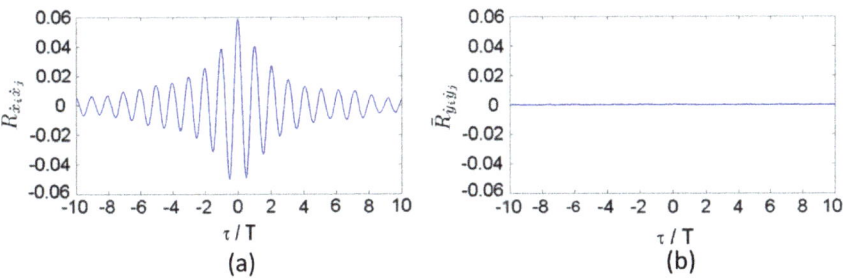

Fig. 4.18 Normalized cross-correlations of velocities of two randomly picked particles in the optimum operating range **a** in the x direction and **b** in the y direction. System parameters are: $\mu = 0.108, e = 0.75, \zeta = 0.004, \mu_s = 0.5, dx/\sigma_{r0} = 5.2, dy/\sigma_{r0} = 3.9, d/\sigma_{r0} = 0.8, N = 16$, under a stationary random excitation in the x direction

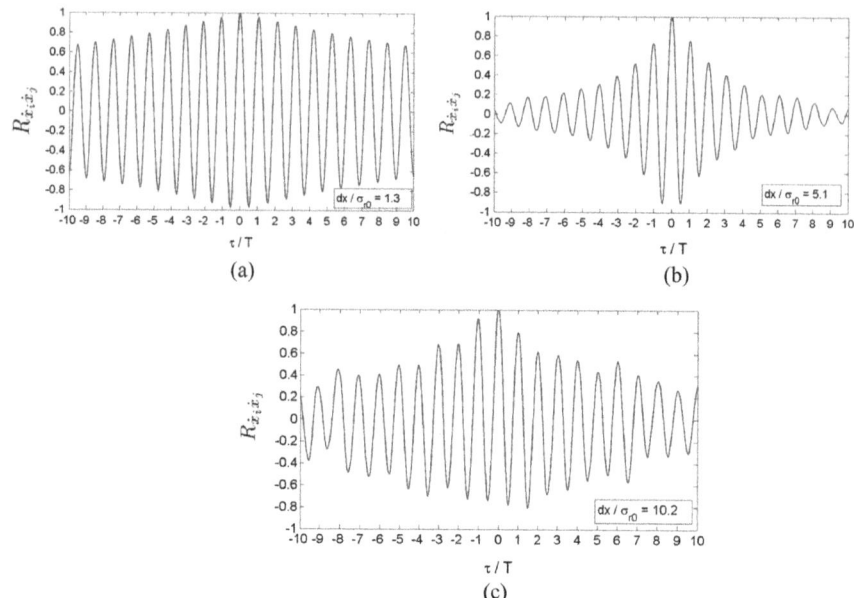

Fig. 4.19 Normalized cross-correlations of velocities in the x direction of two randomly picked particles. **a** Small container size, in the inefficient operating range; **b** medium container size, in the optimum operating range; and **c** large container size, in the inefficient operating range. System parameters are: $\mu = 0.108$, $e = 0.75$, $\zeta = 0.004$, $\mu_s = 0.5$, $dy/\sigma_{r0} = 3.8$, $d/\sigma_{r0} = 0.8$, $N = 16$, under two simultaneous uncorrelated random excitations in the x and y directions

$$R_{\dot{x}_i \dot{x}_j}(\tau) = \frac{E[\dot{x}_i(t)\dot{x}_j(t - \tau)]}{E[\dot{x}_i(t)\dot{x}_j(t)]} = \frac{\frac{1}{T}\int_0^T \dot{x}_i(t)\dot{x}_j(t - \tau)dt}{\frac{1}{T}\int_0^T \dot{x}_i(t)\dot{x}_j(t)dt} \tag{4.11}$$

An interesting observation is that the velocity correlation function decays much faster in the optimum operating range than that in the inefficient operating range (shown in Fig. 4.19a, c). Figure 4.20 draws the three cases in the same figure for comparison. It can be seen that the exponential decay ratio is ≈4.5% for a medium container size in the optimum range, compared to ≈2.2% for a relatively large container size and ≈0.7% for a small container size in the inefficient operating range.

Consequently, it is found that the cross-correlation of velocity of random particles is a good 'global' tool to indicate the performance of the particle damper.

Besides the cross-correlation functions, the autocorrelations of the primary system displacement with or without a particle damper, are also informative. As shown in Fig. 4.21, the former decays quickly in an exponential form, which indicates that the particle damper increases the system damping a lot, while the latter decays in a slower exponential manner, due only to the inherent system damping.

The correlation functions are not the only useful interpreters for optimum behavior of particle dampers; other measures such as the amount of energy dissipation due to impact and friction, and the effective momentum exchange are also useful tools.

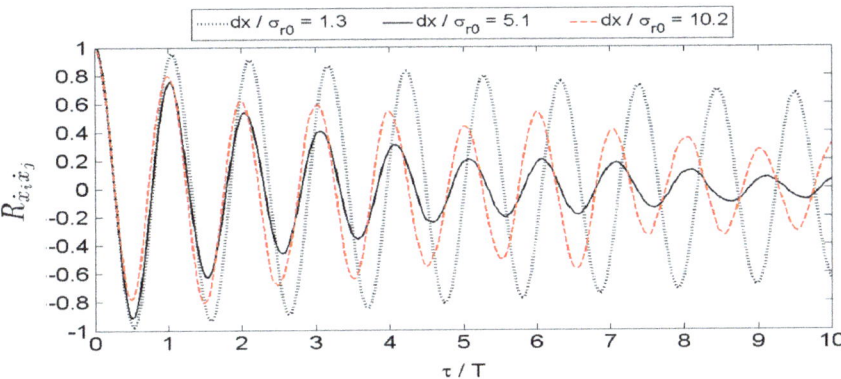

Fig. 4.20 Normalized cross-correlations of velocities in the x direction of two randomly picked particles. $dx/\sigma_{r0} = 1.3$ corresponds to a small container size, in the inefficient operating range; $dx/\sigma_{r0} = 5.1$ corresponds to a medium container size, in the optimum operating range; and $dx/\sigma_{r0} = 10.2$ corresponds to a large container size, in the inefficient operating range

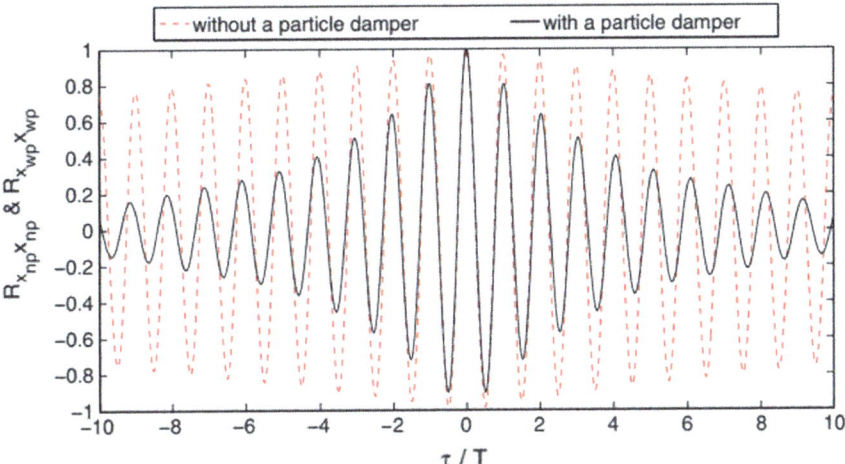

Fig. 4.21 Normalized auto-correlations of the primary system displacement in the x direction, with $\mu = 0.108$, $e = 0.75$, $\zeta = 0.004$, $\mu_s = 0.5$, $dx/\sigma_{r0} = 5.1$, $dy/\sigma_{r0} = 3.8$, $d/\sigma_{r0} = 0.8$, $N = 16$, under two simultaneous uncorrelated random excitations in the x and y directions

According to the contact force model in Chap. 2, the dissipated energy mainly has two sources, which are inelastic impacting and friction, and can be computed thus:

$$E = \begin{cases} \sum \left(2\zeta_2 \sqrt{mk_2}\dot{\delta}_n\dot{\delta}_n dt + \left| F_{ij}^t\dot{\delta}_t dt \right| \right) & (particle - wall) \\ \sum \left(2\zeta_3 \sqrt{\frac{m_i m_j}{m_i + m_j}}k_3\dot{\delta}_n\dot{\delta}_n dt + \left| F_{ij}^t\dot{\delta}_t dt \right| \right) & (particle - particle) \end{cases} \tag{4.12}$$

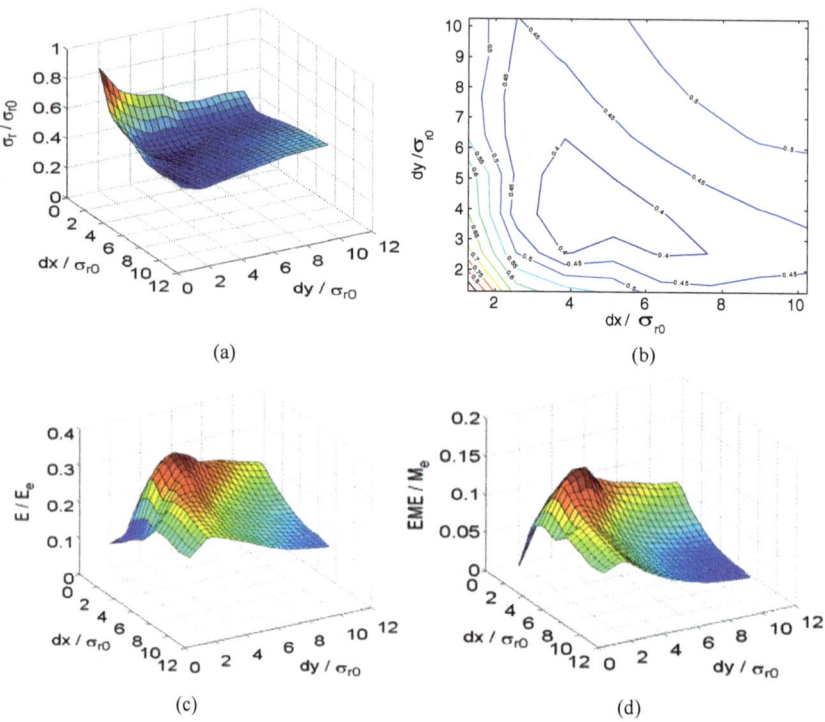

Fig. 4.22 **a** R.M.S. displacement ($\sqrt{x^2 + y^2}$) magnitude response levels; **b** contour plot of σ_r/σ_{r0}; **c** dissipated energy; **d** effective momentum exchange for the primary system with: $\mu = 0.108$, $e = 0.75$, $\zeta = 0.004$, $\mu_s = 0.5$, $d/\sigma_{r0} = 0.8$, $N = 16$, under two simultaneous uncorrelated random excitations in the x and y directions

where dt is the contact time duration, and the dissipated energy is summed over all contact time durations.

Figure 4.22a gives a general impression of how a particle damper operates within a range of container sizes, including lengths and widths (as the height does not influence the behavior very much, it is not included in this study). An optimum region, within which the primary system can get more than a 60% reduction of the R.M.S. response, is clearly shown in the contour plot, Fig. 4.22b. In Fig. 4.22c, the dissipated energy is plotted in dimensionless form by dividing by the energy of the excitation force (E_e). One can find a high ratio of dissipation in the optimum region. Similarly, a normalized version of the effective momentum exchange, which is divided by the momentum exchange of the excitation force (M_e), is plotted in Fig. 4.22d, and gives a relatively high value in the optimum range, too. By comparing the four diagrams, it is found that the amount of energy dissipation and effective momentum exchange correspond well with the response of the system: when the current two quantities are the largest, the response of the system is the smallest, that is, the effect of damper is the best, and when the two quantities become smaller, the effect of damper is correspondingly worse.

In fact, considering two extreme cases, one is that the container is very small, the other is that the container is very large. In the former case, the particles are piled together in many layers, which minimizes the motion of the lower layers, and only creates a vigorous motion in the top-most layers. In the latter, the particles need to spend a long time transiting from one wall to the opposite wall, which leads to fewer effective impacts with the primary system. These two cases will lead to a poor damper performance. When the container size is a moderate value, the damper can play the best role, which is the reason why the particle damper always has an optimal working range.

Through the above discussion, the existence of the optimal performance of particle dampers under different excitations is clearly revealed, and this performance has strong robustness and efficiency, which can provide a good reference for practical engineering. In addition, correlation function, energy dissipation and effective momentum exchange are useful tools to reflect the working characteristics of particle dampers.

4.4.6 Parametric Study

In this section, different parameters, such as the coefficient of restitution (e), excitation levels, container shape, and coefficient of friction (μ_s) are investigated to get a broader view of the performance of particle dampers with a two-degrees-of-freedom system. In order to facilitate the presentation of the results, two-dimensional plots are used hereafter, in which the normalized attenuation level is plotted versus the normalized container size. The width of the container is $dy/\sigma_{r0} = 3.8$ and 16 particles with the same diameter of $d/\sigma_{r0} = 0.8$ are placed in the container. External excitation is a two-way unrelated steady-state random excitation.

1. *Effect of coefficient of restitution*

The coefficient of restitution, e, determines the rebound velocity of the particle and is the ratio between the relative velocity immediately following the impact to the relative velocity just prior to impact. It depends on the type, shape, and surface area of the materials in contact.

Figure 4.23 shows that higher e's lead to a less reduction effect of the primary system's response in small-size containers, and a bigger reduction in large-size containers, compared to lower e's. The reason for this behavior is that a higher e can get a higher relative velocity immediately after the impact, which results in more collisions under a small clearance. In these collisions, much harmful impacts and harmful momentum exchange happen and the effective momentum exchange is reduced. As for the dissipated energy, lower e's will cause the loss of more energy during impacts, and this seems to be dominant when the container sizes are small.

Fig. 4.23 R.M.S. displacement magnitude response with $\mu = 0.108$, $\zeta = 0.004$, $\mu_s = 0.5$ (effect of restitution coefficient)

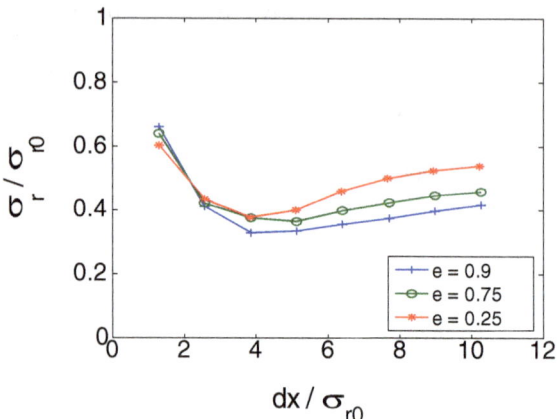

Another significant observation that can be gleaned from Fig. 4.23 is that the sensitivity of a particle damper to changes in dx increases as e decreases, which results in the narrower optimum clearance for smaller e's. Consequently, a particle damper designed with a relatively high value of e can tolerate a broader range of excitation levels, while still performing around the optimum level.

2.   Effect of excitation levels

In this test, five different levels of random excitation are used to investigate the influence of the excitation level on the performance of the particle damper. As shown in Fig. 4.24, the level of excitation plays a very important role. As the level of excitation increases, the efficiency of the damper increases due to the fact that the more energetic motion of the particles increases the exchange of momentum and dissipation of energy. On the other hand, when the excitation is high enough to mobilize all the particles, the response amplitude becomes independent of the intensity of the excitation, provided the dimensionless clearance ratio is kept constant.

Fig. 4.24 R.M.S. displacement magnitude response with $\mu = 0.108$, $e = 0.75$, $\zeta = 0.004$, $\mu_s = 0.5$ (effect of external excitation)

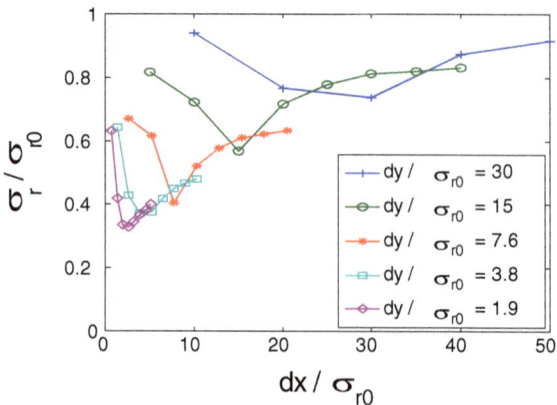

3. *Effect of container shape*

In this test, a cylinder container rather than a cuboid container is used to investigate the effect of container shape. Figure 4.25 displays the R.M.S. response of the primary system with a cylindrical particle damper, which is filled with 16, 64, and 128 particles, respectively. Optimum clearance ranges can easily be found in all cases. The particle damper with 128 particles has a broader range, compared with the 16-particle case. However, for all cases, there is not much difference in the best R.M.S. response levels.

Comparing Fig. 4.25 with Fig. 4.26a–b, which show the behavior of a rectangular-shaped particle damper with the same parameters, a cylindrical particle damper has a better response reduction in the optimum range, that is, there is a 64% R.M.S. response level reduction for a rectangular particle damper compared with a 70% reduction for a cylindrical particle damper. The reason may lie in the symmetric shape, which could cause more beneficial collisions between particles and the container. Additionally, a cylindrically-shaped container is not influenced by the excitation direction.

4. *Effect of coefficient of friction*

Figure 4.26a shows that the damper with small sliding friction coefficient has better vibration reduction effect. The smaller the friction force, the more kinetic energy can be obtained, the more collisions and momentum exchanges can be produced, and the more energy consumed in forward collisions. On the other hand, larger friction coefficient will dissipate more energy in the process of friction and oblique collision. Therefore, the influence of friction coefficient on the performance of damper is very complex.

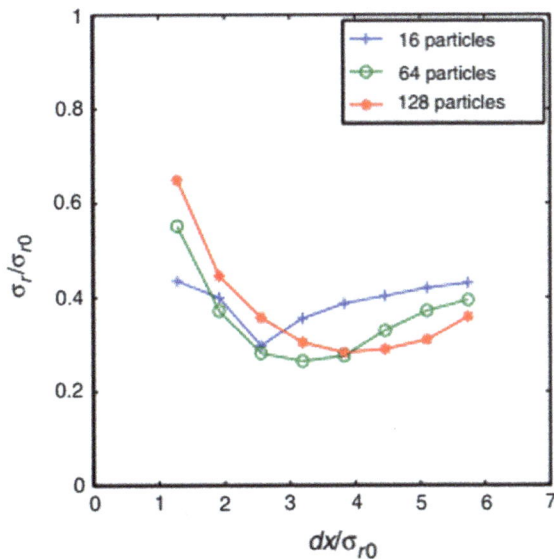

Fig. 4.25 R.M.S. displacement magnitude response levels for the primary system with a cylindrical particle damper, with $\mu = 0.108$, $e = 0.75$, $\zeta = 0.004$, $\mu_s = 0.5$ (effect of container shape)

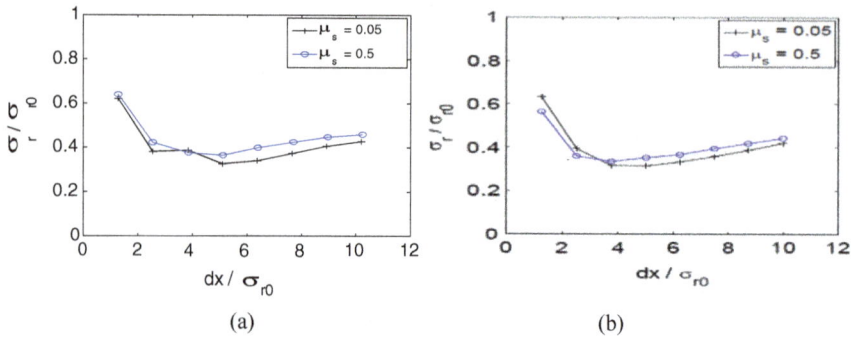

Fig. 4.26 R.M.S. displacement magnitude response levels for the primary system with $\mu = 0.108$, $e = 0.75$, $\zeta = 0.004$ (effect of coefficient of friction): **a** single-unit-multi-particle damper, **b** multi-unit-multi-particle damper

According to the theoretical and numerical work of Bapat [5] for a multi-unit impact damper, the effect of Coulomb friction is generally detrimental. This phenomenon is also shown in Fig. 4.26b, which is for a multi-unit impact damper, introduced in the next section. Both figures show that in large-size containers, less friction leads to a larger response reduction. This indicates that the momentum exchange between particles and the primary system, and the particles' mobility, play very important roles in large-size containers.

4.4.7 Performance Comparison Between Particle Dampers and Multi-unit Impact Dampers

Multi-unit-single-particle dampers have been introduced in the previous sections. The main feature is that there is no collision between particles. For a two-degree-of-freedom system, the particle mass is divided into two parts, one in the x direction and the other in the y direction, so as to reduce the system response caused by the two directions excitation, as shown in Fig. 4.27.

Figure 4.28 shows that the effect of multi-unit-single-particle damper is better than that of single-unit-multi-particle damper when the effective mass ratio is the same. Especially in the case of high coefficient of restitution, the former is insensitive to the change of container size, which shows that it can withstand wider excitation intensity and has good robustness.

On the other hand, considering the two-way vibration reduction, in order to reduce the y-direction vibration, half of the mass is needed in that direction, i.e. $\mu_x = 0.054$, $\mu_y = 0.054$, as shown in Fig. 4.28. Comparing Figs. 4.28 and 4.29, it can be seen that the smaller effective mass ratio can only produce smaller response reduction regardless of the coefficient of restitution. For multi-unit-single-particle dampers, although the maximum amplitude of vibration reduction with low coefficient of

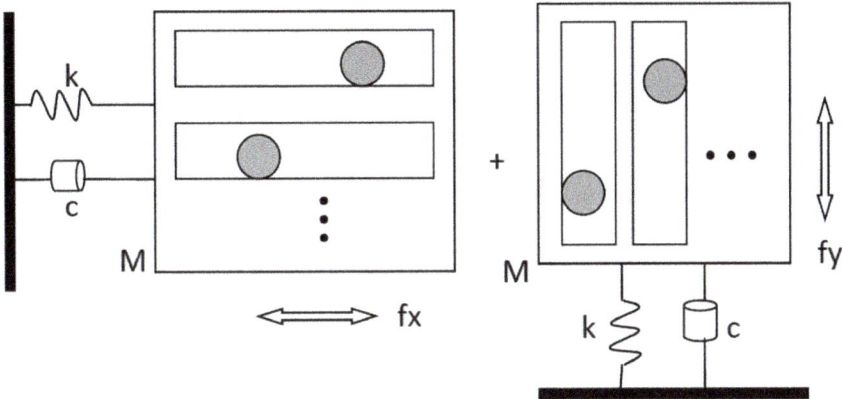

Fig. 4.27 Model of multi-unit impact damper with the total damper mass split in half to control both motion axes

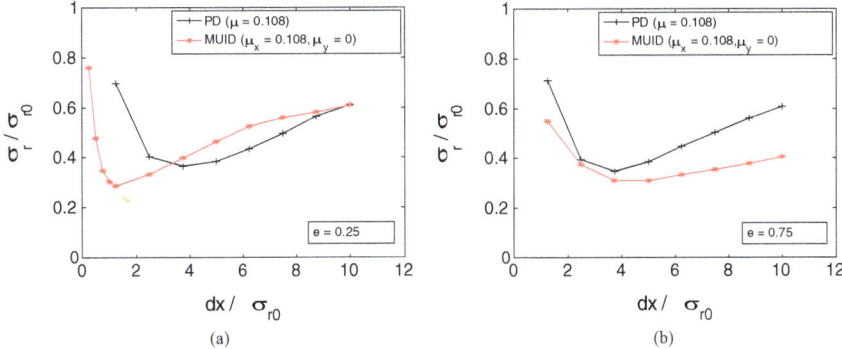

Fig. 4.28 R.M.S. displacement magnitude response levels for the primary system with a particle damper (PD) compared to a multi-unit impact damper (MUID), $\zeta = 0.004$, **a** e $= 0.25$ and **b** e $= 0.75$. The same effective mass ratio is used in both damper systems

restitution is slightly larger than that with high coefficient of restitution, it is more sensitive to the change of container size, indicating that the stability of its optimal operation is not good.

In summary, single-unit-single-particle dampers and multi-unit-single-particle dampers have their own internal characteristics; however, particles with higher coefficient of restitution are better for use in both situations. If the excitation is aligned in the direction where a multi-unit impact damper is installed, this device can provide a larger response reduction compared to a particle damper with the same mass ratio when operating in the optimum range. However, since in practical cases (e.g., a primary system subjected to multi-component ground motion resembling an earthquake) people usually do not know what the excitation direction would be in advance,

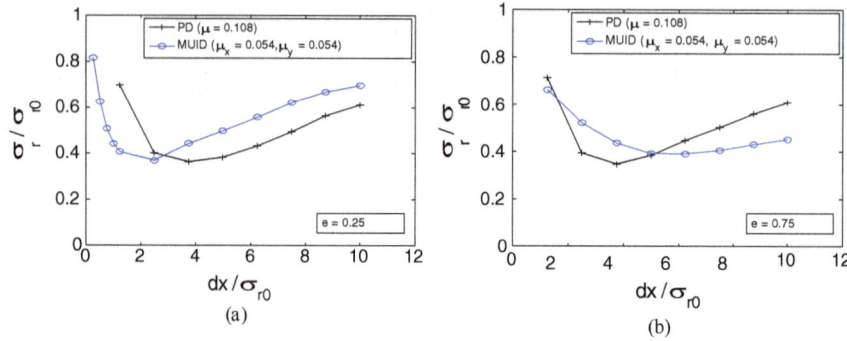

Fig. 4.29 R.M.S. displacement magnitude response levels for the primary system with a particle damper compared to a multi-unit impact damper, $\zeta = 0.004$, **a** e $= 0.25$ and **b** e $= 0.75$. The effective mass ratio of the MUID is one half that of the PD

particle dampers, with their much more robust tolerance of excitation direction, may be a better choice for response control.

References

1. Masri, S.F. 1969. Analytical and experimental studies of multiple-unit impact dampers. *Journal of the Acoustical Society of America* 45 (5): 1111–1117.
2. Friend, R.D., and V.K. Kinra. 2000. Particle impacting damping. *Journal of Sound and Vibration* 233 (1): 93–118.
3. Mao, K.M., M.Y. Wang, and Z.W. Xu, et al. 2004. DEM simulation of particle damping. *Powder Technology* 142(2–3): 154–165.
4. Butt, A.S. 1995. *Dynamics of impact-damped continuous systems, in College of engineering*, 172. Louisiana Tech University: Louisiana.
5. Bapat, C.N., and S. Sankar. 1985. Multiunit impact damper - reexamined. *Journal of Sound and Vibration* 103 (4): 457–469.

Chapter 5
Performance Analysis of Particle Dampers Attached to Multi-degree-of-Freedom (MDOF) Structures

The previous chapter carries out the performance analysis of particle dampers attached to single-degree-of-freedom (SDOF) structures, with an emphasis on the structural dynamic characteristics under correlated stationary random excitations in the x and y directions. Based on such investigation, in this chapter, the main structures are generalized to multi-degree-of-freedom (MDOF) ones, and the damping performance of particle dampers attached to MDOF structures is systematically analyzed. Especially, the external excitations have been expanded from stationary random excitations to nonstationary random excitations, which are more consistent with the actual situation [1–3].

5.1 Analytical Solution to a Particle Damper Attached to a MDOF Structure

5.1.1 Computational Model

Figure 5.1 demonstrates the computational model of an impact damper attached to a MDOF structure. The sinusoidal forcing function is applied to the kth mass and the impact damper is attached to the jth mass. It is assumed that, during a period of the forcing function, two symmetric impacts occur at equal time intervals and at opposite ends of the damper container. This assumed motion is consistent with that which has been found to predominate in the experimental studies of impact dampers as observed by most investigators in this field [4].

© China Machine Press and Springer Nature Singapore Pte Ltd. 2020
Z. Lu et al., *Particle Damping Technology Based Structural Control*,
Springer Tracts in Civil Engineering,
https://doi.org/10.1007/978-981-15-3499-7_5

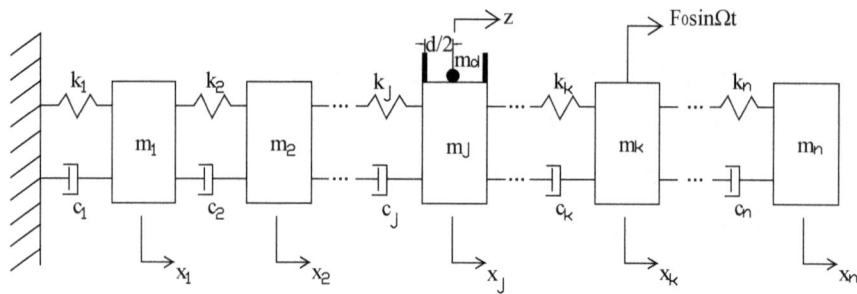

Fig. 5.1 Computational model of an impact damper attached to a MDOF structure

5.1.2   Analytical Solution Method

The equation of model of the system in the absence of the damper is

$$\mathbf{M\ddot{X}} + \mathbf{C\dot{X}} + \mathbf{KX} = \mathbf{F}(t) \tag{5.1}$$

where $\mathbf{M}$, $\mathbf{C}$, $\mathbf{K}$ are the mass, damping, and stiffness matrices, respectively, and $\mathbf{F}(t) = (0, 0, \ldots 0, F_k(t), 0, \ldots, 0)$. Assume that the damping mechanism is the proportional-type satisfying the condition

$$\mathbf{C} = \alpha\mathbf{M} + \beta\mathbf{K} \tag{5.2}$$

where α and β are constants.

Consider the steady-state motion of the system shown in Fig. 5.1 with the origin of the time axis shifted to coincide with the time of occurrence of an impact at $t = t_0$. The result of this shift is to modify the excitation force, giving in the new time scale

$$F_k(t) = F_0 \sin(\Omega t + \alpha_0) \tag{5.3}$$

where a $\alpha_0 = \Omega t_0$ is a phase angle to be determined from the steady-state motion.

Using the normal mode approach, Eq. (5.1) can be transformed into the form

$$\mathbf{M_q\ddot{q}} + C_q\dot{q} + K_q\mathbf{q} = \mathbf{Q_{ex}}(t) \tag{5.4}$$

where the n components of $\mathbf{q}$ are the normal coordinates, M_a, C_a, and K_a are diagonal matrices corresponding to the generalized mass, damping, stiffness matrices, respectively, and where $Q_{ex}(i) = [\varphi]^T F(i)$, and $[\varphi]$ is the modal matrix.

The ith equation of system (5.4) is

$$M_i\ddot{q}_i + C_i\dot{q}_i + K_iq_i = Q_{ex,i} = \varphi_{ki} F_0 \sin(\Omega t + \alpha_0) \tag{5.5}$$

and its solution is

$$
\begin{aligned}
q_i(t) = \exp\left(-\frac{\zeta_i}{r_i}\Omega t\right) \Bigg\{ & \frac{1}{\eta_i}\left(\zeta_i \sin\frac{\eta_i}{r_i}\Omega t + \eta_i \cos\frac{\eta_i}{r_i}\Omega t\right) q_{0i} + \frac{1}{\omega_i \eta_i}\left(\sin\frac{\eta_i}{r_i}\Omega t\right)\dot{q}_{0i} \\
& -\frac{A_i}{\eta_i}\left(\zeta_i \sin\frac{\eta_i}{r_i}\Omega t + \eta_i \cos\frac{\eta_i}{r_i}\Omega t\right)\sin\tau_i - \frac{A_i}{\eta_i}r_i\left(\sin\frac{\eta_i}{r_i}\Omega t\right)\cos\tau_i \Bigg\} \\
& + A_i \sin(\Omega t + \tau_i) \quad i = 1, 2, \ldots, n
\end{aligned}
\tag{5.6}
$$

where

$$
\omega_i = \sqrt{\frac{K_i}{M_i}} \quad \zeta_i = \frac{C_i}{\sqrt{2K_i M_i}} \quad \eta_i = \sqrt{1-\zeta_i^2} \quad r_i = \frac{\Omega}{\omega_i} \quad f_i = \varphi_{ki} F_0
$$

$$
A_i = \frac{f_i/K_i}{\sqrt{(1-r_i^2)^2 + (2\zeta_i r_i)^2}} \quad \psi_i = \tan^{-1}\frac{2\zeta_i r_i}{1-r_i^2} \quad \tau_i = \alpha_0 - \psi_i
$$

$$
q_{0i} = q_i(0) \quad \dot{q}_{ai} = \dot{q}_i(0_+)
$$

and the $+$ subscript indicates conditions immediately after the specified time. Letting the initial displacement and velocity at $t = 0_+$ be

$$
\mathbf{X}(0) = \mathbf{X}_0 = [\varphi]\mathbf{q}_0 \quad \dot{\mathbf{X}}(0_+) = \dot{\mathbf{X}}_a = [\varphi]\dot{\mathbf{q}}_a \tag{5.7}
$$

then

$$
\mathbf{X}(t) = \mathbf{B}_{21}(t)\dot{\mathbf{X}}_a + \mathbf{B}_{22}(t)\mathbf{X}_0 + \mathbf{B}_{23}(t)\mathbf{S}_1 + \mathbf{B}_{24}(t)\mathbf{S}_2 + [\varphi]\mathbf{S}_3(t) \tag{5.8}
$$

$$
\dot{\mathbf{X}}(t) = \mathbf{B}_{31}(t)\dot{\mathbf{X}}_a + \mathbf{B}_{32}(t)\mathbf{X}_0 + \mathbf{B}_{33}(t)\mathbf{S}_1 + \mathbf{B}_{34}(t)\mathbf{S}_2 + [\varphi]\mathbf{S}_4(t) \tag{5.9}
$$

where the undefined matrices and vectors are functions of the system parameters.

Define $y(t)$ as the relative displacement of the damper mass m_d with respect to m_j, then

$$
z(t) = y(t) - X_j(t) \tag{5.10}
$$

Since the time origin coincides with the time of occurrence of an impact,

$$
z(0) = z_0 = y(0) - X_j(0) = \pm\frac{d}{2} \tag{5.11}
$$

During an impact (idealized to be a discontinuous process), the conditions of the system remain unchanged except for the velocities of m_d and m_j, whose instantaneous changes are calculated using the momentum equation and the coefficient of restitution. Noting that in steady state motion

$$\dot{y}(0)_+ = -\frac{2\Omega}{\pi}(X_{0j} + z_0) \tag{5.12}$$

then the velocity vectors before and after an impact are related by

$$\dot{\mathbf{X}}_b = \mathbf{B}_{6q}\dot{\mathbf{X}}_a \tag{5.13}$$

where **B** is a constant diagonal matrix whose elements are equal to unity, except for the jth element which is equal to $(1 - 2e - 2\mu)/(1 - e - 2\mu e)$.

In steady-state motion with two symmetric impacts/cycle of excitation on opposite ends of damper container,

$$\mathbf{X}(t)|_{\Omega t=\pi} = -\mathbf{X}(0) = -\mathbf{X}_0 \tag{5.14}$$

$$\dot{\mathbf{X}}(t)|_{\Omega t=\pi_-} = -\dot{\mathbf{X}}(0)_- = -\dot{\mathbf{X}}_b = -\mathbf{B}_{6q}\dot{\mathbf{X}}_a \tag{5.15}$$

Using (5.14) and (5.15) in conjunction with (5.8) and (5.9)

$$\mathbf{X}_0 = \mathbf{S}_7 \sin \beta_0 + \mathbf{S}_8 \cos \beta_0 \tag{5.16}$$

$$\dot{\mathbf{X}}_a = \mathbf{S}_9 \sin \beta_0 + \mathbf{S}_{10} \cos \beta_0 \tag{5.17}$$

Equations (5.16) and (5.17), together with (5.13), result in

$$\beta_0^{\pm} = \tan^{-1}[(h_1 h_3 \pm h_2 h_4)/(h_2 h_3 \mp h_1 h_4)] \tag{5.18}$$

where the h_1, h_2, h_3, h_4 are functions of $\mathbf{S}_7$, $\mathbf{S}_8$, $\mathbf{S}_9$, $\mathbf{S}_{10}$. With β_0 determined from (5.18), the rest of the unknowns can be found by back substitution.

Figure 5.2 represents an N-storey shear building system equipped with an auxiliary nonlinear particle damper at the top floor, in which a certain number of particles are placed. The MDOF system with a particle damper model is discussed below as an illustration.

The equations of motion of the primary system can be written as

$$\mathbf{M}\ddot{\mathbf{X}} + \mathbf{C}\dot{\mathbf{X}} + \mathbf{K}\mathbf{X} = \mathbf{F} + \mathbf{E}\ddot{x}_g \tag{5.19}$$

$$\mathbf{M} = diag[M_1 \ M_2 \ \dots \ M_N] \tag{5.20}$$

$$\mathbf{C} = \begin{bmatrix} C_1 + C_2 & -C_2 & & \\ -C_2 & C_2 + C_3 & -C_3 & \\ & & \ddots & -C_N \\ & & -C_N & C_N \end{bmatrix} \tag{5.21}$$

Fig. 5.2 Model of a
N-storey shear building
system with a horizontal
particle damper

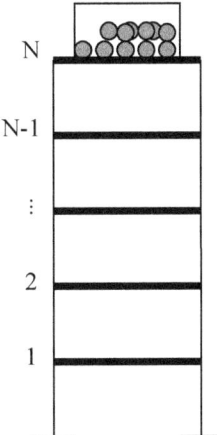

$$
\mathbf{K} = \begin{bmatrix} K_1 + K_2 & -K_2 & & \\ -K_2 & K_2 + K_3 & -K_3 & \\ & & \ddots & -K_N \\ & & -K_N & K_N \end{bmatrix}
$$ (5.22)

$$\mathbf{X} = [X_1 \ X_2 \ \dots \ X_N]^T$$ (5.23)

$$\mathbf{F} = [0 \ 0 \ \dots \ F_N]^T$$ (5.24)

$$\mathbf{E} = [-M_1 - M_2 \ \dots \ - M_N]^T$$ (5.25)

In following section, parametric studies of the performance of a horizontal particle damper attached to a 3DOF system under different dynamic loads are carried out to get a better understanding of the physics involved in particle dampers. The natural frequencies of example primary system are $f_1 = 1.58$ Hz, $f_2 = 4.44$ Hz, and $f_3 = 6.41$ Hz, and the damping ratio is $\zeta = 0.01$.

5.2   A Particle Damper Attached to a MDOF Structure Under Free Vibration

This simulation uses impulses at different storey locations to examine the effect of excitation input location. The duration of the impulse is considered to be very short and an initial velocity is exerted at different storey at the initial instant. A single particle with a mass ratio (the fraction of particle mass and the whole primary system mass) of $\mu = 0.03$, a diameter of 25 mm, and a clearance of 60 mm is applied.

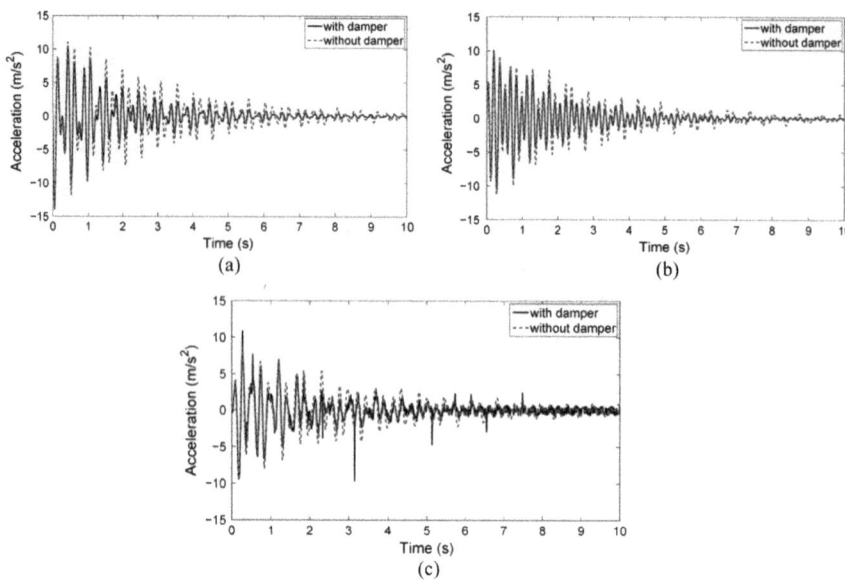

Fig. 5.3 Acceleration time histories of the **a** 1st floor, **b** 2nd floor, and **c** 3rd floor of the primary system, under excitation at the first storey

The response of the primary system with a particle damper decays much faster than that without damper. Figures 5.3 and 5.4 show a sample acceleration and velocity time history of each floor under an impulse at the first floor. One can see that at the third floor, large accelerations are initiated as the impact mass collides with the stops, and then the impacts rapidly reduce the response of the structure at all floor levels. Once the motion of the third floor becomes insufficient for impacts to occur (which depends on the clearance, amplitude of motion and other parameters), the structure oscillates freely, with only internal damping of the structure itself causing the response to decay. Without the particle damper, there is no initial rapid reduction in response and the oscillations decay in a slow exponential form over a longer period of time, as shown by the dashed line. Moreover, it should also be noted that the acceleration of the third floor, where the impact damper is located, can be very high at the moment of collision, although these high accelerations are not evident in the response of the other storeys. As to the velocity response, there are no sudden changes at the instant of collision, and this is because velocity is the integral of accelerations.

Figures 5.5, 5.6, 5.7 shows the corresponding power spectral density (PSD) of the velocity response, under the excitation force acting at the first, second, and third storey, respectively, in which the PSD is plotted in a logarithmic form for better comparison. There is a clear vibration attenuating effect of the impact damper for each storey at the first mode of vibration regardless of the location of the excitation force; however, the vibration control effect upon the higher modes is more clearly

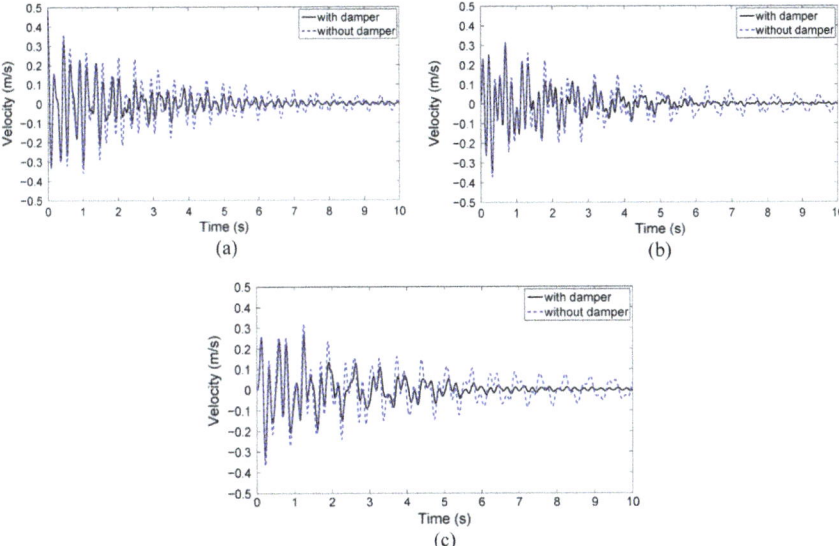

Fig. 5.4   Velocity time histories of the **a** 1st floor, **b** 2nd floor, and **c** 3rd floor of the primary system, under excitation at the first storey

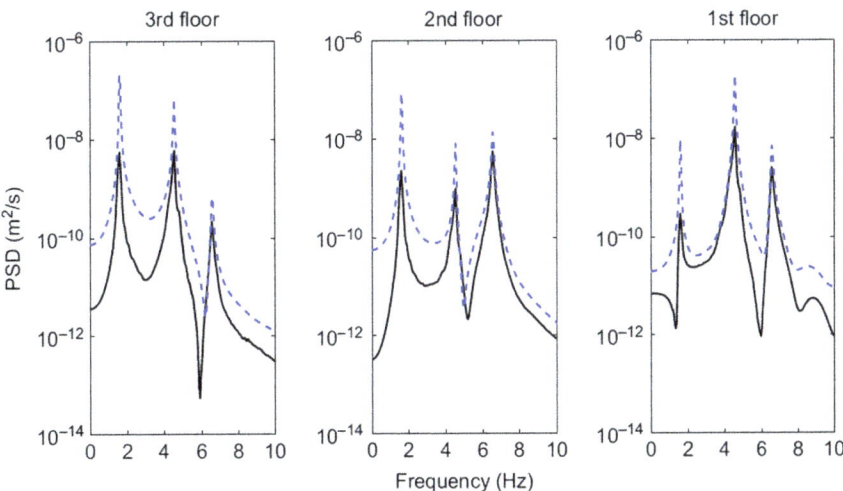

Fig. 5.5   Power spectral density of the primary system velocity response, under excitation at the first storey, in which the dotted line is the without-damper-case, while the solid line is the with-damper-case

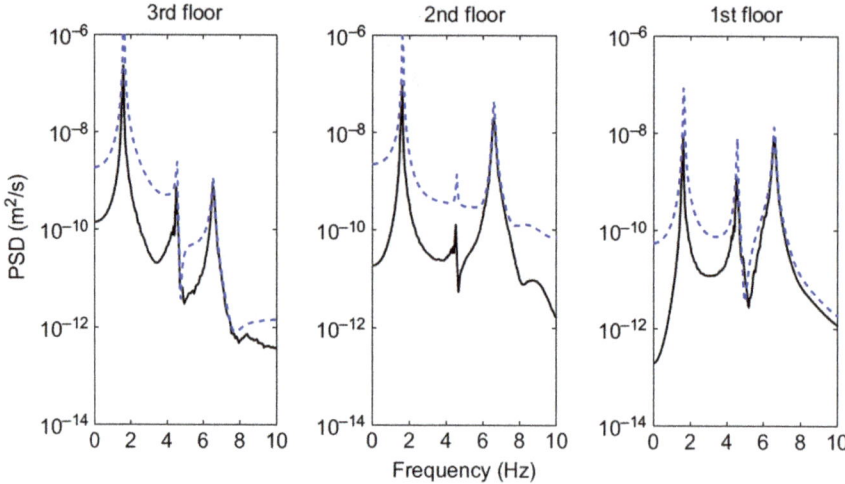

Fig. 5.6 Power spectral density of the primary system velocity response, under excitation at the second storey, in which the dotted line is the without-damper-case, while the solid line is the with-damper-case

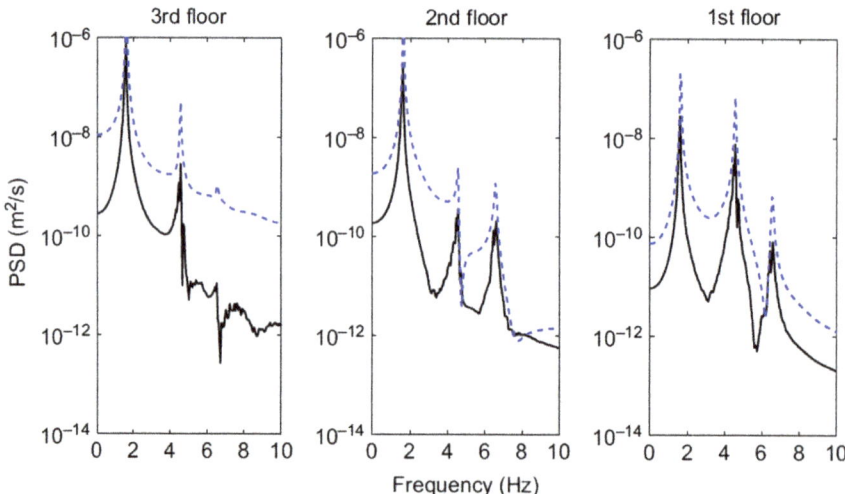

Fig. 5.7 Power spectral density of the primary system velocity response, under excitation at the third storey, in which the dotted line is the without-damper-case, while the solid line is the with-damper-case

affected by location of excitation. In particular, comparing Figs. 5.6 and 5.7, it can be observed that better control of the third mode of each storey occurs when excitation is at the third floor rather than at the second floor. One may also expect that the extent to which certain modes are excited depends on where the structure is excited, leading to a more obvious control effect where the response of the mode is greatly excited. The second mode is a clear example. It is greatly excited when the input force is at the first floor or third floor, as shown in Figs. 5.5 and 5.7; consequently, it is reasonably well controlled by the damper. On the other hand, it is excited very little when the input is at the second floor, as shown in Fig. 5.6. Due to this being the approximate location of an anti-node of the second mode shape; consequently, the controlling effect is not obvious. The above phenomena are also demonstrated in Li's experiments [5].

5.3   A Particle Damper Attached to a MDOF Structure Under Random Vibration

5.3.1   Parametric Study

In this part, different parameters, such as the mass ratio (μ), the coefficient of restitution (e), excitation levels, damping ratio of the primary system (ζ) and damper locations are investigated to get a broader view of the performance of particle dampers with a MDOF primary system. Since the percentage displacement reductions for all storeys have not much differences, to facilitate presentation, only the response of the first floor is demonstrated in the following figures. Other parameters are determined as $dy/\sigma_{x0,1} = 7.5$, $d/\sigma_{x0,1} = 1.4$ and $\mu_s = 0.5$, where dx is length of the container, dy is width of the container, d is diameter of the particle, μ_s is coefficient of friction, $\sigma_{x0,1}$ is root-mean-square value of the displacement of the first floor of the primary system without a particle damper, $\sigma_{x,1}$ is root-mean-square value of the displacement of the first floor of the primary system with a particle damper.

Figure 5.8 shows that increasing the mass ratio of particles can reduce the response of the primary system, but the reduction is not linearly proportional to the increase in the mass ratio; in fact, the effectiveness per unit mass ratio will decrease, in a nonlinear manner, as the mass ratio increases. Another interesting observation is that infinitely increasing the mass of the particles may not reduce the response any further, especially for large container sizes. This phenomenon can be explained by considering the conservation of momentum between the particles and the system. As a specific particle's mass increases, its absolute velocity immediately after the impact decreases which, in turn, reduces its relative velocity, and it takes a longer time to travel towards the other wall of the container. The force of friction also contributes to the reduction in the velocity while the particle is in motion. As the mass is increased beyond a certain value, its relative velocity immediately after the impact does not allow it to overcome the frictional force while in motion, and it comes to rest relative to the system prior to reaching the other wall. At that point, if the system resumes its motion in the same direction and its acceleration is large enough to overcome the force of friction, the particle starts traveling in the opposite direction relative to the

Fig. 5.8 R.M.S. displacement response levels for the first storey of the primary system with damper on the third storey, $e = 0.75$, $\zeta = 0{:}01$, $\mu_s = 0.5$, $dy/\sigma_{x0,1} = 7.5$, $d/\sigma_{x0,1} = 1.4$, and 10 particles (effect of mass ratio)

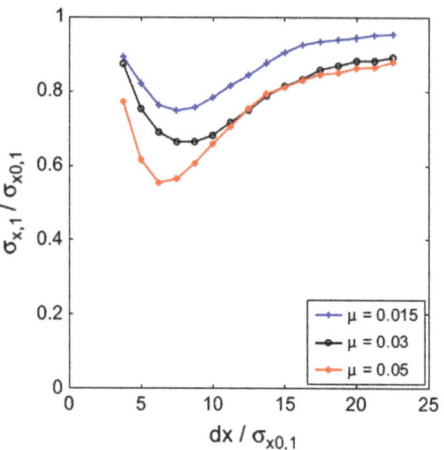

system. If a similar situation arises prior to reaching the other container's boundary, the particle reverses it direction once again. It is possible for the particle to reciprocate between the container's boundaries, while the system goes through several cycles of motion before making the next impact.

The coefficient of restitution e determines the rebound velocity of the particle and is the ratio of the relative velocity immediately following the impact to the relative velocity just prior to impact. It depends on the type, shape, and surface area of the materials coming in contact.

Figure 5.9 shows that higher e's lead to a more reduction in large-size containers, compared to lower e's. The reason for this behavior is that a higher e can result in a higher relative velocity immediately following the impact, which results in more collisions. Another significant observation that can be gleaned from Fig. 5.9 is that the sensitivity of a particle damper to changes in dx (length of the container) increases

Fig. 5.9 R.M.S. displacement response levels for the first storey of the primary system with damper on the third storey, $\mu = 0.05$, $\zeta = 0.01$ and $\mu_s = 0.5$, $dy/\sigma_{x0,1} = 7.5$, $d/\sigma_{x0,1} = 1.4$, and 10 particles (effect of coefficient of restitution)

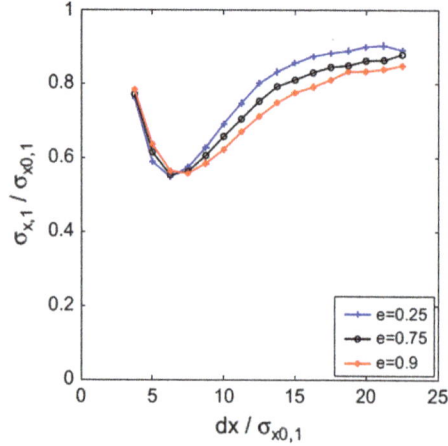

as e decreases, which results in a narrower optimum clearance for smaller e's. Consequently, a particle damper designed with a relatively high value of e can tolerate a broader range of excitation levels, while still performing around the optimum level.

Figure 5.10 shows that the level of excitation plays a very important role. As the level of excitation increases, the efficiency of the damper increases due to the fact that the more energetic motion of the particles increase the exchange of momentum and dissipation of energy. On the other hand, when the excitation is high enough to mobilize all the particles, the response amplitude becomes independent of the intensity of the excitation, provided the dimensionless clearance ratio is maintained constant.

Figure 5.11 shows that the efficiency of the particle damper increases with the damper "elevation" above the ground floor. This is because a higher storey has larger

Fig. 5.10 R.M.S. displacement response levels for the first storey of the primary system with damper on the third storey, $\mu = 0.05$, $e = 0.75$, $\zeta = 0.01$, $\mu_s = 0.5$, and 10 particles (effect of excitation level)

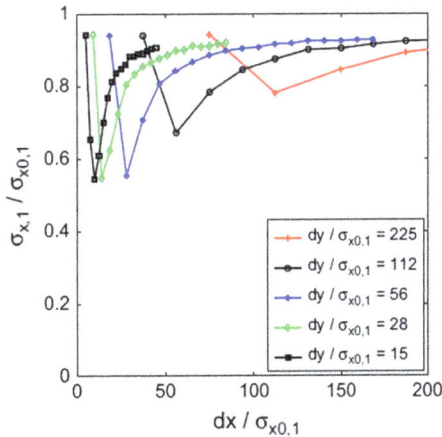

Fig. 5.11 R.M.S. displacement response levels for the first storey of the primary system with $\mu = 0.05$, $e = 0.75$, $\zeta = 0.01$ and $\mu_s = 0.5$, $dy/\sigma_{x0,1} = 7.5$, $d/\sigma_{x0,1} = 1.4$, and 10 particles (effect of damper location)

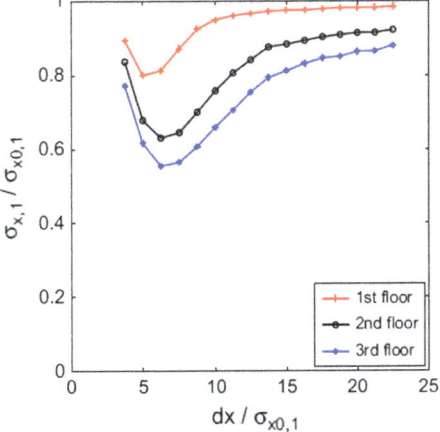

displacement, and it will impart more kinetic energy to the particles, which increases the amount of momentum exchange and energy dissipation.

Figure 5.12 shows that the effectiveness of the particle damper increases as the primary system damping decreases. Consequently, the maximum effect of a particle damper would be achieved for a primary system with a negligible amount of inherent damping.

Figure 5.13 summarizes the effects of mass ratio and the primary system damping on the optimum performance of a particle damper. It is clear that, for a given ζ, the optimum response reduction is not a linear function of the mass ratio. Also, one can conclude that even with very small mass ratios, a properly designed particle damper is capable of substantial attenuation of the R.M.S. response level.

Fig. 5.12 R.M.S. displacement response levels for the first storey of the primary system with damper on the third storey, $\mu = 0.05$, $e = 0.75$, $\mu_s = 0.5$, and 10 particles (effect of primary system damping)

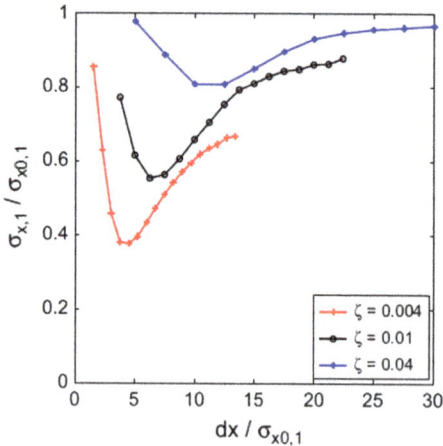

Fig. 5.13 Effect of primary system damping and mass ratio on the performance of particle damper with $e = 0.75$, $\mu_s = 0.5$, and 10 particles

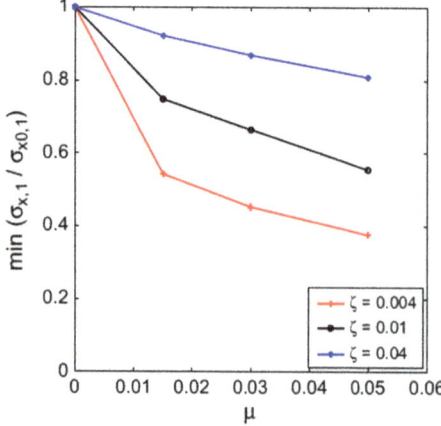

5.3.2   *A Particle Damper Attached to a MDOF Structure Under Nonstationary Random Excitation*

In this part, the performance of particle dampers with a MDOF system under non-stationary excitation $s(t)$ is investigated. One convenient means of generating a synthetic nonstationary random excitation is to modulate a stationary random signal $n(t)$ through multiplication by a deterministic envelope function $g(t)$, as follows [6]:

$$s(t) = g(t)n(t) \tag{5.26}$$

$$g(t) = a_1 \exp(a_2 t) + a_3 \exp(a_4 t) \tag{5.27}$$

where $n(t)$ is the stationary random excitation, and $s(t)$ is the resulting nonstationary part. By a proper choice of a_1, a_2, a_3, and a_4, one can generate a variety of nonstationary excitations, such as earthquake-like excitations.

Simulations were done for the same system discussed in the previous section. The R.M.S. for the resulting non-ergodic process was computed by averaging over a large ensemble of records. For the case studied in this paper, the simulation shows that the resulting R.M.S. will not change after the ensemble size approaches 200 records. Three different envelope functions ($g(t)$) were considered in this study, corresponding to a "fast", "medium'", and "slow" rate of envelope decay.

Figure 5.14 shows the effectiveness of the particle damper in reducing the R.M.S. response for three different envelope functions. For these three cases, the R.M.S. is calculated by averaging over 200 records, and the optimum gap clearance is chosen based on the maximum reduction in the peak value of the R.M.S.

Table 5.1 shows the ratio of the peak R.M.S. ($\sigma_{\max,1}/\sigma_{0\max,1}$) and the ratio of the area under the R.M.S. time history ($\int \sigma_{x,1} dt / \int \sigma_{x0,1} dt$) for the first storey of the primary system, for the three mentioned cases. One can conclude from the displayed results that a particle damper is significantly effective in reducing the area under the R.M.S. time history curve (which is an indication of the response "intensity"). However, the effectiveness of the particle damper in reducing the peak R.M.S. of the nonstationary response is not very significant, especially when the envelope function duration is short. One can find in Table 5.1 that, for $g(t) = g_1(t)$, the nonstationary excitation duration is about 25 natural periods of the system, and the resulting reduction in the peak R.M.S. is only about 15%, while the reduction in the area under the R.M.S. time history is about 46%. The reason for this behavior is due to the particle damper nature: it takes a while for the particles to acquire enough momentum for effective vibration attenuation. By increasing the envelope duration, the behavior of the particle damper improves and gets closer to the stationary one. For example, for

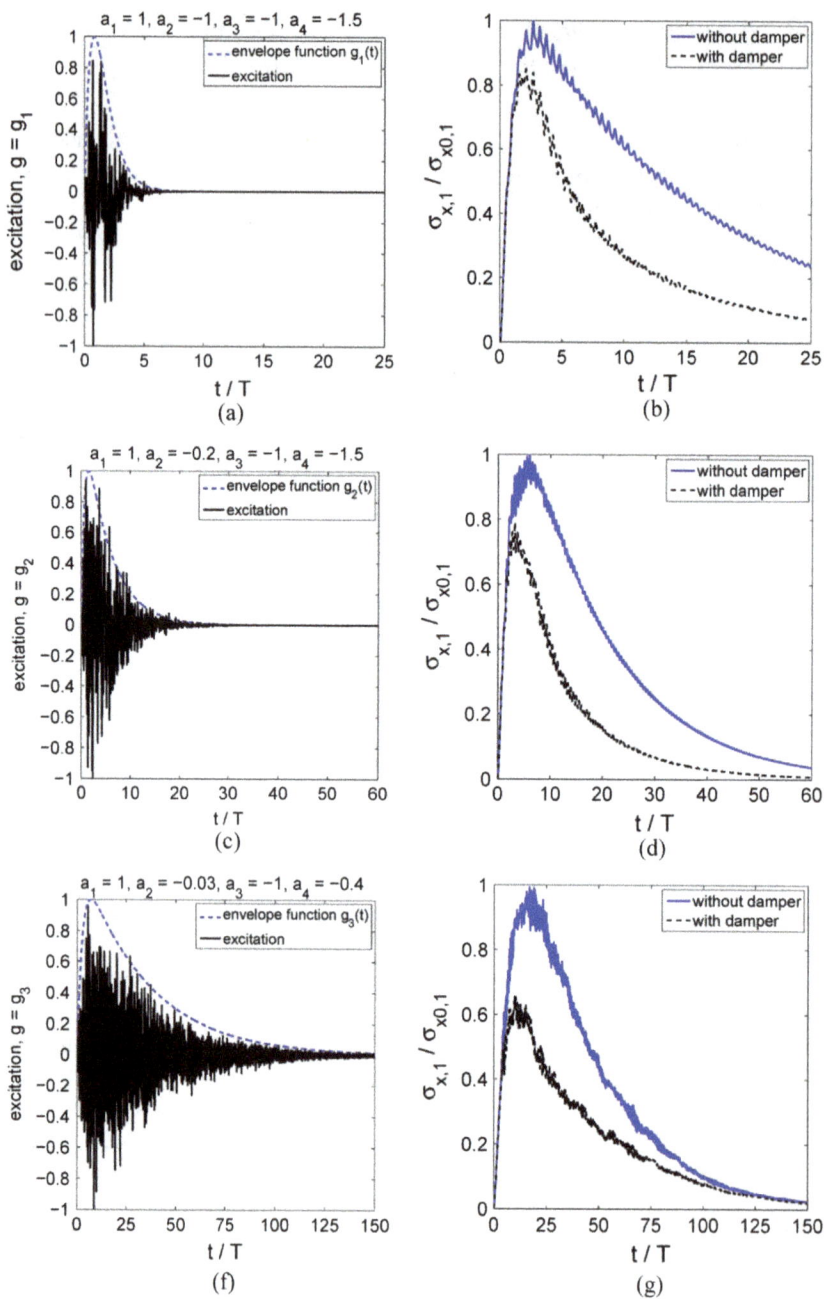

Fig. 5.14 **a, c, e** Exponential envelope function and the resulting nonstationary random excitation; **b, d, f** transient R.M.S. displacement response levels for the first storey of the primary system with damper on the third storey, $\mu = 0.05$, $\zeta = 0.01$, $e = 0.75$, $dx = d_{opt}$, $\mu_s = 0.5$, and 10 particles. **a–b** Excitation envelope: $g(t) = g_1(t) = \exp(-t) - \exp(-1.5t)$, **c–d** Excitation envelope: $g(t) = g_2(t) = \exp(-0.2t) - \exp(-1.5t)$, and **e–f** Excitation envelope: $g(t) = g_3(t) = \exp(-0.03t) - \exp(-0.4t)$

Table 5.1 Summary of the nonstationary simulation results for the first storey of the primary system

Envelope function $g(t)$	Peak R.M.S. ratio $(\sigma_{max,1}/\sigma_{0\,max,1})$	Area under the R.M.S. time history $(\int \sigma_{x,1} dt / \int \sigma_{x0,1} dt)$
$g_1(t)$	0.85	0.54
$g_2(t)$	0.79	0.47
$g_3(t)$	0.66	0.61

Fig. 5.15 Effect of mass ratio on the performance of particle damper in nonstationary random excitation with envelope function $g(t) = g_2(t)$. System parameters are $e = 0.75$, $\zeta = 0.01$, $\mu_s = 0.5$, $dx = d_{opt}$, and 10 particles

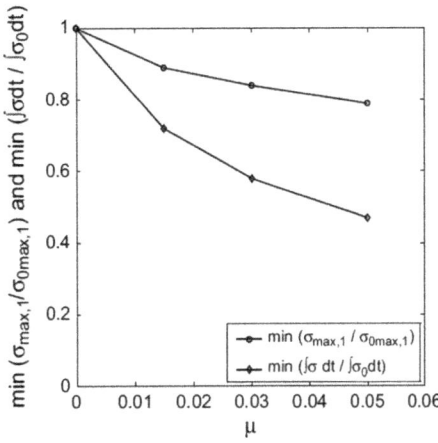

$g(t) = g_3(t)$, the nonstationary excitation duration is about 150 natural periods of the system, and the resulting reduction in the peak R.M.S. is about 34%, which is closer to what is achieved for the same situation in the stationary excitation case (Fig. 5.13 shows about 42% reduction in the stationary R.M.S. with similar system parameters).

Figure 5.15 summarizes the performance of a particle damper in reducing the ratio of the peak R.M.S., and the area under the R.M.S. time history curve, for different mass ratios.

On the other hand, one can relate the peak transient R.M.S. (σ_{max}) to the actual peak response (x_{peak}). Figure 5.16 shows the probability density and cumulative distribution of the actual peak response for more than 200 ensembles. From this figure, one can conclude that with more than 98% confidence, the actual peak response is less than three times the transient R.M.S. peak ($P(x_{1,peak} < 3\sigma_{x1,max}) > 98\%$). Therefore, the information about the statistics of the actual peak response is embedded in the transient R.M.S. plot.

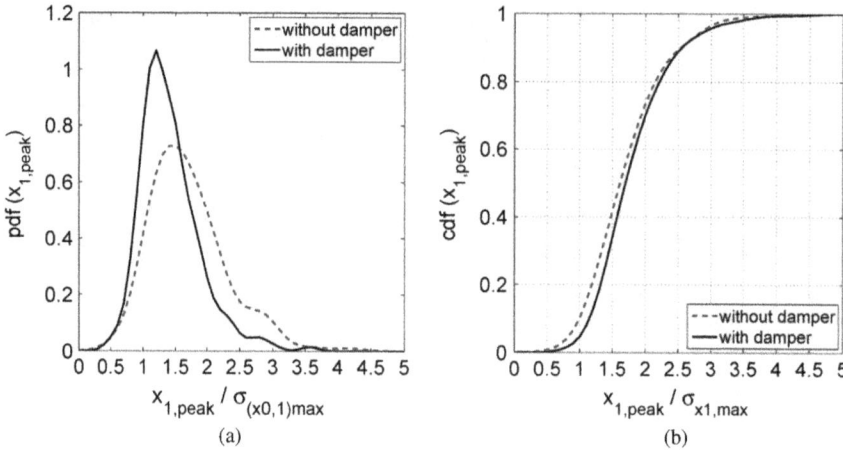

Fig. 5.16 **a** Probability density, and **b** cumulative distribution of the actual peak response for the first storey of the primary system with damper on the third storey, under the nonstationary random excitation with the envelope function $g(t) = g_2(t)$. System parameters are $\mu = 0.05$, $\zeta = 0.01$, $e = 0.75$, $dx = d_{opt}$, $\mu_s = 0.5$, and 10 particles. Based on CDF plot, one can observe that $P(x_{1,\,peak} < 3\,\sigma_{x1,max}) > 98\%$

References

1. Lu, Z., S.F. Masri, and X. Lu. 2011. Studies of the performance of particle dampers attached to a two-degree-of-freedom system under random excitation. *Journal of Vibration and Control* 17 (10): 1454–1471.
2. Lu, Z., S.F. Masri, and X.L. Lu. 2011. Parametric studies of the performance of particle dampers under harmonic excitation. *Sturctural Control and Health Monitoring* 18 (1): 79–98.
3. Lu, Z., X.L. Lu, and S.F. Masri. 2010. Studies of the performance of particle dampers under dynamic loads. *Journal of Sound and Vibration* 329 (26): 5415–5433.
4. Masri, S.F. 1973. Steady-state response of a multidegree system with an impact damper. *Journal of Applied Mechanics* 40 (1): 127–132.
5. Li, K., and A.P. Darby. 2006. Experiments on the effect of an impact damper on a multiple-degree-of- freedom system. *JVC/Journal of Vibration and Control* 12 (5): 445–464.
6. Masri, S.F. 1978. Response of a multidegree-of-freedom system to nonstationary random excitation. *Journal of Applied Mechanics* 45 (3): 649–656.

Chapter 6
Shaking Table Test Study on Particle Damping Technology

In the former five chapters, from shallow to deep, from simple to complex, the performances of single-degree-of-freedom system, double-degree-of-freedom system and multi-degree-of-freedom system with additional particle dampers were analyzed in detail. The effects of different parameters on the performances of the system were discussed. Some state parameters and motion characteristics characterizing the optimal operation of particle dampers were found, and some meaningful results were obtained. As particle damper is a relatively new structural control technology, especially for its application in civil engineering field, and its structural characteristics make the whole system a highly non-linear system, so the shaking table test of the structural model of the system is very important. Based on the theoretical research of the former five chapters, the particle dampers (multi-unit-multi-particle dampers) are installed in the three-storey single-span steel frame model and the five-storey single-span steel frame model respectively, and the shaking table tests are carried out to further study the vibration reduction performance of the device. Compared with the results of numerical analysis, the accuracy of the numerical analysis model of particle damper proposed in Chap. 3 is further verified.

6.1 Shaking Table Test of a Three-Story Steel Frame Attached with a Particle Damper

6.1.1 Experiment Design

The structural model used in the test is a three-storey steel frame model, and its plane and elevation are shown in Fig. 6.1. The height of frame storey is 2 m and the plane dimension is 1.95 m × 1.9 m. The frame column is made of No. 10 I-beam, the main beam of the frame adopts No. 12.6 channel steel, and the secondary beam adopts No. 10 channel steel, members are welded together.

© China Machine Press and Springer Nature Singapore Pte Ltd. 2020
Z. Lu et al., *Particle Damping Technology Based Structural Control*,
Springer Tracts in Civil Engineering,
https://doi.org/10.1007/978-981-15-3499-7_6

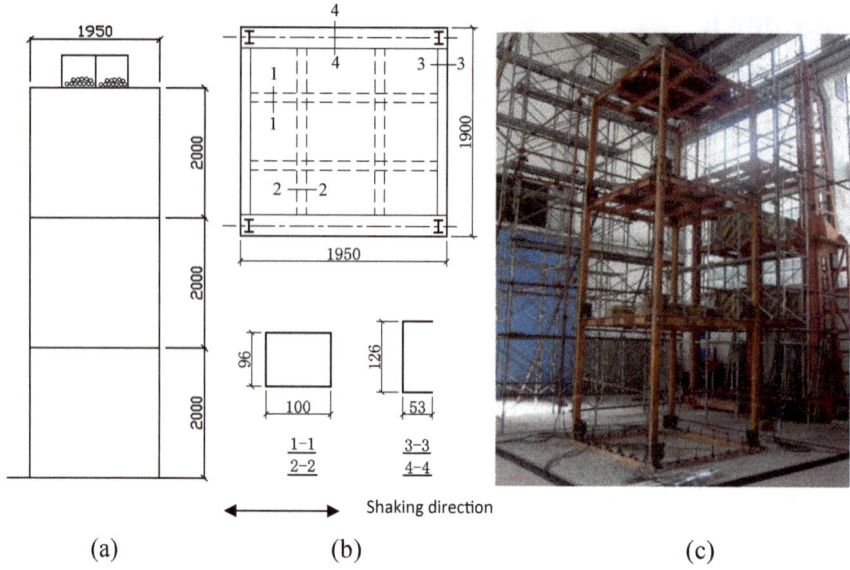

Fig. 6.1 Configuration of frame model with multi-unit particle dampers (unit: mm). **a** Elevation; **b** Floor plan; and **c** Photo of the experimental model

In order to make the fundamental frequency of the testing frame equal to around 1.0 Hz, which was close to the frequency of a typical high-rise building, additional masses were applied to each floor by bolting them to the floor beam. The total masses from the first floor to the roof, including the frame self-weight during testing, were 1915 kg, 1915 kg, and 2124 kg, respectively. The primary system had a damping ratio of 0.013. The natural frequencies of the primary system are $f_1 = 1.07$ Hz, $f_2 = 3.2$ Hz, and $f_3 = 4.8$ Hz.

The multi-unit particle dampers were made of steel plates consisting of four rectangular containers, with the dimensions of length 0.49 m × width 0.49 m × height 0.5 m. A total of 63 steel ball bearings were put into each container, with the total mass of 135 kg, which was 2.25% of the primary system mass, the diameter of the ball is 50.8 mm.

In order to evaluate the performance of the multi-unit particle damper system under different seismic actions, four earthquake time histories of acceleration were selected as the input data during the shaking table testing. The four time histories of acceleration were Kobe (1995, NS), El Centro (1940, NS), Wenchuan (2008, NS) and Shanghai design code specified artificial earthquake accelerogram (SHW2, 1996). The peak value of the acceleration was increased gradually from 0.05 to 0.2 g (g is the acceleration due to gravity), and the time interval was 0.02 s. The time-domain and frequency-domain characteristics of seismic waves are shown in Fig. 6.2. In the whole test, the shaking table only vibrates in the direction of weaker stiffness of frame column. Accelerometers and displacement meters are arranged on each floor of the frame model to monitor its vibration response.

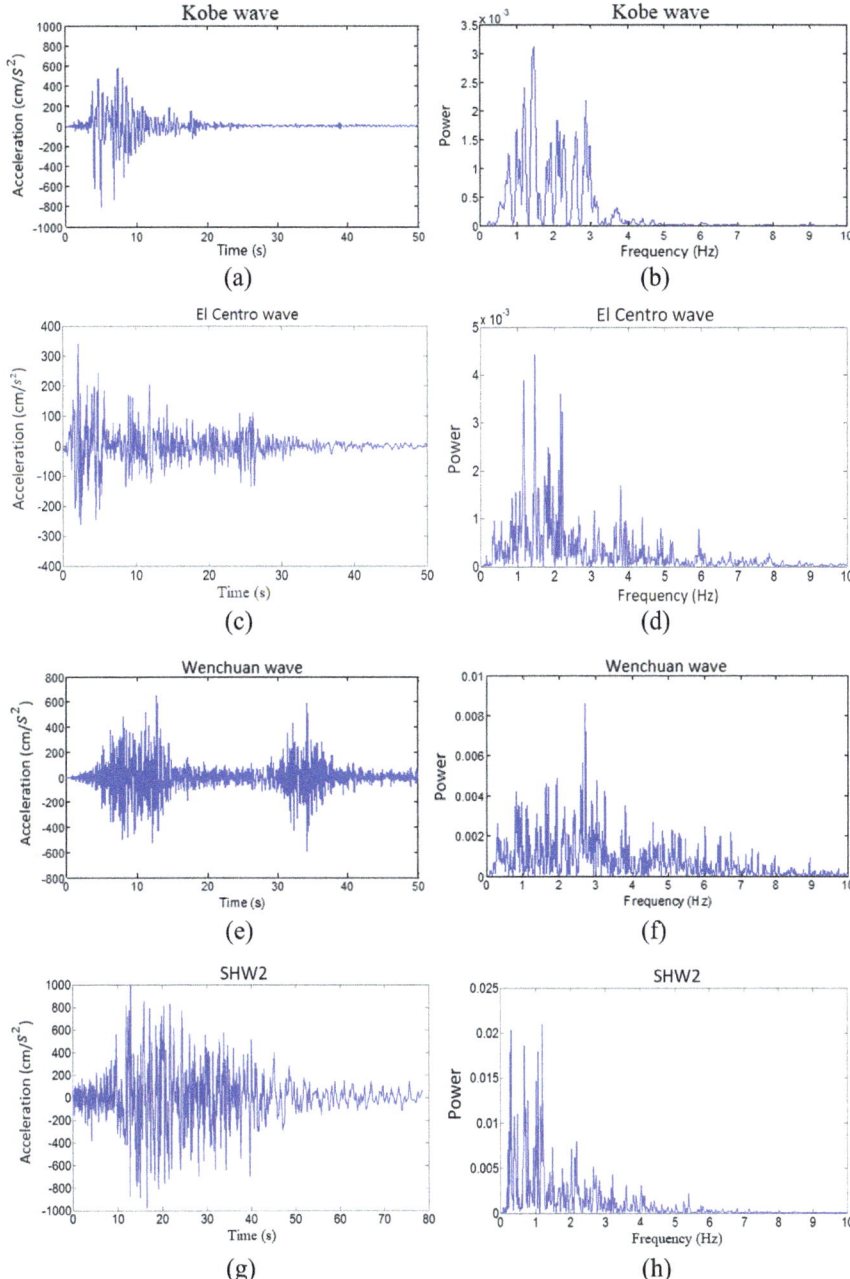

Fig. 6.2 Characteristic curves of seismic waves used in the test, left column is time history curve, right column is spectrum characteristic curve **a–b** Kobe wave, **c–d** El Centro wave, **e–f** Wenchuan wave and **g–h** SHW2 wave

Kobe wave was recorded by Kobe Oceanographic Observatory near the epicenter of the Hanshin earthquake (M7.2) on January 17, 1995. This earthquake is a typical downward earthquake in the city. The epicenter distance of Kobe Ocean Meteorological Station is 0.4 km. The time interval of discrete acceleration was 0.02 s, and the peak acceleration values of N-S component, E-W component and U-D component were 818.02 gal, 617.29 gal and 332.24 gal, respectively. The N-S component is selected as the X-direction input in the experiment.

El Centro wave is the acceleration time history recorded at El Centro station by the IMPERIAL Valley earthquake (M7.1) on May 18, 1940. It is a classical seismic record widely used in structural test and seismic response analysis. The time interval of discrete acceleration was 0.02 s, and the peak acceleration values of N-S component, E-W component and U-D component were 341.7 gal, 210.1 gal and 206.3 gal, respectively. The N-S component is selected as the X-direction input in the experiment.

Wenchuan wave is the acceleration time history recorded at Wolong Town Station in Wenchuan County, Sichuan Province, China (M8.0) on May 12, 2008. The time interval of discrete acceleration was 0.005 s, and the peak acceleration values of N-S and U-D components were 652.851 gal and 948.103 gal, respectively. The N-S component is selected as the X-direction input in the experiment.

The main strong earthquakes of Shanghai Artificial Wave (SHW2) lasted about 50 s, and all the excitations were 78 s long, and the time interval of discrete acceleration excitations was 0.02 s.

6.1.2   Results of Shaking Table Test

1. *Top displacement response of model*

Maximum displacement of the top story of frame is an important parameter in seismic design, and it is not enough to give the peak value of structural displacement when evaluating structural damage. The characteristics of the response in the whole time history need to be studied. In random vibration, the energy level of random variables is usually expressed by the R.M.S. response. The expression of R.M.S. is as follows:

$$R.M.S. = sqrt\left(\frac{1}{n}\sum_{i=1}^{n}x_i^2\right) \tag{6.1}$$

The maximum displacement responses at the roof of the test frame and their root mean square. (R.M.S.) value are shown in Table 6.1 for all the test runs (The response of the uncontrolled frame under SHW2 (0.2 g) wave was so large that it may cause the frame to collapse, consequently, this test was not carried out in the experiment). It can be seen that:

Table 6.1 Maximum displacement and their R.M.S. value (mm) at the roof of the test frame

Seismic input	Peak input value (g)	Test frame with dampers		Test frame without dampers		Reduction effect (%)	
		X_3	σ_3	X_3	σ_3	X_3	σ_3
Kobe	0.05	38.335	7.385	42.727	12.401	10.3	40.4
	0.1	66.665	12.899	73.984	19.882	9.9	35.1
	0.2	110.979	17.356	116.063	21.807	4.4	20.4
El Centro	0.05	30.366	6.552	33.131	10.525	8.3	37.7
	0.1	49.319	11.044	53.936	18.095	8.6	39.0
	0.2	81.416	15.308	92.143	24.672	11.6	38.0
Wenchuan	0.05	23.118	5.915	26.073	6.699	11.3	11.7
	0.1	43.994	10.991	47.435	12.470	7.3	11.9
	0.2	75.354	18.063	78.938	20.889	4.5	13.5
SHW2	0.05	70.774	18.337	83.027	29.306	14.8	37.4
	0.1	96.420	23.228	118.393	29.656	18.6	21.7
	0.2	–	–	–	–	–	–

(1) The frame with multi-unit particle dampers has smaller response of displacement compared with that of the frame without particle dampers.

(2) The vibration reduction effect (reduction effect = (response of system without dampers − response of system with dampers)/response of system without dampers) of the R.M.S. of displacement is much better than that of the peak displacement, in which the former is 11.7–40.4%, and the latter is 4.4–18.6%. This means multi-unit particle dampers can help the primary system to dissipate a lot of input earthquake energy. Additionally, the displacement can also be effectively reduced.

(3) The vibration reduction effect is different under different seismic inputs. In the experiment, the system under Wenchuan excitation resulted in the worst reduction effect. The reason may lie in the frequency characterization of the input earthquake excitations. Figure 6.2 shows the excitation of Kobe wave and Wenchuan wave in the time domain and frequency domain, respectively. One can see that the main frequency of Kobe wave is around 1.4, 1.5 and 1.1 Hz, which is near the fundamental frequency of the primary system, while that of Wenchuan wave is around 2.7 Hz.

Figure 6.3 shows the displacement time history at the roof level of the test frame, one can see that the particle dampers system not only reduces the maximum response of the displacement, but also makes the whole time history attenuate quickly, so that the response in most of the time is reduced. This is also an additional evidence that the R.M.S. of the displacement reduction effect is better than the maximum displacement reduction effect. Another interesting phenomenon in Fig. 6.3 is that the responses of the controlled and uncontrolled system are the same at the very beginning time period, after a while, the controlled curve begins to decay quickly. This is a similar

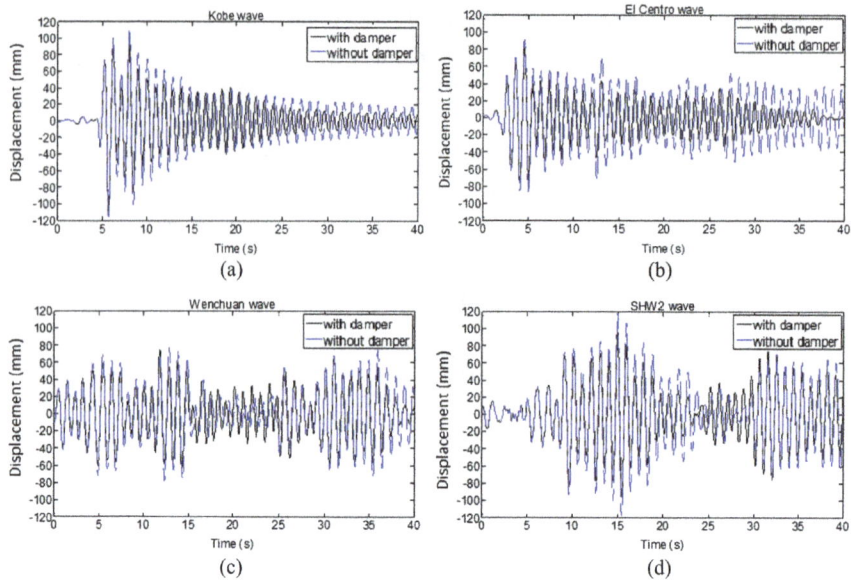

Fig. 6.3 Displacement time history at roof level of the test frame. **a** Kobe wave (0.2 g); **b** El Centro wave (0.2 g); **c** Wenchuan wave (0.2 g); and **d** SHW2 wave (0.1 g)

phenomenon encountered in the operation of Tuned Mass Damper. The vibration reduction effect is not good at the very beginning and becomes better as time goes by. The reason is that it takes some time for particles in the container to impact the wall of the container. After certain impacts, the particle damper system starts to dissipate the input energy by momentum transfer.

2. *Maximum acceleration response and interlayer displacement response at the top of the model*

Not only the displacement can be reduced, the acceleration and interstory drift can also be attenuated in most cases. Table 6.2 shows the maximum acceleration responses at the roof (A3) and the maximum interstory responses at the first floor (X1) of the test frame. One can see that the acceleration and interstory drift of the controlled frame are smaller than that of the uncontrolled frame in most cases (except for the Wenchuan wave (0.2 g) case); however, the reduction effect of interstory drift (0.1–6.4%) is not as good as that of acceleration (2.3–19.1%). The reason may lie on the position of the damper, which is on the top floor of the frame. From Table 6.2, the response reduction effects of the frame with particle dampers under Wenchuan wave are also the worst, especially for the top floor's acceleration in 0.2 g case, which is enlarged. This is also an evidence of the complex influence of input excitations.

Figure 6.4 shows the acceleration time history curve of the top floor of the frame model under different types of seismic excitation. It can be found that, similar to the displacement time history curve given in Fig. 6.3, the multi-unit-multi-particle damper not only reduces the maximum acceleration response in most cases, but

Table 6.2 Maximum acceleration at the roof (g) and maximum interstory drift at the first floor (mm) of the test frame

Seismic input	Peak input value (g)	Test frame with dampers		Test frame without dampers		Reduction effect (%)	
		A_3 (g)	X_1 (mm)	A_3 (g)	X_1 (mm)	A_3	X_1
Kobe	0.05	0.213	19.185	0.240	20.498	11.3	6.4
	0.1	0.366	33.713	0.398	33.749	8.0	0.1
	0.2	0.591	58.178	0.637	59.025	7.2	1.4
El Centro	0.05	0.178	18.080	0.198	18.419	10.1	1.8
	0.1	0.296	29.627	0.311	30.703	4.8	3.5
	0.2	0.501	52.471	0.567	55.743	11.6	5.9
Wenchuan	0.05	0.168	14.335	0.172	14.757	2.3	2.9
	0.1	0.318	26.947	0.345	28.479	7.8	5.4
	0.2	0.474	60.269	0.452	60.833	−4.9	0.9
SHW2	0.05	0.362	35.587	0.430	37.155	15.8	4.2
	0.1	0.473	58.534	0.586	60.075	19.1	2.6
	0.2	–	–	–	–	–	–

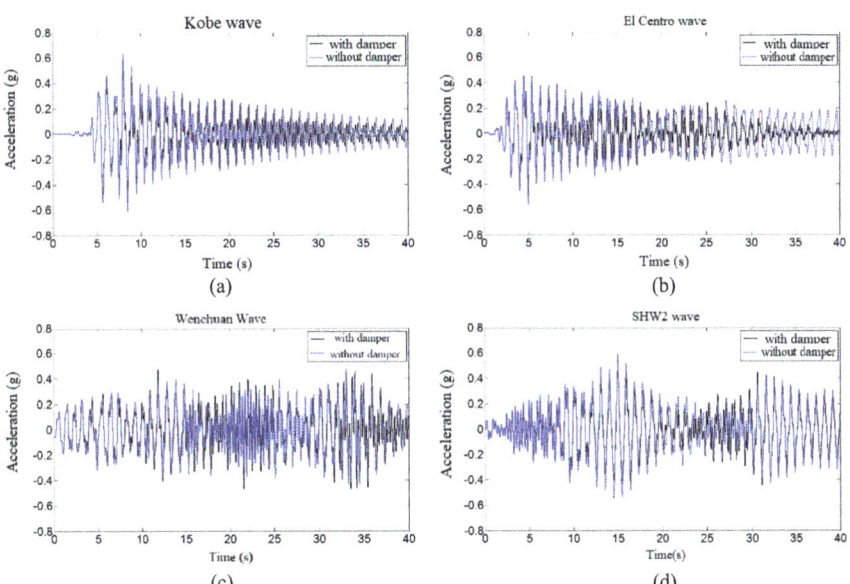

Fig. 6.4 Acceleration time history at roof level of the test frame. **a** Kobe wave (0.2 g); **b** El Centro wave (0.2 g); **c** Wenchuan wave (0.2 g); and **d** SHW2 wave (0.1 g)

also reduces its response in the whole time history. However, the effect of vibration reduction under different excitations is different, especially under Wenchuan wave excitation. In fact, the displacement response of steel frame structure under Wenchuan wave excitation is smaller than that under other excitations, which is one of the reasons for the poor vibration reduction effect of the system (Larger frame response can make the particles in the container collide more intensely, thus consuming more energy through the momentum exchange and energy dissipation between particles and the main structure to enhance the vibration reduction effect).

3. *Maximum displacement and acceleration response curves of each layer of the model*

Figure 6.5 shows the maximum displacement and maximum acceleration at every floor of the test frame under different seismic inputs. One can find that generally each floor of the frame can achieve vibration attenuation in almost all test inputs, although the reduction effects have slight differences. As a structure is like a filter, which can remove the high frequency component of the excitation when the earthquake wave is transmitting from the ground to the upper structure, and finally the fundamental frequency dominates the vibration of the structure. However, in the lower floor of the structure, there may still be a lot of high frequency components. For the fact that the acceleration is related to the square of the frequency, the acceleration response at the bottom contains high frequency components. Consequently, for the first floor, it may

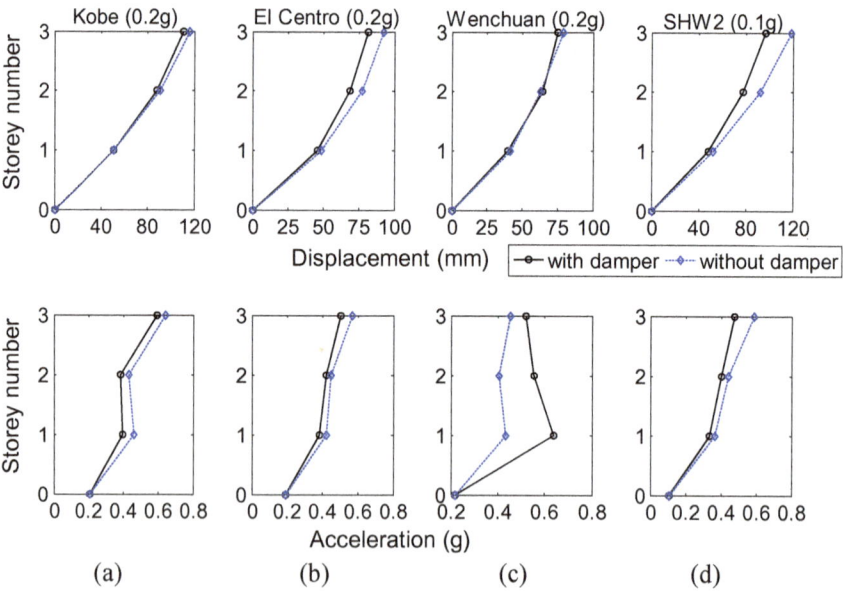

Fig. 6.5 Maximum displacement (the upper line) and maximum acceleration (the lower line) at every floor of the test frame. The circle legend shows the response of the test frame without damper, while the diamond legend shows that with damper. **a** Kobe wave (0.2 g); **b** El Centro wave (0.2 g); **c** Wenchuan wave (0.2 g); and **d** SHW2 wave (0.1 g)

exist a situation that although the displacement is small, the maximum acceleration is larger than the top floor, as shown in Fig. 6.5c.

4. *Typical experiment of motion process*

Figure 6.6 gives a series of snapshots from a shot video to show a typical time period of the motion of the system with particles. It could be found that in a certain period, plug flow motions of the particles, rather than unorderly movements, can be observed. That is to say, the particles gather together and move together in one direction. After the collision with the container wall, they move together in the opposite direction, instead of unorderly movement in all directions. The plug flow phenomenon is consistent with the theoretical analysis of Chap. 4, and it is a similar qualitative behavior to what occurs in a single-particle impact damper when operating with two-impacts-per-cycle [1].

5. *Discussion about experiment results*

Through the analysis of the above test results, it can be seen that the steel frame with multi-unit-multi-particle dampers can obtain better vibration control effect (including displacement response of roof, R.M.S. displacement response and maximum acceleration response of roof) under the input of Kobe wave, El Centro wave and SHW2 wave, especially the R.M.S. response reflecting vibration energy is the best. It can also be seen from the time-history curve of the response. Under Wenchuan wave input, the effect of vibration reduction is the worst, especially at 0.2 g condition, the acceleration response is still enlarged. On the one hand, it is related to the characteristics of input excitation (the frequency spectrum of Wenchuan wave indicates that the main frequency of Wenchuan wave is around 2.7 Hz, while the main frequency of other waves is close to the natural frequency of main frame, i.e. 1 Hz); on the other hand, it is also related to the smaller displacement response of steel frame under Wenchuan wave input. The small response of the steel frame results in less severe collision between particles and the main structure, thus the momentum exchange and energy dissipation between the two are relatively less, and the effect of vibration reduction is also poor.

Similar to tuned mass dampers, multi-unit-multi-particle dampers have poor vibration reduction effect in the early stage and better vibration reduction effect in the later stage. This is due to the collision of particles with the wall of the container and the formation of particle flow by these colliding particles, which takes a certain amount of time.

6.1.3 Verification of Discrete Element Model

In order to verify the feasibility and correctness of the spherical discrete element numerical model of particle damper proposed in Chap. 3, some structural parameters of the experimental model are input into the program to observe whether the calculated results are in good agreement with the experimental results.

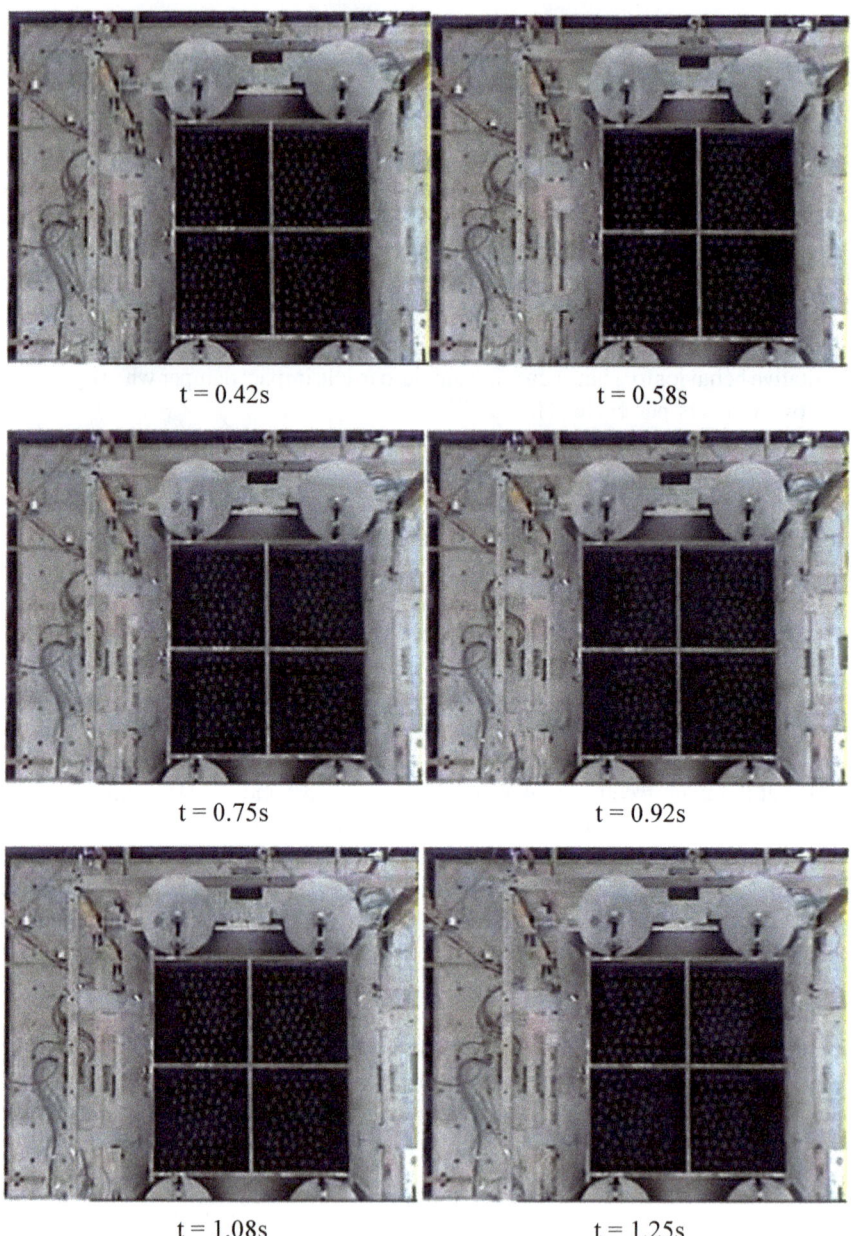

t = 0.42s   t = 0.58s

t = 0.75s   t = 0.92s

t = 1.08s   t = 1.25s

Fig. 6.6 Snapshots of the motion of the test frame with multi-unit particle dampers under El Centro (0.2 g) input

Table 6.3 Values of system parameters

Parameter	Value
Container number	4
Total particle number	63×4 ($\mu = 2.25\%$)
Diameter of the particle (mm)	50.4
Density of the particle (kg/m^3)	7800
Coefficient of friction	0.5
Critical damping ratio of the damper	0.1
Stiffness of the spring between particle and wall (N/m)	100,000
Stiffness of the spring between particle and particle (N/m)	100,000

According to the test records, a complete wave is intercepted from the acceleration time history curve of the collected platform as the input wave. The calculated results are compared with the displacement time history curve of the top layer of the model collected at the same time. The correctness of the proposed numerical model and algorithm is verified by the coincidence degree of the theoretical and measured curves. In numerical simulation, for convenience, the initial values of displacement and velocity of each particle are assigned to zero. Therefore, the calculated results are different from the experimental records in the initial stage. With the gradual progress of the calculation process, the influence of initial value selection on the structural system response will gradually decrease.

The numerical models, algorithms and computational models are detailed in Chapter.

The mass, stiffness and damping of the test frame model, the system parameters used for calculation are shown in Table 6.3.

$$\mathbf{M} = \begin{bmatrix} 1915 & 0 & 0 \\ 0 & 1915 & 0 \\ 0 & 0 & 2124 \end{bmatrix} \text{kg} \quad \mathbf{K} = \begin{bmatrix} 933000 & -466500 & 0 \\ -466500 & 933000 & -466500 \\ 0 & -466500 & 466500 \end{bmatrix} \text{N/m}$$

$$\zeta_1 = 0.013$$

Figures 6.7, 6.8, 6.9 and 6.10 respectively draw the calculated and experimental curves of the displacement and acceleration time histories of the top floor of the frame model with multi-unit-multi-particle dampers under 0.1 and 0.2 g seismic excitations, which can be found to be in good agreement with each other. Table 6.4 illustrates the comparison of simulation results and experimental results for the maximum displacement of the roof of the test frame with the particle damper system, which illustrates that there is a good agreement between the test and analysis results. The results all demonstrate that the proposed analytical method can estimate the response of particle damper system under earthquake excitations in an acceptable accuracy.

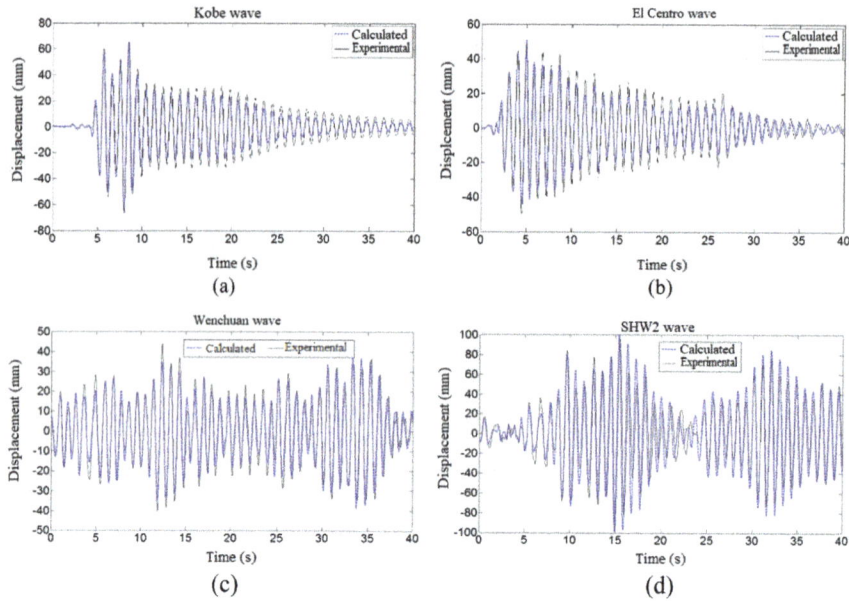

Fig. 6.7 Displacement time histories at roof level of the test frame with multi-unit particle dampers under 0.1 g earthquake excitations. **a** Kobe wave, **b** El Centro wave, **c** Wenchuan wave and **d** SHW2 wave

6.2   Shaking Table Test of a Five-Story Steel Frame Attached with a Particle Damper

Particle dampers can be used in vibration control of engineering structures in many ways. In the previous section, particle dampers are fixed on the roof of structures. In this section, another method based on tuned mass dampers is introduced, which can be regarded as an improved application of tuned mass dampers.

Tuned mass damper (TMD) is an effective and commonly used passive control device. It is generally composed of solid mass, spring element and damping element. Viscous damper is usually used to provide additional damping and dissipate energy. However, its frequency band is narrow and viscous damper is vulnerable to temperature and poor durability. Particle Tuned Mass Damper (PTMD) is a method of using TMD to suspend a container containing particles of metal or other materials in a place where structural vibration is greater. It uses tuned mass damper, inelastic collision and friction between particles and container wall to consume vibration energy [2–4]. The technology has the advantages of simple concept, wide frequency bandwidth, temperature insensitivity, good durability and easy to use in harsh environment [5–7]. It also combines the mature implementation method of tuned mass damper, so it has good application prospects in civil engineering [8–10].

In this section, a five-storey steel frame is taken as the research object, and the particle tuned mass damper (PTMD) is applied to the field of seismic control in

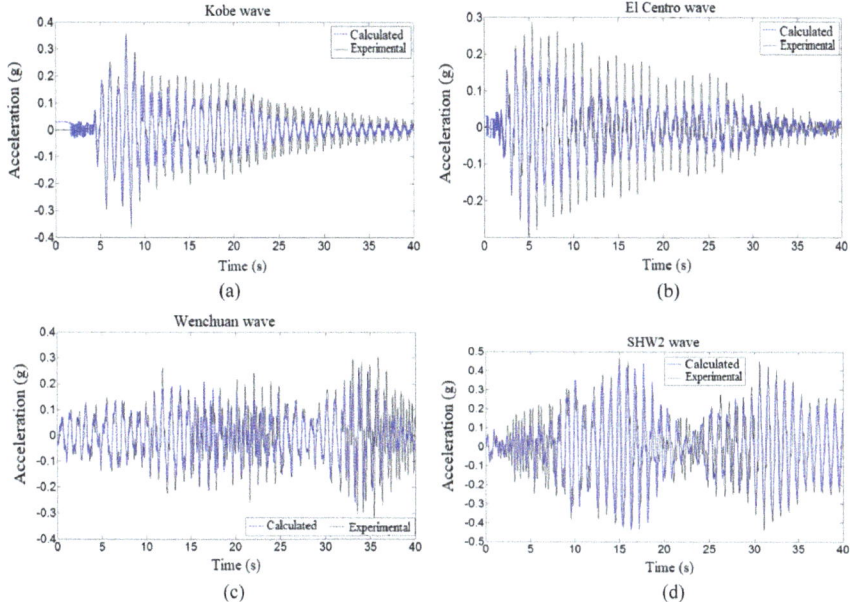

Fig. 6.8 Acceleration time histories at roof level of the test frame with multi-unit particle dampers under 0.1 g earthquake excitations. **a** Kobe wave, **b** El Centro wave, **c** Wenchuan wave and **d** SHW2 wave

civil engineering. The response of structures with and without particle tuned mass dampers is compared and analyzed through the natural earthquake wave and Shanghai artificial wave excitation test, and the effects of different seismic wave excitation, suspension length (frequency ratio), mass ratio and gap clearance on damping are studied. The influence law of the damper's vibration reduction performance provides a basis for further research and application of particle damping technology.

6.2.1 Experiment Design

The experimental model consists of a five-storey steel frame as the primary structure and a particle tuned mass damper. The height of single layer is 1.06 m and the total height is 5.30 m. The frame columns consist high-strength steel plates (Q690) with width × length × height dimensions of 15 mm × 180 mm × 1060 mm. The slabs consist of steel plates (Q345) with plane dimensions of 2 m × 2 m and a thickness of 30 mm. The total weight of frame structure is 6000 kg. The first frequency and the second and third frequencies of the primary structure are 1 Hz, 3 Hz and 5 Hz, respectively, and the damping ratio is 2%. The particle tuned mass damper is suspended on the top of the primary structure during the experiment as shown in Fig. 6.11. The damper particles are made of steel balls with a diameter of 51 mm,

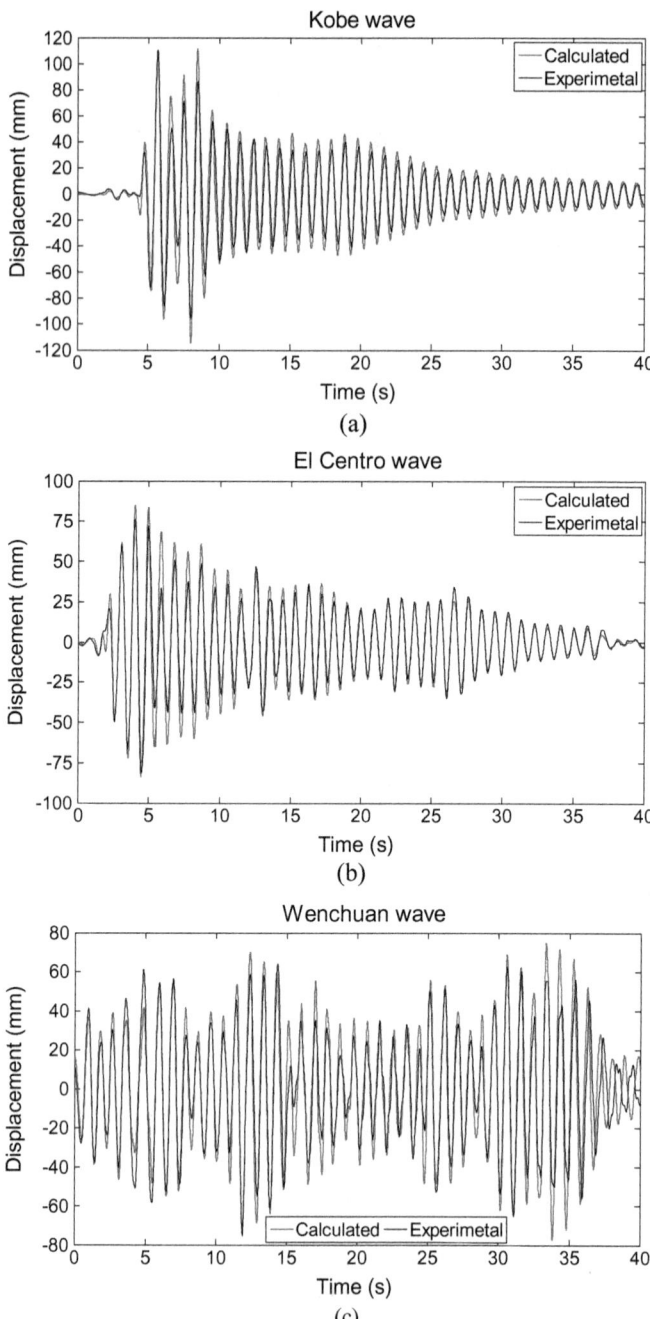

Fig. 6.9 Displacement time histories at roof level of the test frame with multi-unit particle dampers under 0.2 g earthquake excitations. **a** Kobe wave, **b** El Centro wave and **c** Wenchuan wave

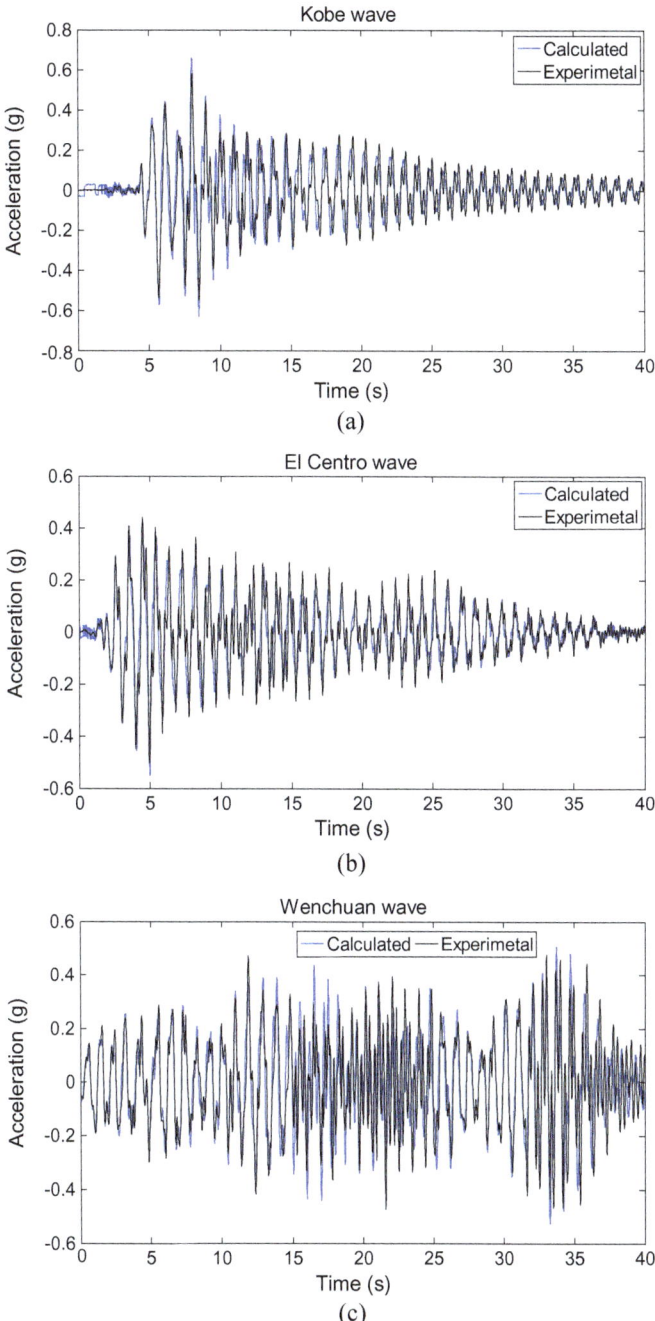

Fig. 6.10 Acceleration time histories at roof level of the test frame with multi-unit particle dampers under 0.2 g earthquake excitations. **a** Kobe wave, **b** El Centro wave and **c** Wenchuan wave

Table 6.4 Comparison of calculated results and experimental results for the maximum displacement of the roof of the test frame with multi-unit particle dampers

Seismic input	Peak input value (g)	Calculated value (mm)	Experimental value (mm)	Error (%)
Kobe	0.05	37.726	38.335	−1.6
	0.1	67.638	66.665	1.5
	0.2	114.519	110.979	3.2
El Centro	0.05	29.713	30.366	−2.2
	0.1	49.472	49.319	0.3
	0.2	84.206	81.416	3.4
Wenchuan	0.05	22.418	23.118	−3.0
	0.1	42.113	43.994	−4.3
	0.2	77.174	75.354	2.4
SHW2	0.05	69.821	70.774	−1.3
	0.1	98.465	96.420	2.1
	0.2	–	–	–

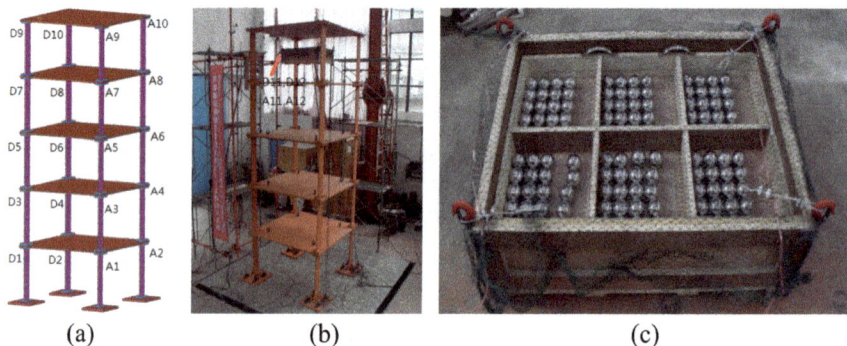

      (a)   (b)   (c)

Fig. 6.11 Configuration of the test specimen: **a** primary structure; **b** photo of the experimental model; **c** photo of the particle tuned mass damper

totaling 180 balls, which are evenly placed in the container. The container consists of wooden plates with a thickness of 4 cm, and the outer dimensions are 1000 mm × 640 mm × 300 mm. Three mutually clamped and consolidated boards are used to separate the middle of the rectangular container to form a container with 12 inner diameters of 288 × 283 × 120 mm in the upper and lower layers, thus forming a multi-unit damper. The total mass of the container is 39.345 kg. The top of the damper is covered with a protective net. The total mass ratio of the damper to the primary system is 2.26%.

In order to study the damping effect of particle tuned mass damper, shaking table tests were carried out on steel frame models with and without particle tuned mass damper under actual seismic wave input and Shanghai artificial wave input. Four ground motions (El Centro wave (1940, NS), Wenchuan wave (1995, NS), Japan 311 wave (2011, NS) and Shanghai artificial wave (SHW2, 1996)) are utilized in the shaking table tests to investigate the vibration control effects of the PTMD under different seismic actions. For tuned mass damper and particle damper, parameters such as suspension length, mass ratio and gap clearance are designed to investigate the relationship between the shock absorption performance of the new damper and the key parameters, so as to provide design basis for further application.

6.2.2 Results of Shaking Table Test

1. *Damping effect of PTMD*

(1) *Responses to steady-state and unsteady-state random excitations*

Shaking table test and steady-state random excitation test are often used to study the vibration control effect of PTMD. Figures 6.12 and 6.13 are the comparison of displacement and acceleration responses of structures with and without dampers under shaking table test, respectively. It can be seen from the graphics that PTMD can rapidly reduce the displacement and acceleration response of the main structure in a short period of time.

In addition, Table 6.5 lists the effect of peak displacement and acceleration, R.M.S. displacement and acceleration on the top floor of the main structure under EI Centro wave (0.05 g) and white noise. The reduction rate is defined as follows: the reduction rate = (response of structure without PTMD-response of structure with PTMD)/response of structure without PTMD × 100%. The table shows that not only the peak displacement and peak acceleration response are reduced, but also the R.M.S. displacement and R.M.S. acceleration response are significantly reduced. The best R.M.S. response reduction rate can reach 54.06%. At the same time, it

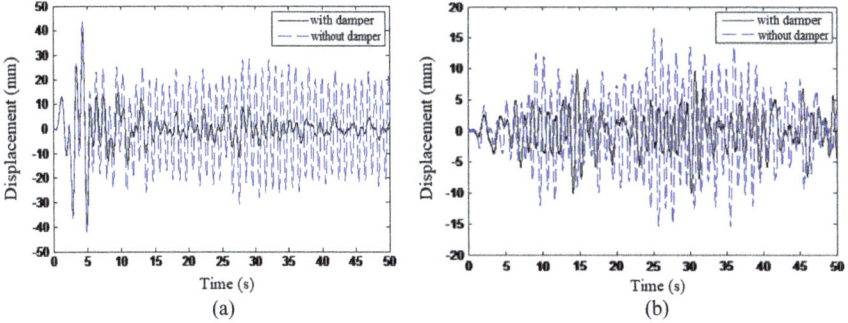

Fig. 6.12 Time history curve of top floor displacement of steel frame structure: **a** El Centro wave (0.05 g); **b** white noise

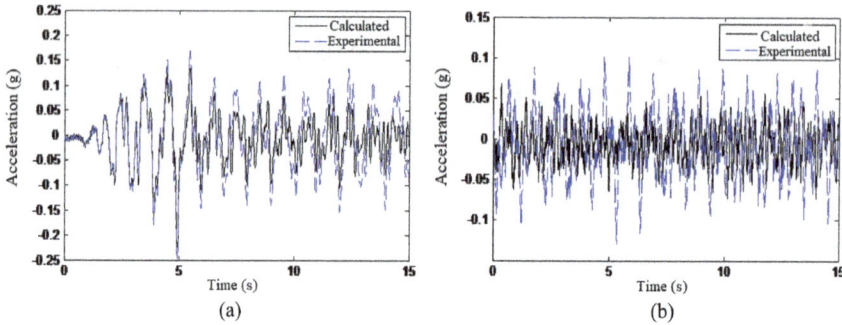

Fig. 6.13 Time history curve of top floor acceleration of steel frame structure: **a** El Centro wave (0.05 g); **b** white noise

Table 6.5 Displacement and acceleration damping rate of top story of steel frame structures

Seismic input	El Centro wave			
	Peak Disp (mm)	Peak Acc (g)	R.M.S. Disp (mm)	R.M.S. Acc (g)
Uncontrolled	44.02	0.2467	15.78	0.0674
Controlled	39.29	0.2243	7.25	0.0491
Reduction effects (%)	10.75	9.08	54.06	27.15
Seismic input	White noise			
	Peak Disp (mm)	Peak Acc (g)	R.M.S. Disp (mm)	R.M.S. Acc (g)
Uncontrolled	16.52	0.1294	6.22	0.0406
Controlled	10.12	0.0740	3.14	0.0224
Reduction effects (%)	38.74	42.81	49.52	44.83

also shows that PTMD can reduce the response of the main structure in the whole earthquake cycle. In a word, PTMD has better shock absorption effect under both steady-state and unsteady-state random excitation.

(2) *Response of different seismic waves*

In shaking table test, accelerometers and laser displacement meters are installed on the model floor to measure acceleration and displacement. The effect of damper can be preliminarily judged by the change of peak value of displacement and acceleration. R.M.S. response is often used to represent the energy level of random variables in random vibration. The damping performance of dampers is measured by calculating the R.M.S. response of acceleration and displacement.

Tables 6.6 and 6.7 list the displacement and acceleration responses at the roof of the test frame under different seismic wave intensities, respectively, including the peak value and the R.M.S. value. For safety reasons, experiments with the uncontrolled structure under the SHW2 wave (0.1 g), El Centro wave (0.2 g), Wenchuan wave

Table 6.6 Displacement responses at the roof of the test frame (mm)

Seismic input		El Centro		Wenchuan		Japan 311		SHW2	
		Peak	R.M.S.	Peak	R.M.S.	Peak	R.M.S.	Peak	R.M.S.
0.05 g	Uncontrolled	44.04	13.42	10.16	2.00	7.00	1.40	91.73	41.97
	Controlled	39.29	5.07	9.67	1.78	5.83	0.90	55.94	11.68
	Reduction effects (%)	10.79	62.22	4.82	11.00	16.71	35.71	**39.01**	**72.17**
0.1 g	Uncontrolled	102.13	30.49	26.00	6.56	13.81	4.19	–	–
	Controlled	79.00	11.16	20.07	4.12	13.51	2.48	112.04	19.08
	Reduction effects (%)	22.65	63.40	22.82	37.20	2.20	40.81	–	–
0.2 g	Uncontrolled	–	–	–	–	31.92	12.47	–	–
	Controlled	148.16	19.94	35.00	7.36	25.60	4.57	147.60	26.30
	Reduction effects (%)	–	–	–	–	18.56	63.35	–	–

Table 6.7 Acceleration responses at the roof of the test frame (g)

Seismic input		El Centro		Wenchuan		Japan 311		SHW2	
		Peak	R.M.S.	Peak	R.M.S.	Peak	R.M.S.	Peak	R.M.S.
0.05 g	Uncontrolled	0.2447	0.0577	0.2140	0.0555	0.1146	0.0285	0.4142	0.1741
	Controlled	0.2253	0.0253	0.2015	0.0471	0.1034	0.0250	0.2752	0.0505
	Reduction effects (%)	7.93	56.15	5.84	15.14	9.77	12.28	33.56	70.99
0.1 g	Uncontrolled	0.5542	0.1302	0.4540	0.1240	0.2782	0.0720	–	–
	Controlled	0.4796	0.0600	0.4457	0.1148	0.2662	0.0665	0.6312	0.0775
	Reduction effects (%)	13.46	53.92	1.83	7.42	4.31	7.64	–	–
0.2 g	Uncontrolled	–	–	–	–	0.5722	0.1442	–	–
	Controlled	0.9935	0.1367	0.8340	0.1937	0.5446	0.1324	1.5493	0.1593
	Reduction effects (%)	–	–	–	–	4.78	8.11	–	–

(0.2 g) and SHW2 wave (0.2 g) were not conducted. When observing the results under the SHW2 wave (0.05 g) and those under the other three waves (0.1 g), the primary structure vibrated violently, so we extrapolated that these waves may cause the test frame to collapse. However, the experiments with the controlled structure were conducted under these high-level seismic inputs by estimating the reduction effects of the PTMD.

The responses of the test frame with the PTMD attached were smaller than most of the responses of the uncontrolled structure, which demonstrates stable and efficient attenuation effects from the PTMD. In addition, the vibration control effects for the R.M.S. response were generally more obvious than those for the peak response, which indicates that the PTMD efficiently attenuated the entire response of the primary structure over a period of time. Additionally, the vibration reduction effects for the acceleration response generally were not as good as those for the displacement response because collisions could cause abrupt changes in acceleration.

It could be seen from the tables that the reduction effects were favorable, especially under the SHW2 and El Centro waves. The best vibration control effects for the peak and R.M.S. values of the displacement responses were 39.01% and 72.17% respectively; the values for the acceleration responses were 33.56% and 70.99%, respectively, which means that the particle tuned mass damper can help the main structure absorb and dissipate a large part of the seismic input energy.

Under four different seismic inputs, the effects of damper are different, which is related to the frequency characteristics of input excitation. Among them, the R.M.S. response of displacement of SHW2 is more effective than that of all seismic excitations, which shows that the performance of the PTMD under SHW2 is better than that of the actual seismic excitation. On the one hand, the main frequency of SHW2 (about 1.1 Hz) is similar to the frequency of the main structure. On the other hand, the energy of SHW2 itself is relatively large. In the case of small earthquake input, the main structure can still drive the damper to sway sufficiently to dissipate energy, thus reducing the main structure vibration. Because the magnitude and characteristics of seismic waves are different, the reduction rates of PTMD are different under different seismic waves.

Figure 6.14 consists of the acceleration time history and displacement time history curve of the roof of steel frame under actual seismic wave and artificial wave. It could be found that the frame structure with additional dampers not only reduces the response peak value obviously, but also makes the response time history curve decay rapidly in the whole time period. In addition, as observed in the displacement time history curve of Fig. 6.14b, the effect of vibration reduction is not obvious at the initial stage of response, which is similar to that of tuned mass dampers. This is because the damper oscillation is small and the particles do not fully collide to dissipate energy; however, with the passage of time, the particles collide with the wall of the container continuously and the energy dissipation occurs, and the damping effect begins to appear.

Figure 6.15 shows the power spectral density curves of the displacement responses at the 5th floor, the 3rd floor and the 1st floor of the test frame, respectively. The reduction effects were greater for higher floors than for lower floors. Moreover, the

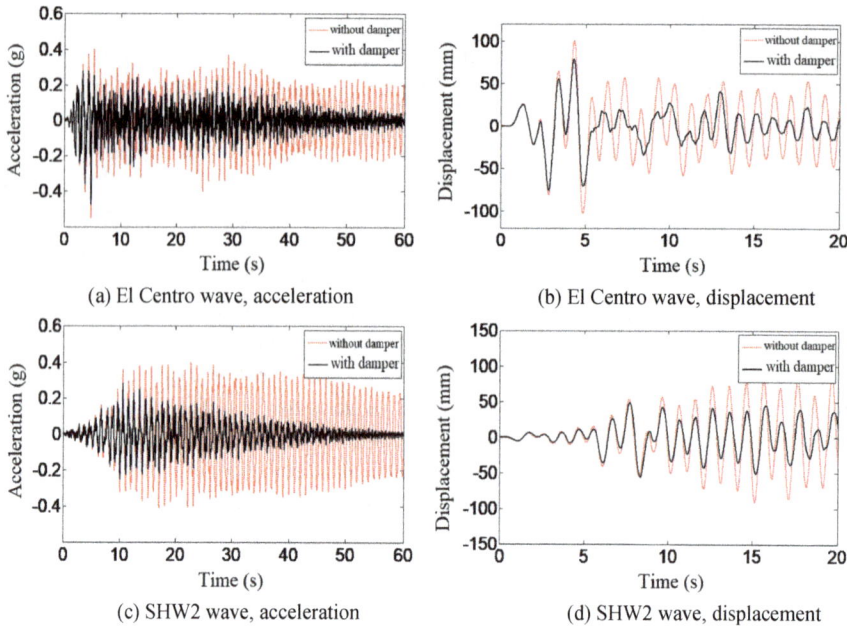

Fig. 6.14 Time-history curve of top-story dynamic response of rigid frame model under seismic excitation

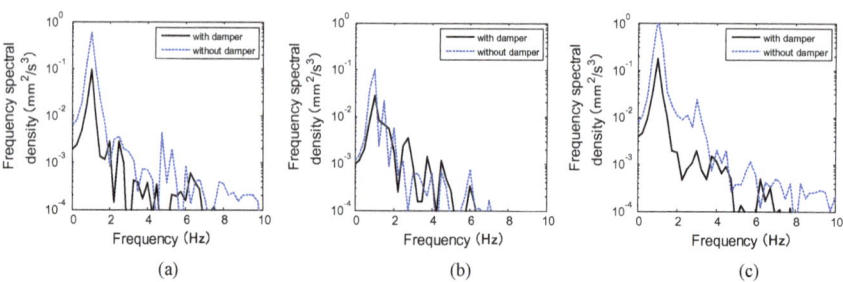

Fig. 6.15 Power spectral density curves of the displacement responses at **a** the 5th floor, **b** the 3rd floor, and **c** the 1st floor

PTMD significantly attenuated the first mode of vibration because of its installation position, whereas the reduction effects for higher modes were unstable, which is related to the suspension position of the damper. The additional damper is suspended on the top floor, which is the place where the displacement of the first mode corresponds to the maximum, so the control effect is good, but the effect of the higher mode control is not obvious, because it is not suspended at the position where the maximum displacement is located in the higher mode. However, considering that the first mode has the greatest participation in the structural response, the preferred position of the damper should be on the top floor with the greatest displacement. Li also found the same experimental phenomenon.

(3) *Peak responses at every floor*

Figures 6.16 and 6.17 show the maximum displacement and maximum acceleration at every floor of the test frame under different seismic inputs, respectively. Different vibration reduction effects occurred under different seismic inputs. Specifically, the responses at every floor were effectively reduced under the El Centro wave and SHW2 wave, whereas the PTMD exhibited little improvement in terms of vibration control and the responses at some floors even increased under the Wenchuan wave and Japan 311 wave. These results indicate that the vibration control effects of the PTMD had a complicated relationship with the characteristics of the seismic inputs, which will be further discussed in next part of this section.

Moreover, the vibration reduction effects under these seismic waves generally improved as the floor increased, and the vibration control effects of the top floor were the best. These results mainly occurred because the response at the top floor

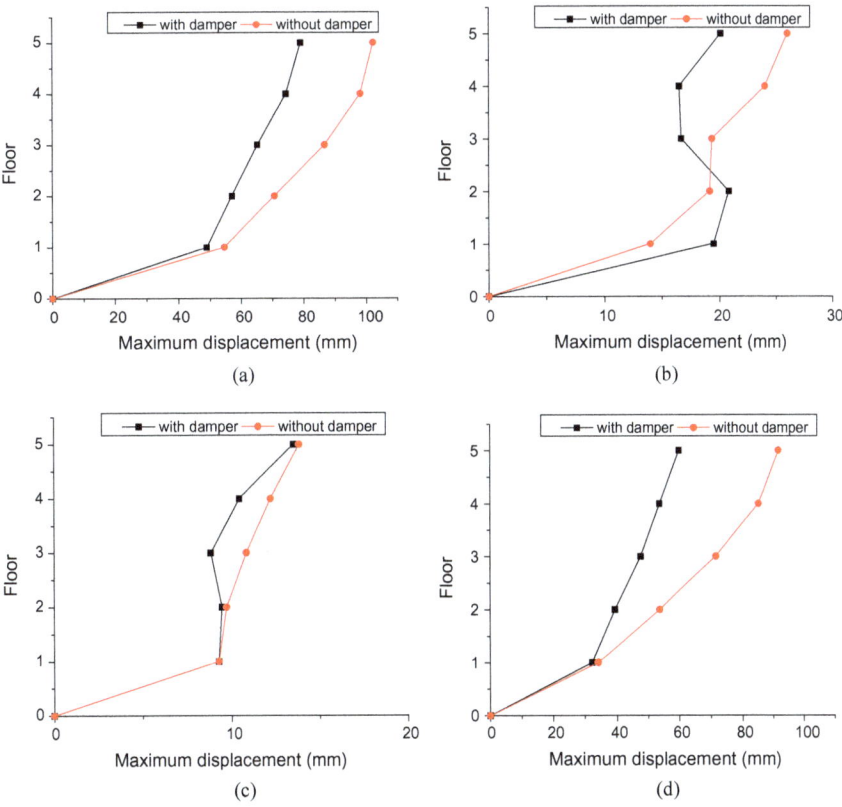

Fig. 6.16 Maximum displacements at every floor of the test frame under the **a** El Centro (0.1 g), **b** Wenchuan (0.1 g), **c** Japan 311 (0.1 g), and **d** SHW2 waves (0.05 g)

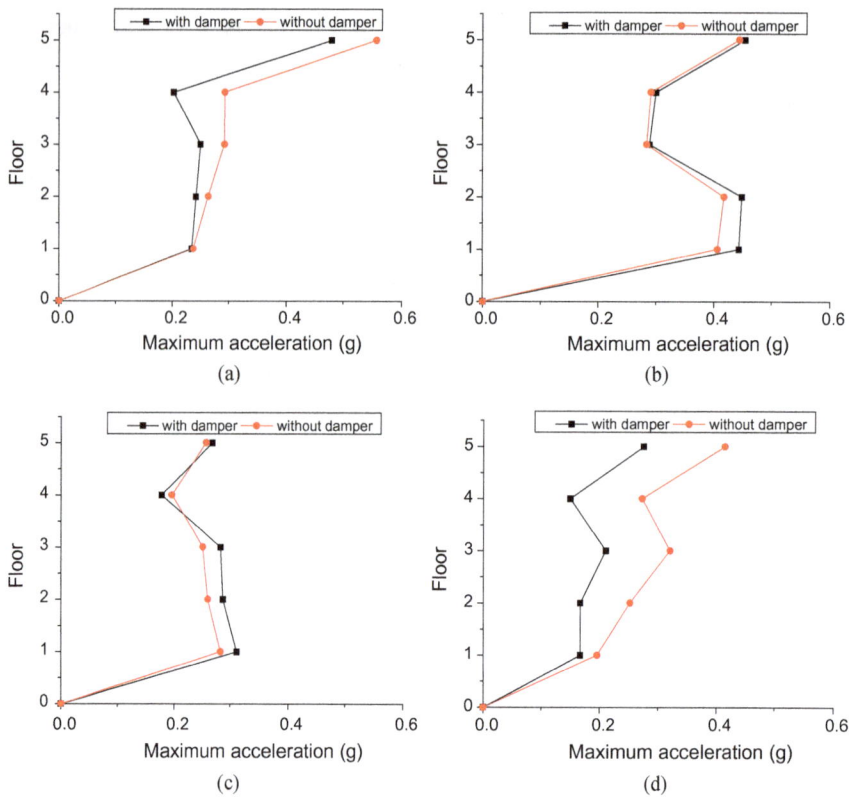

Fig. 6.17 Maximum accelerations at every floor of the test frame under the **a** El Centro (0.1 g), **b** Wenchuan (0.1 g), **c** Japan 311 (0.1 g), and **d** SHW2 waves (0.05 g)

was the largest and because the PTMD was installed on the top floor; consequently, the response at the top floor could be suppressed to a large degree.

Interestingly, the maximum acceleration of the first floor could be larger than that of the upper floors, especially under the Wenchuan wave and Japan 311 wave. On the one hand, the building could be regarded as a filter, and the high-frequency components of the excitation were gradually reduced as the earthquake waves were transmitted from the ground to the upper parts of the structure. Finally, the fundamental frequency dominated the vibration of the structure. However, the high-frequency components of the excitation still acted on the lower floors of the structure. On the other hand, the acceleration is directly proportional to the square of the frequency. Therefore, the acceleration responses on the lower floors, especially the first floor, were larger than those on the upper floors, although the displacements were small.

(4) *Effects of the frequency characteristics and amplitude level of the input*

The frequency characteristics and amplitude level of the seismic input play very important roles in the vibration control of PTMDs. Therefore, the reduction effects

Fig. 6.18 Acceleration response spectra of the input motions and the design acceleration response spectra

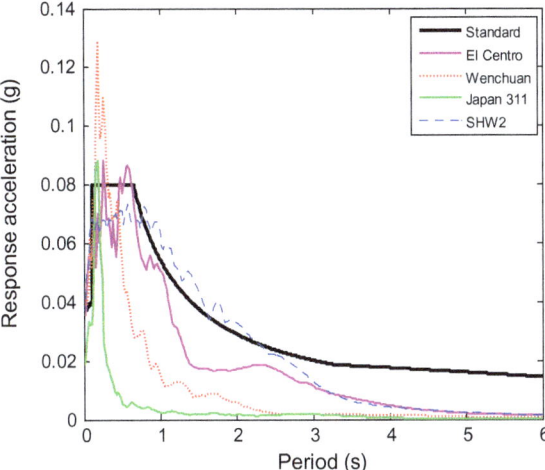

of PTMDs when applying different types of input with the same amplitude and the effects under a certain input with increasing amplitudes are investigated.

As listed in Tables 6.6 and 6.7, the reduction effects of the displacement and acceleration responses of the PTMD were different under four types of excitations, and the effects for both the maximum and R.M.S. responses were the best under the SHW2 wave. This result was mainly caused by the frequency characteristics of the input. The acceleration response spectra of the seismic input and design acceleration response spectra are plotted in Fig. 6.18. The fundamental frequency of the SHW2 wave was approximately 1.1 Hz, which was the closest to the natural frequency of the primary system (1 Hz). In this situation, even if the amplitude of seismic wave is small, the main structure vibrates violently, and the particle collision is also violent. Thus, the PTMD significantly dissipated the energy under the SHW2 input. However, the strongest components of the El Centro wave, Wenchuan wave and Japan 311 wave were concentrated at approximately 2 Hz, 6 Hz and 6 Hz, respectively, which were significantly different from the natural frequency of the primary structure.

Although the frequency characteristics of the seismic input influenced the reduction effects of the PTMD, Tables 6.6 and 6.7 show that the system responses were generally decreased by the PTMD to a certain degree under different types of earthquakes, which indicates the robustness of the PTMD.

Tables 6.6 and 6.7 show that the reduction effects improved as the amplitude level of the input increased. Thus, the larger the response of the primary structure, the better the reduction effects of the PTMD. For example, the reduction effects of the maximum displacement and acceleration under the El Centro wave (0.05 g) were only 10.79% and 4.17%, respectively; however, these values under the El Centro wave (0.1 g) increased to 22.65% and 12.73%, respectively, because the response of the primary structure was mild and the damper was not sufficiently shaken when the excitation energy was small. Under these conditions, the particles moved chaotically rather than in a plug flow pattern, while the plug flow, which has been verified by many experiments and theories, is a phenomenal characterization of the optimal vibration reduction effect of the particle damper. However, violent collisions occurred among the particles and between the particles and the container as the input energy increased, which quickly dissipated the input energy. This observation can also explain why the reduction effects under the Wenchuan wave and Japan 311 wave did not improve as the seismic amplitude increased. Tables 6.6 and 6.7 show that the responses under these two seismic waves were really small, and some responses (under 0.1 g) may have been smaller than those under the El Centro wave (0.05 g). Thus, PTMDs may have more favorable potential applications in major or even larger earthquakes.

2. *Parametric study of PTMD*

The influence of some system parameters, including the auxiliary mass ratio, gap clearance, mass ratio of particles to the total auxiliary mass, frequency characteristic and amplitude level of input, and buffered material, on the vibration control effects of the PTMD was experimentally investigated to understand the physical working mechanisms of the PTMD and optimize its damping performance.

(1) *Effects of the auxiliary mass ratio*

The auxiliary mass ratio is a very important parameter to the vibration control effects of TMDs because this factor indirectly influences the optimal frequency and the optimal damping of TMDs. Similarly, the auxiliary mass ratio significantly affects the energy dissipation of particle dampers. Therefore, the influence of the auxiliary mass ratio on the damping performance of PTMDs was examined by adding particles and keeping the containers the same, thus adjusting the auxiliary mass ratio to 0.66, 1.19, 1.73, 2.26 and 2.8%. To maintain the same gap clearance, the number of particles in each container remained 20, and the number of containers with particles was changed. The containers that included particles were symmetrically positioned. The reduction effects of the maximum responses and R.M.S. responses at the roof of the test frame under the El Centro wave (0.1 g) when using different damper masses are shown in Fig. 6.19a, b, respectively.

When the auxiliary mass ratio ranged from 0 to 2.8%, larger auxiliary mass ratios could achieve better reduction effects in terms of the maximum responses. The collision of particles constitutes a large portion of the energy dissipation that is achieved by PTMDs, so increasing the mass of particles can increase the energy dissipation for a certain range of auxiliary mass ratios. When the auxiliary mass ratio ranged from 0.66 to 2.8%, the reduction effects improved slightly in terms of the R.M.S. responses as the auxiliary mass ratio increased.

(2) *Effects of gap clearance*

The gap clearance of the particles to the wall of the container is a very important parameter that influences the vibration control effects of particle dampers. In the experiment, the influence of the gap clearance on the vibration reduction was examined by setting this factor as 0 D, 1.64 D, 2.64 D, 3.64 D and 5.64 D (D is the diameter of the particle).

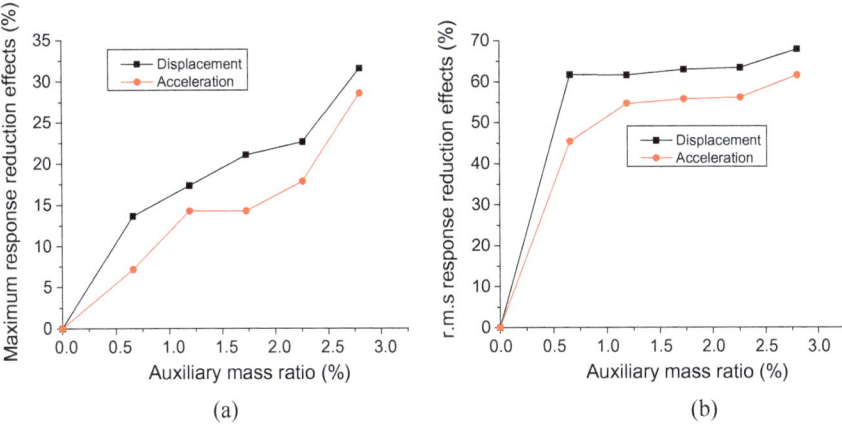

Fig. 6.19 Effects of the auxiliary mass ratio on vibration reduction: **a** maximum response, **b** R.M.S. response

Figure 6.20a shows the reduction effects of the R.M.S. acceleration responses at the roof of the test frame under different gap clearances. It is found that the damping ratio of R.M.S. of acceleration is more obvious under different gap clearances. With the increase of gap clearances, there will be a trend of first increasing and then decreasing. This shows that the nearer gap clearance between particles and containers is not the better, but there is an optimal gap clearance. This is because the energy dissipation of the particle damper is mainly through the collision friction between particles and the wall of the container. If the gap clearance was too small, the particles were very likely to move together with the container, which led to few collisions. However, if the gap clearance was too large, a long time was required for the particles to move from one end of the container to the other, and only limited

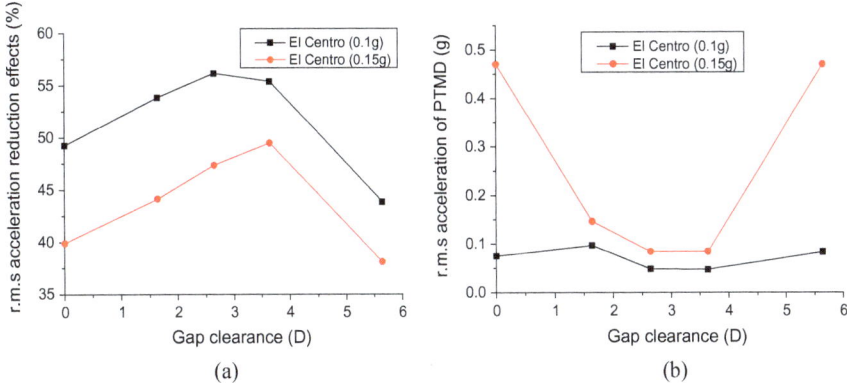

Fig. 6.20 Effects of gap clearance on the vibration reduction: **a** reduction effects of the R.M.S. acceleration at the roof of the test frame; **b** R.M.S. acceleration of the PTMD

collisions occurred, which affected the suppression efficiency. Therefore, the gap clearance should be kept as the optimal value to achieve more efficient collisions among particles and between particles and the container.

In addition, the R.M.S. acceleration responses of the PTMD varies with gap clearance are plotted in Fig. 6.20b. As the gap clearance increased, the R.M.S. responses of the PTMD first decreased and then increased. Moreover, by comparing Fig. 6.20a, b, it is shown that the reduction effects of the primary structure were more favorable when the responses of the PTMD were relatively small. Additionally, the best damping performance can be achieved with a proper gap clearance, in which the damper experiences stable and efficient movement. The reduction effects were satisfactory and the responses of the PTMD were relatively mild when the gap clearance ranged from 1.6 D to 3.6 D.

For 0 D and 5.64 D (approximately equal to the length of the container) case, the PTMD could be considered a TMD by increasing the mass of the container and eliminating the particles to keep the total auxiliary mass ratio the same. Additionally, the PTMD could dissipate extra energy through collisions between particles compared to traditional TMDs.

(3) *Effects of the mass ratio of particles to the total auxiliary mass*

Increasing the auxiliary mass ratio can improve the vibration control effects in a certain range. Furthermore, the influence of higher particle mass on the vibration control effects when the total auxiliary mass ratio is kept constant is an intriguing topic. On the one hand, the collisions among particles and the collisions between particles and the container constitute a large portion of the energy dissipation, so increasing the particle mass can increase the energy dissipation. On the other hand, the damping ratio of the PTMD can be indirectly influenced by changing the particle number.

In the experiment, the mass ratio of the particles to the total auxiliary mass was adjusted to be 0, 0.2, 0.38, 0.64 and 0.76, while the total auxiliary mass ratio was maintained as 1.73%. Figure 6.21 shows the reduction effects of the peak acceleration and R.M.S. acceleration responses at the roof of the test frame under different particle masses. The response reduction effects improved as the mass ratio of particles to the auxiliary mass ratio increased. In terms of an optimal auxiliary mass ratio, increasing the proportion of the particle mass can improve the vibration control effects to a certain extent.

(4) *Effect of suspension length (frequency ratio)*

The suspension tuned mass damper mainly uses resonance principle to change the suspension length so that the frequency of the damper is the same as that of the main structure, so that the main structure can transfer the most energy to the damper, thus reducing the vibration of the main structure; the frequency of the particle damper is variable, and the frequency and the intensity of the oscillation of the damper box, the elastic modulus of the particles and the damper cavity are related to many factors. At present, there is no corresponding calculation formula.

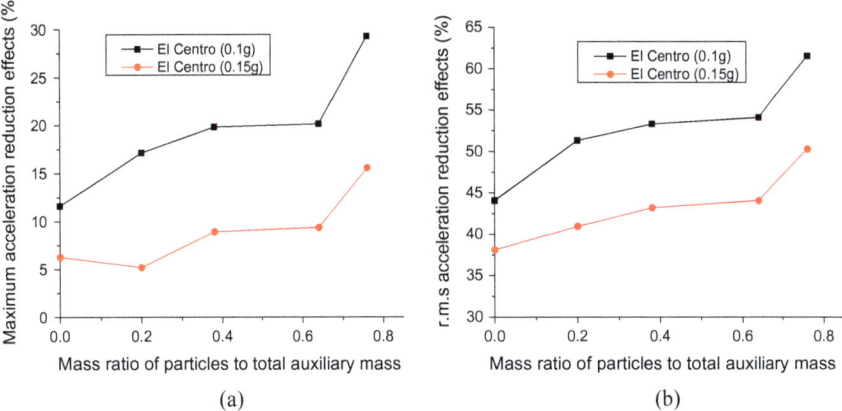

Fig. 6.21 Effects of the mass ratio of particles to the total auxiliary mass on the vibration reduction: **a** Maximum acceleration; **b** R.M.S. acceleration

In order to investigate the frequency effect of particle tuned mass damper, the suspension length parameters are analyzed from the angle of tuned mass damper. The calculation formula of single pendulum is used to calculate the suspension length parameters: $T = 2\pi\sqrt{(L/g)}$, $T = 1/f \Rightarrow L = g \times (1/2\pi f)^2$, which can be used to calculate the suspension length of damper. In the course of the test, the influence of the parameters of different frequencies on the shock absorption characteristics of dampers is investigated by changing different suspension lengths. Figure 6.22 shows the seismic reduction effect curve of particle tuned mass damper under El Centro seismic wave with different suspension lengths. The results show that:

Fig. 6.22 Comparisons of seismic reduction effects of particle tuned mass dampers with different frequency ratios

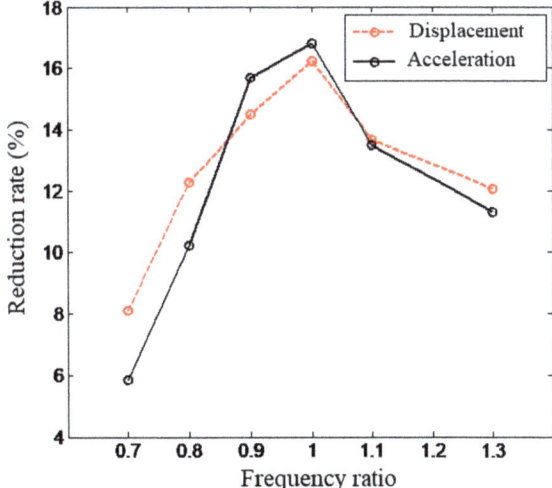

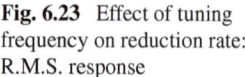
Fig. 6.23 Effect of tuning
frequency on reduction rate:
R.M.S. response

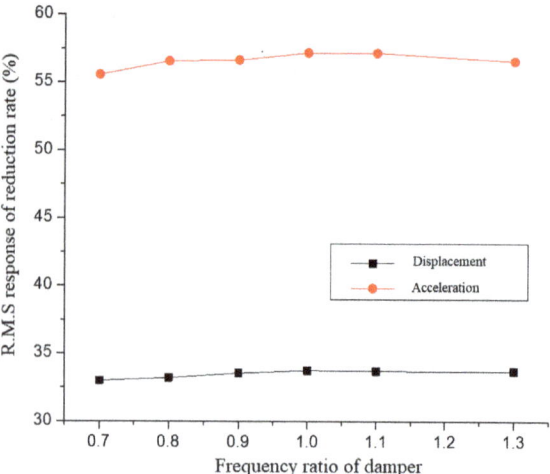

(1) When tuning (i.e. 1.0f), the damping effect is the best, and the maximum value
     of displacement and acceleration can reach 16–18%. When the frequency of
     the damper deviates from 1.0f, the effect of the damper decreases, which is the
     same as that of the suspended tuned mass damper.
(2) Without tuning, i.e. 1.3f, the maximum displacement damping rate of particle
     tuned mass damper can still reach 12%. This shows that the new damper has a
     wider frequency band and has a certain robustness.

From Fig. 6.23, it could be seen that the R.M.S. reduction of displacement and
acceleration responses of PTMD is better, and the reduction of response is not susceptible to the influence of suspension length. Compared with the maximum response,
the R.M.S. displacement response of the damper is reduced by 34% and the R.M.S.
acceleration response by 57%. Even if the frequency of the damper deviates from the
frequency of the main structure, the R.M.S. response of the damper can be reduced
effectively. It can be seen that PTMD can increase the frequency band of the damper.

(5) *Effects of the buffered material*

Rigid container walls are usually used in traditional particle dampers, which may
generate relatively large impact forces during collisions. Thus, several methods are
proposed to decrease the noise level and alleviate material degradation and local
deformation from violent impacts. For example, packaging the particles inside a soft
bag to form a bean bag damper [11] and sticking the buffered material onto the inner
surface of the container walls are both efficient approaches.

In this experiment, rubber spacers with 10 mm width were stuck on the vertical
inner walls of the container, and the performance of the PTMD with/without buffered
material was compared. Tables 6.8 and 6.9 list the vibration reduction effects of the
displacement and acceleration responses at the roof of the test frame under different
seismic inputs (0.1 g), including both the peak and R.M.S. responses. The improvements in the vibration control effects from the buffered material can be defined as

Table 6.8 Comparison of the vibration reduction effects of the PTMD for the displacement responses (mm) with/without buffered material under 0.1 g seismic input

Seismic input	El Centro		Wenchuan		Japan 311		SHW2	
	Peak	R.M.S.	Peak	R.M.S.	Peak	R.M.S.	Peak	R.M.S.
Uncontrolled	102.13	30.49	26.00	6.56	13.81	4.19	–	–
Rigid	79.00	11.16	20.07	4.12	13.51	2.48	112.04	19.08
Reduction effects (%)	22.65	63.40	22.82	37.20	2.20	40.81	–	–
Buffered	76.72	10.82	18.32	3.58	12.10	1.89	107.45	18.21
Reduction effects (%)	24.88	64.51	29.54	45.43	12.38	54.89	–	–
Improvement (%)	2.89	3.05	8.72	13.11	10.44	23.79	4.10	4.56

Table 6.9 Comparison of the vibration reduction effects of the PTMD for the acceleration responses (g) with/without buffered material under 0.1 g seismic input

Seismic input	El Centro		Wenchuan		Japan 311		SHW2	
	Peak	R.M.S.	Peak	R.M.S.	Peak	R.M.S.	Peak	R.M.S.
Uncontrolled	0.5542	0.1302	0.4540	0.1240	0.2782	0.0720	–	–
Rigid	0.4796	0.0600	0.4457	0.1148	0.2662	0.0665	0.6312	0.0775
Reduction effects (%)	13.46	53.92	1.83	7.42	4.31	7.64	–	–
Buffered	0.4411	0.0546	0.3986	0.0984	0.2357	0.0585	0.5955	0.0693
Reduction effects (%)	20.41	58.06	12.21	20.65	15.27	18.75	–	–
Improvement (%)	8.03	9.00	10.57	14.29	11.46	12.03	5.66	10.58

follows: improvement = (response of controlled rigid structure − response of controlled structure with buffered material)/(response of controlled rigid structure) × 100%. The reduction effects of the buffered PTMD were improved compared to conventional rigid PTMDs, with the greatest improvement reaching 23.79%.

Li and Darby [12] investigated the influence of different types of buffered materials on the vibration reduction effects of impact dampers and preliminarily found that the effects significantly depended on the magnitude of the contact force and the contact time. In conclusion, buffered material can help dissipate more energy than rigid PTMD, but this improvement is variable, and soft material with a high coefficient of restitution is more preferable.

3. *Comparison of PTMD and TMD*

Some hybrid non-linear energy dissipation devices were originally designed to reduce the response of the system by different damping methods, while increasing the robustness of vibration control. For this purpose, as a passive control device, particle tuned mass damper (PTMD) introduces the concept of particle damping into traditional dampers. The PTMD can not only tune the frequency of the main structure, but also dissipate energy by collision within the particle damper. On the other hand, the narrow frequency band is a disadvantage of TMD control. This multi-channel energy dissipation method is introduced into the design of PTMD. This non-linear mechanism makes PTMD not particularly sensitive to tuning frequency. Therefore, the robustness of vibration control is improved.

In order to compare the damping character of PTMD and TMD, some experiment cases have been proposed.

(1) *Improvement of vibration control effect*

A lot of research and engineering practice of TMD show that TMD has better vibration control effect under steady excitation, such as under wind vibration, but the effect of vibration reduction under unsteady excitation is not satisfactory, for example, under earthquake. Unlike TMD, many studies have shown that particle dampers can effectively mitigate earthquake action. Therefore, in the next part, we preliminarily compare the shock absorption effects of PTMD and TMD.

PTMD is a damper consisting of 240 steel balls with 51 mm diameter evenly distributed in the container. Its mass ratio is 2.8% of the total mass. The TMD with the same mass ratio is realized by removing the ball and replacing it with a mass block fixed to the container of the same mass.

Table 6.10 lists the response ratios of displacement to acceleration of steel frames roof with PTMD or TMD under El Centro (0.1 g) seismic waves, including peak and R.M.S. responses. The improvement of vibration control effect of PTMD is calculated by following formulas: improvement rate = (response of additional TMD structure − response of additional PTMD structure)/response of additional TMD structure × 100%.

It can be seen from Table 6.10 that PTMD plays a very good role in controlling displacement and acceleration responses under earthquakes. Especially for R.M.S. response control, PTMD is much better than TMD. PTMD uses a variety of methods to dissipate energy. Through frequency modulation, it can absorb more energy from

Table 6.10 Comparisons of Top Floor Responses of Steel Frame Structures with additional PTMD and TMD		TMD	PTMD	Improvement rate (%)
	Peak displacement (mm)	74.34	69.87	6.01
	Peak acceleration (g)	0.4945	0.3955	19.98
	R.M.S. displacement (mm)	10.62	9.79	7.82
	R.M.S. acceleration (g)	0.0499	0.0727	31.36

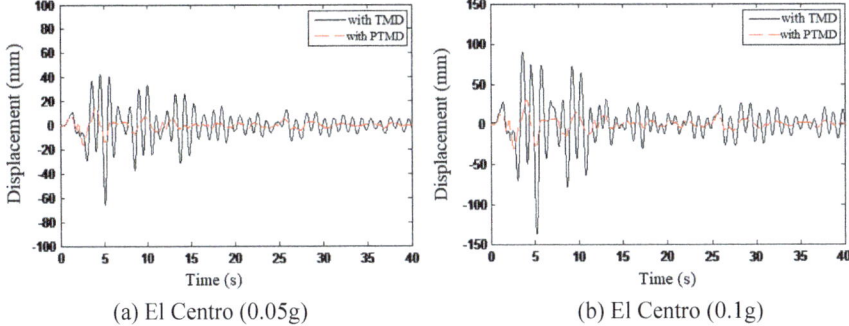

(a) El Centro (0.05g)   (b) El Centro (0.1g)

Fig. 6.24 Displacement of PTMD and TMD

the main structure and dissipate more energy at the same time. TMD can reduce the first-order modal response of the main structure under earthquake by adjusting the frequency and damping, but the higher-order modal response of the structure under earthquake is also very strong. In a word, in most cases, the effect of PTMD is better than that of TMD.

(2) *Attenuation of damper response*

In addition to studying the vibration control effect of the main structure with additional dampers, we should also pay attention to the response of the damper itself. In fact, when the damper has better damping effect, the response of the damper itself is smaller, which is also an important factor to be considered in the design. This section mainly compares the displacement of PTMD and TMD.

The displacement of the damper under El Centro waves (0.05 and 0.1 g) is shown in Fig. 6.24. It can be found that under the same conditions, the displacement of PTMD is obviously smaller than that of TMD. The response of the main structure is obviously reduced, and the response of the damper is relatively small, which is a good phenomenon. This means that the motion of damper is a stable and effective motion, and the space required for motion is small, which is very beneficial to commercial buildings. From this point of view, PTMD is a very good choice in engineering applications.

6.2.3 Verification of Simplified Equivalent Model

The experimental results discussed above have certain reference value for the structural engineering application of particle damping technology. Considering the complexity of seismic excitation and the strong nonlinearity of particle collision, a simple and effective numerical simulation method must be found in engineering application. On the one hand, this method needs to grasp the most essential mechanism of particle damping, on the other hand, it needs to be simple and practical enough to facilitate the design and use of practical projects.

In this section, the equivalent simplification algorithm proposed in Chap. 3 is adopted, that is, the particle swarm is equivalent to a single particle based on certain equivalence principle to establish a simplified model. The contribution of the collision force between the particle and the container wall to the structural vibration reduction is considered. The numerical analysis of shaking table test is carried out to verify the reliability and feasibility of the simplified algorithm.

1. *Equivalent simplification method for MDOF system*

In Chap. 3, the simplified algorithm for single degree of freedom is introduced in detail. Similarly, if the primary structure is modeled in the solution as an *n*-DOF vertical linear cantilever beam, then the suspended container of the PTMD can be considered a single pendulum and as the $(n + 1)$th DOF. The impact force from collisions between the simplified single particle and the container walls can be seen as an external force F_p that acts on the container. The simplified computational model is shown in Fig. 6.25, in which the subscript i represents the ith floor and F_i is the seismic force that acts on that floor. The governing equation of the primary system can be written as Eq. (6.2):

$$\begin{cases} \mathbf{M\ddot{X}} + \mathbf{C\dot{X}} + \mathbf{KX} = \mathbf{E}\ddot{x}g + \varphi(c_c\dot{y}_1 + k_c y_1) \\ m_c\ddot{x}_c + c_c\dot{y}_1 + k_c y_1 - c_p H(y_2, \dot{y}_2) - k_p G(y_2) = 0 \\ m_p\ddot{x}_p + c_p H(y_2, \dot{y}_2) + k_p G(y_2) = 0 \end{cases} \tag{6.2}$$

$$\mathbf{M} = \text{diag}[\, m_1 \ m_2 \ \ldots \ m_n \,] \tag{6.3}$$

Fig. 6.25 Simplified computational model

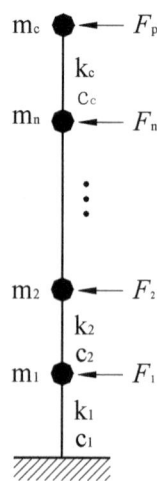

$$C = \begin{bmatrix} c_1 + c_2 & -c_2 & & & \\ -c_2 & c_2 + c_3 & -c_3 & & \\ & -c_3 & \cdots & & \\ & & & \cdots & -c_n \\ & & & -c_n & c_n \end{bmatrix} \tag{6.4}$$

$$K = \begin{bmatrix} k_1 + k_2 & -k_2 & & & \\ -k_2 & k_2 + k_3 & -k_3 & & \\ & -k_3 & \cdots & & \\ & & & \cdots & -k_n \\ & & & -k_n & k_n \end{bmatrix} \tag{6.5}$$

$$X = [\, x_1 \; x_2 \; \ldots \; x_n \,]^T \tag{6.6}$$

$$E = [\, -m_1 \; -m_2 \; \ldots \; -m_n \,]^T \tag{6.7}$$

$$\varphi = [\, 0 \; 0 \; \ldots \; 0 \; 1 \,]^T \tag{6.8}$$

where subscripts c and p represent the cavity and particle of the damper respectively, M, C and K are the mass, damping, and stiffness matrices of the primary structure, respectively; X is an n-dimensional displacement vector of the structure; E and $\ddot{x}g$ are the matrix-induced ground acceleration and ground acceleration, respectively; and φ is an n-dimensional location vector of the control force, whose nth component is 1 and the other components are 0. y_1 is the relative displacement of the container with respect to the primary system, i.e., $y_1 = x_c - x_5$, and y_2 is the relative displacement of the simplified particle with respect to the container, i.e., $y_2 = x_p - x_c$. $G(y_2)$ and $H(y_2, \dot{y}_2)$ are nonlinear functions (See Chap. 3, Sect. 3.1.2).

2. *Parameter determination*

In this simulation, 180 steel balls with 51 mm diameter were equally placed into 12 containers in two layers. For convenience, the behavior of the particles in each container was assumed to be the same. Therefore, the impact forces on all the containers from collisions between the particles and the containers could be calculated as the product of the number of containers and the impact force of one container.

The parameters that were used in the numerical simulation are listed in Table 6.11. The damping ratio of the primary structure is 0.02, and the mass matrix and stiffness

Table 6.11 System parameter values

	Mass (kg)	Circular frequency (rad/s)	Damping ratio
Container	39.345	6.28	0.10
Equivalent single particle	7.96	125.60	0.20

matrix of the primary structure is as follows:

$$
\mathbf{M} = \begin{bmatrix} 1200 & & & & \\ & 1200 & & & \\ & & 1200 & & \\ & & & 1200 & \\ & & & & 1200 \end{bmatrix} \text{kg}
$$

$$
\mathbf{K} = \begin{bmatrix} 1063050 & -558390 & & & \\ -558390 & 1137820 & -579530 & & \\ & -579430 & 1158850 & -579420 & \\ & & -579420 & 1185850 & -606430 \\ & & & -606430 & 606430 \end{bmatrix} \text{N/m}
$$

For the calculation of the circular frequency and damping ratio of the cavity and the equivalent single particle, see Chap. 3, Sect. 3.1.1.

3. *Programming and verification*

A MATLAB program has been conducted to show the performance of the PTMD under four seismic excitations, the results are compared with that in experiments.

Figure 6.26 shows the calculated and experimental results of the acceleration time histories at the top of the test structure with the PTMD under the El Centro wave (0.2 g) and SHW2 wave (0.05 g). These results generally matched, which means that the proposed equivalent simplification method can simulate the trend of system motion more effectively in a reasonable range.

In addition, a comparison of the calculated and experimental results of the peak acceleration and peak displacement responses at the top of the test structure with the PTMD under El Centro wave is listed in Table 6.12. The errors were limited within an acceptable range, which indicates that the simplified numerical simulation method that was proposed in this paper can yield relatively accurate estimates of the

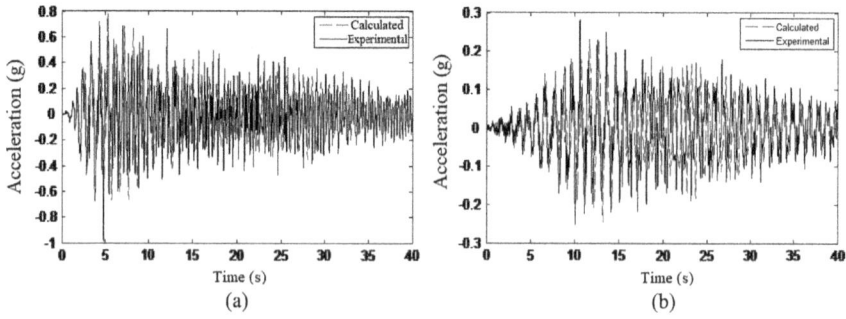

Fig. 6.26 Acceleration time histories at the top of the test structure with the PTMD under the **a** El Centro wave (0.2 g) and **b** SHW2 wave (0.05 g)

Table 6.12 Comparison of the calculated and experimental results for the peak acceleration and displacement response on top of the test structure with the PTMD

Input	Acceleration (m/s^2)			Displacement (mm)		
	Calculation	Experiment	Error (%)	Calculation	Experiment	Error (%)
El Centro (0.1 g)	0.4899	0.4796	2.15	80.9500	79.0030	2.46
Wenchuan (0.1 g)	0.4280	0.4540	5.73	21.9297	20.0658	−9.29
Japan 311 (0.1 g)	0.2548	0.2682	4.99	12.7873	13.5102	5.35
SHW2 (0.05 g)	0.2749	0.2751	0.07	54.2773	55.9410	8.84

peak response, especially for the acceleration responses. This is also a very important control index of comfort in structural design.

Based on the above results, although the proposed equivalent calculation method simplifies the highly non-linear characteristics of particle dampers to a certain extent, such as ignoring the collision force between particles, the method still grasps the core point of using particle dampers to reduce vibration, that is, the collision force between particles and the wall of the container. As this method can simulate the key control force which affects the damper's vibration reduction effect, it can achieve the goal of satisfying the accuracy requirement by simple calculation, and can be further used in practical engineering design.

References

1. Masri, S.F. 1969. Analytical and experimental studies of multiple-unit impact dampers. *Journal of the Acoustical Society of America* 45 (5): 1111–1117.
2. Panossian, H.V. 1992. Structural damping enhancement via non-obstructive particle damping technique. *Journal of Vibration and Acoustics* 114 (1): 101–105.
3. Zhao, L., P. liu, and Y. Lu. 2009. Experimental study on damping characteristics of non-obstructive micro-particle damping column. *Journal of Vibration and Shock* 28 (8): 1–5.
4. Panossian, H.V. 1991. Non-obstructive particle damping tests on aluminum beams. In *Proceedings of the damping 91 conference*. San Diego California.
5. Panossian, H.V. 1989. Non-obstructive impact damping applications for cryogenic environments. In *Proceedings of damping*.
6. Bhatti, R.A., and Y. Wang. 2009. Simulation of particle damping under centrifugal loads. World Academy of Science Engineering & Technology 57.
7. Hollkamp, J.J., R.W. Gordon. 1998. Experiments with particle damping. In *Smart structures and materials 1998: Passive damping and isolation*. San Diego: SPIE.
8. Xia, Z., Y. Shan, and X. Liu. 2007. Particle damping experiment based on cantilever beam. *Journal of Aerospace Power* 10: 1737–1741.
9. Yan W., Y. Huang, Haoxiang He, et al. 2010. Particle damping technology and its application prospect in civil engineering. *World Earthquake Engineering* 26 (4): 18–24.
10. Lu, Z., X. lv, and W. Yan. 2012 Experimental study on damping control of particle damper. *China Civil Engineering Journal* (S1): 243–247.

11. Popplewell, N., S.E. Semercigil. 1989. Performance of the bean bag impact damper for a sinusoidal external force. *Journal of Sound and Vibration* 133 (2): 193–223.
12. Li, K., A.P. Darby. 2006. An experimental investigation into the use of a buffered impact damper. *Journal of Sound and Vibration* 291: 844–860.

Chapter 7
Wind Tunnel Test Study on Particle Damping Technology

In the previous chapter, the shaking table test of particle damping technology was systematically studied. This chapter will carry out a deeper analysis of its damping effect under wind excitations. Earthquake and wind are the dynamic excitations that must be considered in the design of civil engineering structures. The excitation properties of them are different. Earthquakes tend to be short-duration with multiple frequencies, and wind excitations are more similar to white noise and last longer. Therefore, it is more challenging to use the same structural vibration control method to have better damping effect on the two kinds of different excitations. Common linear dampers, such as tuned mass dampers, tend to have better control effects on wind excitations and poor damping effects on earthquakes. Particle damping technology can broaden the damping band of the damper by the nonlinear action of particle collision. Its damping characteristics under earthquake action have been verified. This section will continue to explore its damping characteristics under wind excitations, especially the influence of different design parameters, and finally introduces the simplified design method and its realization way for engineering application.

7.1 Experiment Design

In order to study the influence of PTMD on wind-induced vibration control of high-rise buildings, the accelerometer and the laser displacement meter are used to measure the acceleration and displacement of the selected models under uncontrolled and controlled state in the TJ-2 atmospheric boundary layer wind tunnel of the State Key Laboratory of Disaster Reduction in Civil Engineering in Tongji University.

1. *Model parameter selection*

The test model is selected based on the Benchmark (Phase III) model which is always used in wind-induced vibration control in civil engineering. The Benchmark problem

© China Machine Press and Springer Nature Singapore Pte Ltd. 2020
Z. Lu et al., *Particle Damping Technology Based Structural Control*,
Springer Tracts in Civil Engineering,
https://doi.org/10.1007/978-981-15-3499-7_7

of vibration control of civil engineering structures is to establish a complete structural vibration control system inspection and evaluation system under the same structural model, environmental interference and performance indicators, so as to provide a common platform for comparing different control methods and strategies.

The Benchmark problem has gone through three stages. The first phase of the study is based on two three-layer framework models proposed by Professor Spencer. They are studied under the action of earthquakes using active tendon (ATS) control and active mass driver (AMD) control. Various control algorithms and control effects are compared [1]. The second phase is based on the problems of earthquake and wind excitations. Two representative practical projects are selected as the research object: one is a 20-story seismic steel structure designed according to the California State Code (proposed by Professor Spencer) [2], the other is a 76-story and 306 m-high reinforced concrete wind-resistant tower to be built in Melbourne, Australia (proposed by Professor Yang) [3]. The effectiveness and applicability of various control schemes are studied and compared. The wind load at this phase uses the random wind time history generated by the Davenport transverse wind direction and speed spectrum. Based on the researches of the second stage, the third-phase vibration control Benchmark problem is studied to get more realistic structure's response under real dynamic loads. The wind load model for structural vibration control analysis is established based on the wind load obtained from the structural model wind tunnel experiment and considering the structure and its components entering a nonlinear state under earthquake, which emphasizes the significance of Benchmark research from a more practical perspective. The representative research objects at this stage are: three 3-, 9-, and 20-story seismic steel structures designed according to the California State Code, and the 76-story reinforced concrete wind-resistant tower built according to the wind tunnel test [4].

The structure selected for this wind tunnel test is a frame-tube structure composed of a concrete core tube and an outer frame. The core tube is designed to bear wind loads and the outer frame is used to bear the main gravity loads. The structural plane is square and chamfered at two opposite corners. The total mass of the structure is 153×103 t, the total volume is 510×103 m^3, and the mass density is about 300 kg/m^3. The structure has a width of 42 m and a structural aspect ratio of 7.3, which is a wind-sensitive structure [5]. The inner core tube has a size of 21×21 m and the outer frame column has a distance of 6.5 m. Each layer has 24 pillars, which are evenly distributed around the cylinder, and are fixedly connected to the box-section beam with a height of 0.9 m and a width of 0.4 m at each floor. The concrete compressive strength is 60 Mpa and the elastic modulus is 40 Gpa. The column size, the core tube thickness and the quality of each floor vary with height. The structural plan and elevation are shown in Fig. 7.1.

Since the structure is substantially symmetrical in two horizontal directions, the stiffness and the center of the mass coincide, thereby avoiding a significant horizontal torsional coupled vibration [6, 7].

The top three natural vibration frequency of the structure is 0.160, 0.765, 1.992 Hz, respectively, and the corresponding vibration mode is shown in Fig. 7.2.

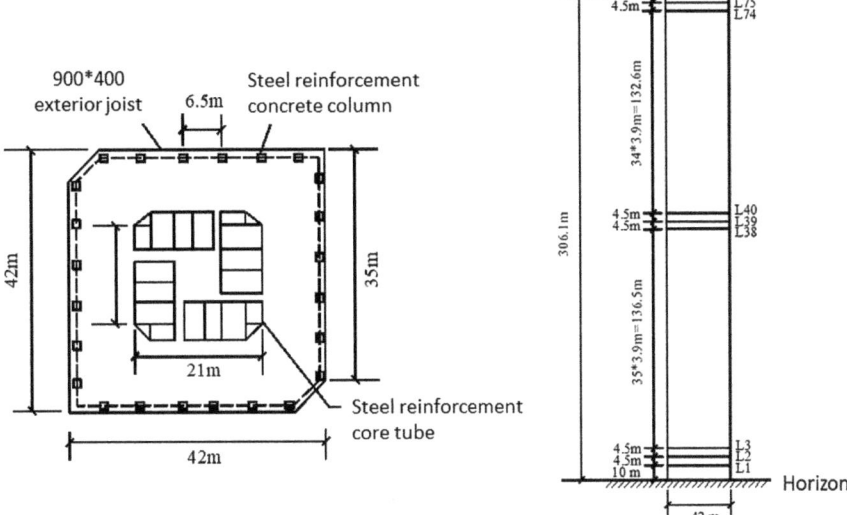

Fig. 7.1 Plane and elevation of Benchmark model

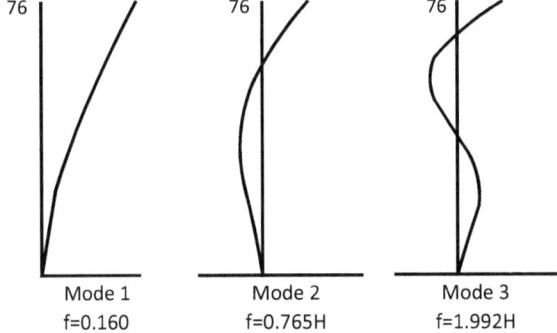

Fig. 7.2 The top three mode diagram of Benchmark model

2. *Scale determination*

In general, pneumatic similarities include structural length, density, elastic similarity, and similarity in airflow density, viscosity, velocity, and gravitational acceleration. These physical quantities can be represented by several dimensionless parameters, such as Reynolds number, Froude number, Strouhal number, Cauchy number, density ratio, damping ratio, etc. The Pneumatic similar parameters are summarized in Table 7.1. Wind tunnel tests for buildings should try to meet these similar conditions. If the Reynolds numbers of the two flows are equal, the viscous forces of the fluid are similar. For turbulent flow with a large Reynolds number, the inertial force plays

Table 7.1 Pneumatic similar parameter

Dimensionless parameter	Expression	Physical meaning	Similar requirements
Reynolds number	$\rho DU/\mu$	Air inertial force/air viscous force	Bluff body is allowed not to be simulated
Froude number	U^2/Dg	Flow inertial force/gravity	Strictly similar
Strouhal number	n_s/DU	Unsteady motion inertial force/inflow inertial force	Strictly similar
Cauchy number	$E/\rho U^2$	Structural elastic force/pneumatic inertial force	Comprehensive consideration
Density ratio	ρ_s/ρ	Structural elastic force/pneumatic inertial force	Strictly similar
Damping ratio	δ	Energy consumption per cycle/vibration total energy	Strictly similar

a leading role, and the viscous force is relatively small. The requirement for equal Reynolds number can be relatively low.

In Table 7.1, the parameter ρ represents the air mass density, generally $\rho = 1.225$ kg/m^3; U represents the average wind speed; D represents the structural feature size, and μ represents the air motion viscosity coefficient; g represents the gravitational acceleration; f represents the structural vibration frequency; E represents Structural material elastic modulus; ρs represents the structural material mass density; δ represents the structural damping logarithmic decay rate.

(1) *Model length scale*

The wind tunnel test was carried out in the TJ-2 wind tunnel in State Key Laboratory of Disaster Reduction in Civil Engineering in Tongji University. The TJ-2 wind tunnel is a typical horizontally arranged closed-flow boundary layer wind tunnel, which is a steel-concrete hybrid structure. The experimental section is 15 m long and adopts a rectangular section with a width of 3 m and a height of 2.5 m. The controllable wind speed of the air tunnel is 0.5–60 m/s, which can be continuously adjusted [8]. Due to the experiment of the high-rise building model with additional particle damper, in order to ensure the same proportion of the particles and the building, the proportion of the model should be enlarged as much as possible, and the blocking ratio should be less than 8%. The length scale of the selected model is 1/200. At this ratio, the size of the model is 21 cm × 21 cm × 153 cm. The blocking ratio of the model is $\sqrt{21^2 + 21^2} \times 153/(300 \times 250) = 6\% < 8\%$, which can satisfy the requirements. At the same time, the experimental model is relatively regular, and located in the middle of the wind tunnel. Therefore, it will not be affected by the wall effect.

(2) *Structural density ratio*

Since the density of the air flowing in the wind tunnel is the same as that in the atmosphere, the air mass density ratio is $\lambda \rho = \rho_m / \rho_p = 1$. Similarly, the density of the model must be the same as that of the supertall building structure because the mass density of the prototype is about 300 kg/m³. Therefore, the mass density of the model is also 300 kg/m³.

(3) *Model scale*

By determining the length scale and density scale of the model, other dimensions of the model can be derived using dimensional analysis, as shown in Table 7.2.

3. *Model design parameter determination*
(1) *Model mass*

The similarity ratio of the model density is 1. The density of the model is 300 kg/m³ the mass of the prototype is 153×103 t. The length scale of the model is 1/200. The

Table 7.2 Model scale

Parameter	Symbol	Unit	Similarity ratio	Similar requirements
Length	L	m	$\lambda L = 1{:}200$	Geometric similarity ratio
Density	ρ	Kg/m³	$\lambda \rho = 1$	Same
Velocity	u	m/s	$\lambda v = 1/\sqrt{\lambda L} = 1/\sqrt{200}$	Froude number
Time	t	s	$\lambda t = 1 {:} \lambda f = 1/\sqrt{200}$	Strouhal number
Frequency	f	Hz	$\lambda f = \lambda v / \lambda L = \sqrt{200}{:}1$	Strouhal number
Gravity acceleration	g	m/s²	$\lambda g = 1$	Same
Displacement response	d	m	$\lambda d = \lambda L = 1/200$	Dimension
Acceleration response	a	m/s²	$\lambda_a = \lambda_L \lambda_f^2 = 1$	Dimension
Unit length quality	m	Kg/m	$\lambda_m = \lambda_\rho \lambda_L^2 = 1 : 200^2$	Dimension
Unit mass moment	Jm	Kgm²/m	$\lambda_{J_m} = \lambda_\rho \lambda_L^4 = 1 : 200^4$	Dimension
Bending stiffness	EI	Nm²	$\lambda_{EI} = \lambda_E \lambda_L^4 = 1 : 200^5$	Dimension
Torsional stiffness	GJd	Nm²	$\lambda_{GJ_d} = \lambda_G \lambda_L^4 = 1 : 200^5$	Dimension
Axial stiffness	EA	N	$\lambda_{EA} = \lambda_E \lambda_L^2 = 1 : 200^3$	Dimension
Damping ratio	ξ	–	$\lambda \xi = 1$	Same

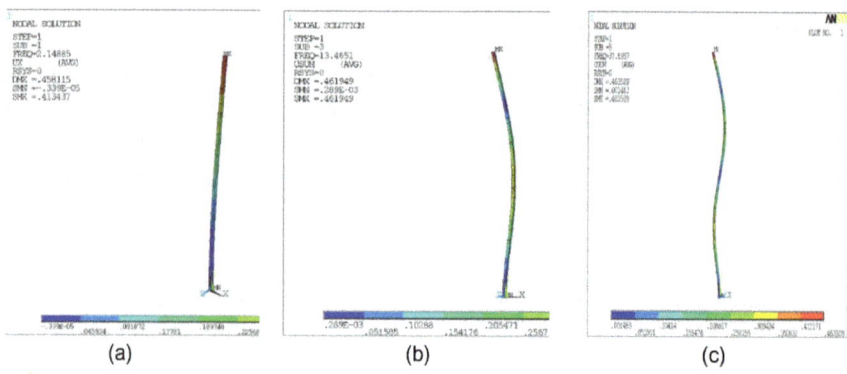

Fig. 7.3 Mode analysis chart of core beam: **a** The first mode; **b** The second mode; **c** The third mode

quality of the model can be determined as: $m = M \times \lambda_l^3 = 153 \times 10^6 \times (1/200)^3 = 19.125$ Kg.

(2)   *Model fundamental frequency*

The fundamental frequency of the building prototype is: $fs = 0.16$ Hz. According to the frequency similarity ratio, the vibration frequency of the model is: $fm = \sqrt{200} \times 0.16 = 2.26$ Hz.

(3)   *Implementation of model dynamic characteristics*

Before the wind tunnel test, the finite element analysis is performed on the fabricated model using ANSYS finite element software, and the size of the core beam is adjusted to simulate the model shape. The final design of the core beam has a cross-sectional dimension of 1.55 cm × 1.55 cm. The vibration mode analyzed by ANSYS is shown in Fig. 7.3.

The model design and physical comparison are shown in Fig. 7.4.

The installation position of the sensor on the model is shown in Fig. 7.5. The model is divided into three sections of equal height. The laser displacement meters and the accelerometers are installed at the same height in the across-wind and along-wind direction. A total of 6 laser displacement meters and 6 accelerometers are installed.

4.   *The design of particle tuned mass damper*
(1)   *Damper cavity*

The purpose of this test is to investigate the damping effect of the particle damper. In order to reduce the proportion of the damper cavity mass to the total mass of the damper, a hardwood with a lower density is selected as the damper cavity, and the thickness of the board is 0.1 cm. The PTMD reduces vibration by tuning, which is realized by adjusting the suspension line at the top of the damper. Considering the swing of the suspension damper during the movement and the area occupied by the particles, the size of the inner cavity of the damper has three types: 7.5 cm × 7.5 cm, 8.5 cm × 8.5 cm, 9 cm × 9 cm. The cavity type includes single layer, double

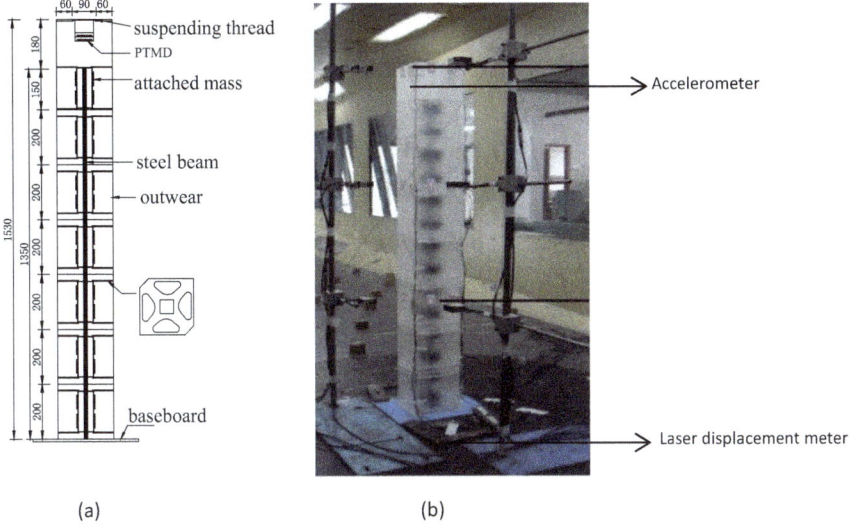

Fig. 7.4 Test model design and photo: **a** Model design; **b** Photo

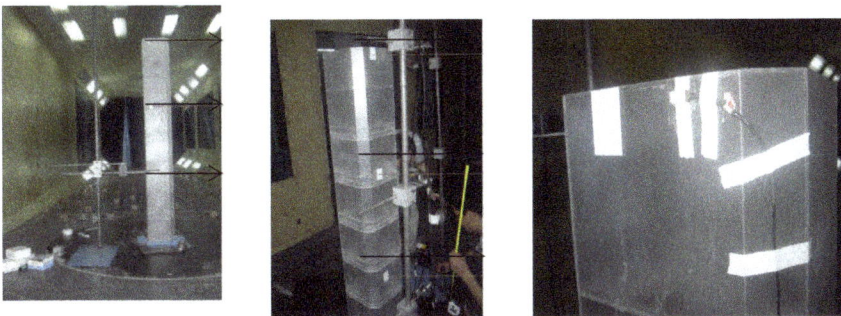

Fig. 7.5 Distribution of accelerometers and laser displacement meters

layer and three-layer. Each layer has the same height, which are 1.5 cm, as shown in Fig. 7.6.

In order to compare the damping effect of the particle damper under different interval modes of the cavity, different intervals are set inside the cavity. In this experiment, two different intervals are set: 4-square division and 9-square division. The interval form is shown in Fig. 7.7.

(2) *Damper particles*

In order to compare the effects of particles' density on the damping effect of the particle damper, steel particles, copper particles, lead particles and tungsten carbide particles are selected as the particle material in this test. In order to study the effect of particle diameter on the damping effect of the damper, considering the number of particles accommodating the damper cavity at different mass ratios and the difficulty

Fig. 7.6 Three-layer damper cavity

(a)   (b)

Fig. 7.7 Different damper cavities with different division modes: **a** 4-square division; **b** 9-square division

of obtaining the material, the diameter selected for the lead particles is 2, 3, 4, 6, 8 and 10 mm. Particles of the other three materials have diameters of 6 mm. The selected particle material for the test is shown in Fig. 7.8.

(3) *Suspension line*

In order to realize the tuning function of the damper, the suspension is used to adjust the frequency of the damper. Four equal length cycloids are set, and the four corners of the damper cavity are respectively connected with the prefabricated small holes at the top of the model. The connection mode is simplified to the hinge. In this experiment, the suspension line was made of ordinary nylon thread.

Because the natural vibration frequency of the model is high, to make the vibration frequency of the damper the same as the vibration frequency of the model, the length of the cycloid is 4.86 cm, which is taken as 5 cm.

At the same time, in order to investigate the effect of the particle damper on the vibration control of the main structure without tuning, the frequency of the damper is adjusted by setting the length of different cycloids in this test, and three suspension lengths of 4, 6 and 7 cm are selected. Finally, the particle tuned mass damper can be obtained, shown in Fig. 7.9.

Fig. 7.8 Particles for testing: **a** 6 mm tungsten carbide particles; **b** 6 mm copper particles; **c** 6 mm steel particles; **d** 2 mm lead particles; **e** 3 mm lead particles; **f** 4 mm lead particles; **g** 6 mm lead particles; **h** 10 mm lead particles

5. *TMD design*

In order to compare the damping performance of the particle damper and TMD under the same mass ratio, the TMD condition is added in this test, and the mass of the TMD and the particle damper are ensured to be the same by adding a mass in the damper cavity. The tuned mass damper is shown in Fig. 7.10.

Table 7.3 Summary of design wind pressures of major cities in China

Design wind pressure	50 years (kN/m^2)	100 years (kN/m^2)
Taitung	0.9	1.05
Sanya	0.85	1.05
Hong Kong	0.9	0.95
Xiamen	0.8	0.9
Shenzhen	0.75	0.9
Dalian	0.65	0.75
Shanghai	0.55	0.6
Beijing	0.45	0.5

6. *Wind environment simulation*

(1) *Wind field*

In view of the limitations of the experimental conditions, only A, B and C class wind field are simulated in this experiment. Since the high-rise buildings are mostly located in the wind field of class C, most of the tests are completed in the wind field of class C. The wind field layout is shown in Fig. 7.11.

(2) *Wind speed*

According to different design wind pressures in major cities across the country in 'Load code for the design of building structures', the wind pressure variation range of different design lives are summarized. The variation range of basic wind pressure in one hundred years is 0.5–1.1 kN/m^2, corresponding to the gradient wind speed under Class B wind field, which is shown in Table 7.3. For the class C wind field (400 m) and the class D wind field (450 m), the gradient wind height is smaller than the wind tunnel height after the scaling (1/200). Therefore, the wind speed is the same as the wind speed in the class B wind field in the two different wind fields. The comparison of design wind pressure and wind tunnel wind speed is shown in Table 7.4.

Considering the feasibility of the wind speed setting device in the wind tunnel and the test operation, the wind speeds in the actual test conditions are 3.5, 4.0, 4.5, 5.0 and 5.5 m/s.

Table 7.4 Comparison of design wind pressure and wind tunnel wind speed

Basic wind pressure (KN/m^2)	Basic wind speed (m/s)	Gradient wind speed (m/s)	Wind tunnel wind speed (m/s)
0.50	28.28	49.96	3.53
0.65	32.25	56.96	4.03
0.80	35.78	63.19	4.47
1.00	40.00	70.65	5.00

Fig. 7.9   Particle tuned mass damper

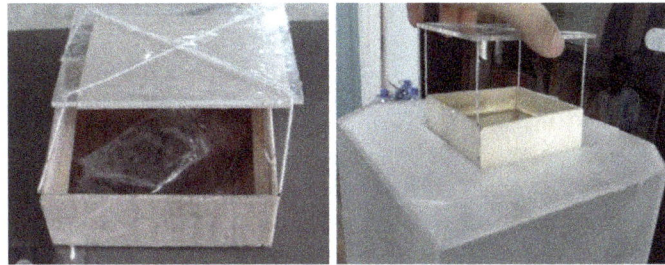

Fig. 7.10   Tuned mass damper

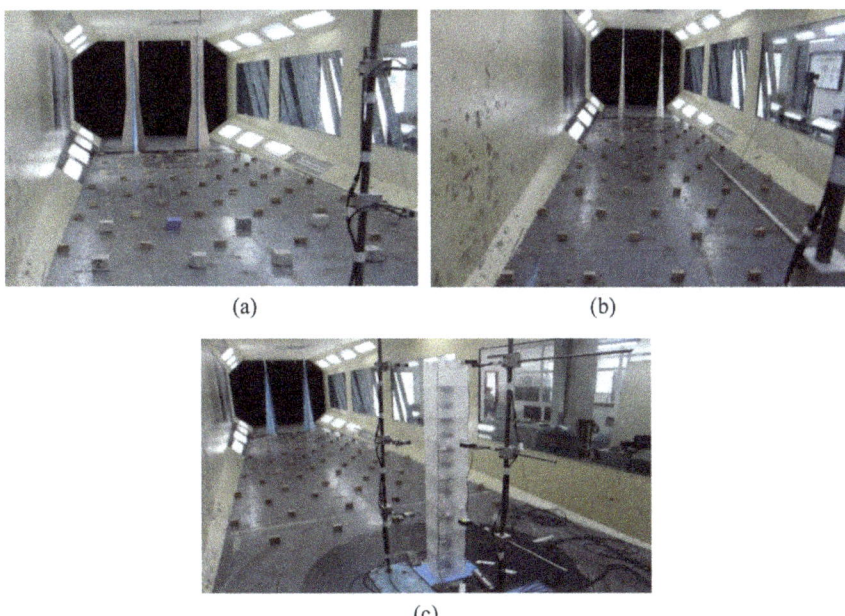

Fig. 7.11   Wind field layout: **a** Class A wind field; **b** Class B wind field; **c** Class C wind field

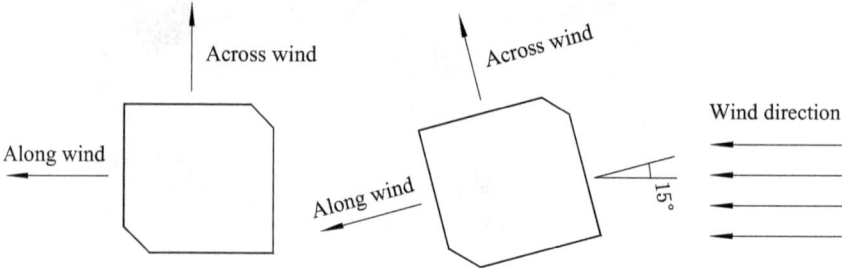

Fig. 7.12 The relationship between wind inclination angle and model

(3) *Wind inclination angle*

The model section selected in this test is a square with a chamfer angle. The Wind inclination angle varies from 0 to 90° to traverse all possible inclination angles of the model. Therefore, the selected wind inclination angles are 0°, 15°, 30°. 45°, 60°, 75°, and 90°. The wind inclination angle and model relationship are shown in Fig. 7.12.

(4) *Sampling time*

In order to realistically simulate the working state of PTMD, the sampling time of 10 min average maximum wind speed in real environment is simulated according to the requirements of basic wind pressure sampling in "Load code for the design of building structures" GB50009-2012, and the final sampling time is determined according to the time scale ratio, which is $10 \times 60 \times 1/14.14 = 42$ s. The wind speed is added to the ideal wind speed from zero. The wind speed is unstable in the first 80 s, and the sampling begins when the wind speed is stable.

7. *Test conditions*

The test conditions are designed from the perspective of PTMD and wind field environment. A summary of the variables for all operating conditions is shown in Table 7.5.

(1) *Mass ratio*

The mass ratio is defined as the total mass of the PTMD (damper cavity and total mass of the particles) to the mass of the model. Different mass ratios will have different effects on the damping effect of the damper, and the mass ratio directly affects the damper's promotion and application in engineering. In order to investigate the damping effect of the damper under different mass ratios, the working conditions of different mass ratios are set. Only the overall mass ratio of the damper to the model is changed (The damper cavity is unchanged, and the particle mass changes. Therefore the total mass of the damper is changed), other parameters remain unchanged. The parameters are summarized as shown in Table 7.6.

Table 7.5 Test variable summary table

1	Mass ratio (%)	0.5	0.75	1	1.25	1.5	–	–
2	Particle material	Steel	Copper	Lead	Tungsten carbide	–	–	–
3	Particle diameter (mm)	2	3	4	6	8	10	–
4	Mass ratio of container to particles	Original	0.25	0.5	1	2	4	–
5	Suspension length (cm)	4	5	6	7	–	–	–
6	Cavity type	No interval	4-square division	9-square division	–	–	–	–
7	Cavity size	7.5 cm × 7.5 cm	8.5 cm × 8.5 cm	9 cm × 9 cm	–	–	–	–
8	Wind speed (m/s)	3.5	4	4.5	5	6	7	–
9	Wind inclination angle (degrees)	0	15	30	45	60	75	90
10	Wind field	B	C	D	–	–	–	–

Table 7.6 Summary of test conditions under different mass ratios

Condition	Mass ratio (%)	Particle	Wind speed (m/s)	Wind field	Wind inclination angle	Suspension length (cm)	Damper cavity
1–1	0.5	2 mm lead	4	C	0	5	9 cm × 9 cm No interval
1–2	0.75						
1–3	1						
1–4	1.25						
1–5	1.5						

(2) *Particle materials*

Different particle materials have different properties (such as stiffness, density and friction coefficient). Therefore, different material types will have different effects on the damping effect of the damper. Four common particle materials are selected according to the material density. The material and working condition settings are summarized in Table 7.7.

Table 7.7 Summary of test conditions under different particle densities

Condition	Particle material	Mass ratio (%)	Diameter (mm)	Wind speed	Wind field	Wind inclination angle	Suspension length	Damper cavity
2–1	Steel	1	6	4	C	0	5 cm	9 cm × 9 cm No interval
2–2	Copper							
2–3	Lead							
2–4	Tungsten carbide							

Table 7.8 Summary of test conditions with different particle diameters

Condition	Diameter (mm)	Mass ratio (%)	Particle material	Wind speed (m/s)	Wind field	Wind inclination angle	Suspension length	Damper cavity
3–1	2	1	Lead	5	C	0	5 cm	9 cm × 9 cm No interval
3–2	3							
3–3	4							
3–4	6							
3–5	8							
3–6	10							

(3) *Particle diameter*

Different particle diameters in the particle damper may affect the damping effect of the damper. To explore the effect particle diameters on the damping performance of PTMD, particles of different diameters are set under the same material conditions. Other parameters remain the same. The summary of working conditions of different diameters is shown in Table 7.8.

(4) *Damper cavity mass and particle mass ratio*

The PTMD is composed of the damper cavity and the particles. The difference between the cavity and the particle mass will affect the motion state and motion form of the particles, which will have different effects on the damping effect of the damper. In order to investigate the damping effect of PTMD on the damper under different damper cavity and particle mass ratio, the response conditions under different ratios are tested. The summary is shown in Table 7.9.

(5) *Suspension length*

The PTMD in this test is suspended, and the different suspension lengths correspond to different frequencies of the damper. The tuning frequency of the damper is changed by adjusting the length of the suspension line. The vibration frequency of actual

Table 7.9 Summary of test conditions under different mass ratio of container to particles

Condition	Cavity mass and particle mass ratio	Total mass ratio (%)	Particle material	Wind speed (m/s)	Wind field	Wind inclination angle	Suspension length	Damper cavity
4–1	original	1	2 mm lead	4	C	0	5 cm	9 cm × 9 cm No interval
4–2	0.25							
4–3	0.5							
4–4	1							
4–5	2							
4–6	4							

Table 7.10 Summary of test conditions under different suspension lengths

Condition	Suspension length	Wind inclination angle	Mass ratio (%)	Particle material	Wind speed	Wind field	Damper cavity
5–1	4 cm	0	1	6 mm tungsten carbide	5	C	9 cm × 9 cm No interval
5–2	5 cm						
5–3	6 cm						
5–4	7 cm						

structural buildings tends to change greatly in different environments. Therefore, it is of great significance to examine the damping effect of PTMD under different suspension lengths. In this test, the working conditions under different suspension lengths are considered. The summary is shown in Table 7.10.

(6) *Different types of damper cavity*

The different interval modes inside the damper cavity will have different effects on the collision between the particles and the damper cavity wall, and thus have different effects on the internal energy dissipation mode of PTMD. To qualitatively investigate the effect of different interval forms on the damping effect of dampers, two different cavity interval modes are set. The test conditions are summarized in Table 7.11.

(7) *Wind speed*

The actual buildings have different vibration responses at different wind speeds, and the structural response tends to increase as the wind speed increases. In the case of increased structural response, PTMD should also exhibit different damping effects. In order to investigate the effect of different wind speed on the damping effect of dampers, conditions of different wind speed are set. The summary is shown in Table 7.12.

Table 7.11 Summary of test conditions under different damper cavity type

Condition	Damper cavity	Mass ratio (%)	Particle material	Diameter (mm)	Wind speed	Wind field	Wind inclination angle	Suspension length (cm)
6–1	1 (No interval)	0.7	steel	6	5	C	0	5
6–2	2 (4-square division)							
6–3	3 (9-square division)							

Table 7.12 Summary of test conditions under different wind speed excitations

Condition	Wind speed (m/s)	Wind field	Mass ratio (%)	Particle material	Wind load inclination	Suspension length (cm)	Damper cavity
7–1	3.5	C	1	6 mm steel	0	5	7.5 cm × 7.5 cm No interval
7–2	4						
7–3	4.5						
7–4	5						

(8) *Inclination angle*

In the actual structure, the direction of the wind flow changes randomly, and the vibration response of the structure will show different states with the change of the flow direction. Therefore, investigating the effect of PTMD on the vibration control of the structure under different inclination angles is significant. Therefore, the test conditions are shown in Table 7.13.

Table 7.13 Summary of test conditions under different inclination angles

Condition	Inclination angle	Mass ratio (%)	Particle material	Wind speed	Wind field	Suspension length (cm)	Damper cavity type
8–1	0	1	2 mm lead	5	C	5	9 cm × 9 cm No interval
8–2	15						
8–3	30						
8–4	45						
8–5	60						
8–6	75						
8–7	90						

7.2 Results of Wind Tunnel Test

7.2.1 Dynamic Characteristics of Test Model

Structural dynamic characteristics are inherent characteristics of the structure, including natural frequency, damping, and vibration mode. These parameters are determined by factors such as structural composition, stiffness, mass distribution, material properties, and structural connections, and are independent of external loads.

The methods for measuring the dynamic characteristics of the structure include free vibration method, forced vibration method and pulsation method. Among them, the free vibration method is widely used because of its convenient implementation [9]. In this test, the initial displacement method in the free vibration method is used, that is, the pre-added initial displacement is suddenly released, and the structure is freely vibrated, and the free attenuation curve of the top response of the model is measured.

1. *Self-vibration characteristics*

The natural frequency of the across-wind direction and along-wind direction of the model structure can be identified by Fourier spectrum obtained by the Fourier transform of the free attenuation curve. The Fourier spectrums in both directions are shown in Fig. 7.13. It could be seen that the natural frequency of the along-wind direction is 2.16 Hz, and the natural frequency of the across-wind direction is 2.17 Hz.

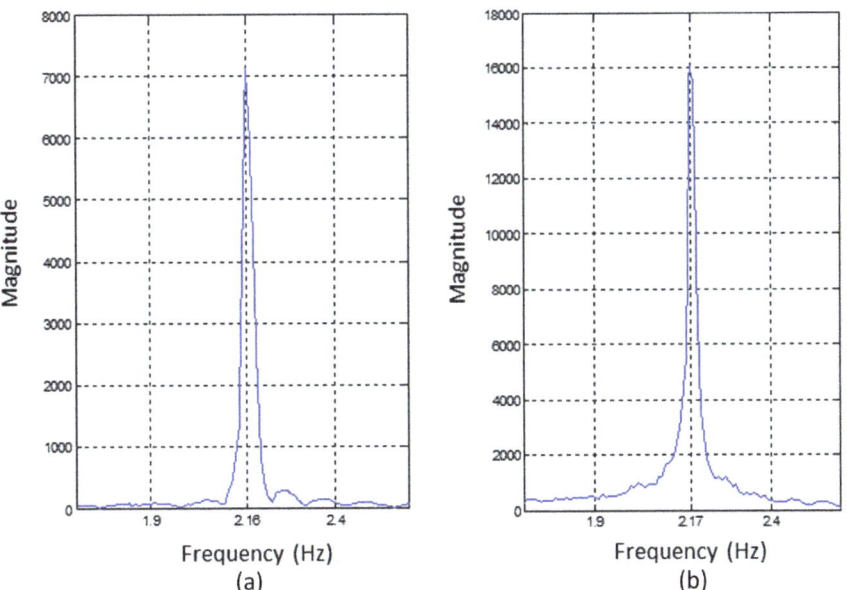

Fig. 7.13 Fourier spectrum of the free attenuation curve of the main structure in two directions: **a** Along-wind direction; **b** Across-wind direction

2. *Damping ratio*

The structural damping is determined (measured without wind) using the method shown by Eq. (7.1) [10] :

$$\zeta_s = \frac{\ln(y_m/y_n)}{2(n-m)\pi} \tag{7.1}$$

where y_i is the amplitude of the ith cycle of free attenuating vibration of the model.

When measuring the structural damping ratio, the damping ratio of the structure can be obtained if any two peaks of the free attenuation curve are known.

In the course of this experiment, the damping measurement of the aeroelastic model of a square-section super high-rise building is not so simple. Since the dynamic characteristics of the model in the two orthogonal lateral bending directions are completely consistent, when the initial displacement is applied in the along-wind direction of the model, the across-wind direction also vibrates, since the self-vibration frequencies in the two directions are almost the same, resulting in coupling of the vibration in two directions (called beat phenomenon), which brings great difficulty to the damping measurement. The vibration curves in both directions are shown in Fig. 7.14, where Δ is the displacement.

It could be seen from Fig. 7.14 that the initial displacement is applied in the along-wind direction, and the structure is freely attenuated in the along-wind direction at the initial moment. As time passes, part of the energy is converted into vibration in the across-wind direction. In order to eliminate the influence of energy transfer on the damping measurement, it is assumed that the along-wind direction (x direction) and the across-wind direction (y direction) have the same stiffness, frequency and damping ratio [11]. Its initial vibrational energy can be expressed as:

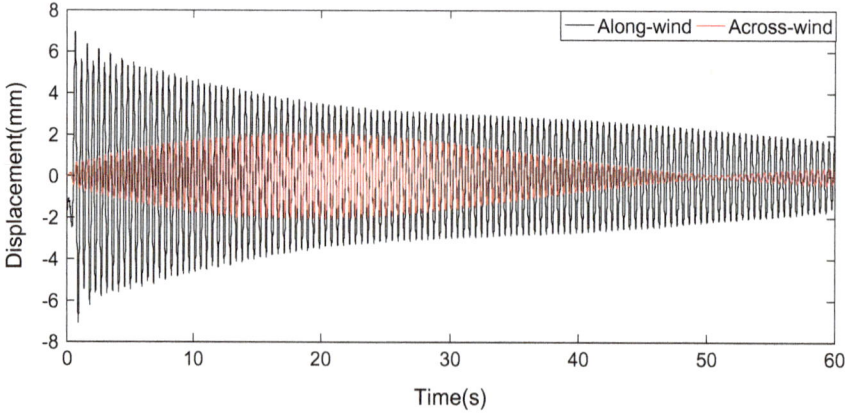

Fig. 7.14 Free attenuation curves under the initial displacement condition of the along-wind direction

$$E_{0x} = 0.5k A_{0x}^2 \tag{7.2}$$

$$E_{0y} = 0.5k A_{0y}^2 \tag{7.3}$$

Among them, A_{0x}, A_{0y} are the vibration amplitudes in two directions, respectively, k is the stiffness of the model in both directions. If there is no energy transfer in both directions, the vibrational energy in each direction after n-cycles is:

$$E_{nx} = 0.5k A_{nx}^2 \tag{7.4}$$

$$E_{ny} = 0.5k A_{ny}^2 \tag{7.5}$$

Therefore, the sum of the vibrational energy consumed by the damping can be calculated:

$$\Delta E_\zeta = (E_{0x} - E_{nx}) + (E_{0y} - E_{ny}) = 0.5k[(A_{0x}^2 + A_{0y}^2) - (A_{nx}^2 + A_{ny}^2)] \tag{7.6}$$

Define the vibration vector sum r as:

$$r^2 = x^2 + y^2 \tag{7.7}$$

Due to the effect of "beat", the vibrations of the along-wind and the across-wind are in phase, and:

$$A_r^2 = A_x^2 + A_y^2 \tag{7.8}$$

Bring into Eq. (7.6) to get:

$$\Delta E_\zeta = 0.5\,k(A_{0r}^2 - A_{nr}^2) \tag{7.9}$$

Assuming that after n cycles of vibration, the energy from the along-wind to the across-wind is ΔE_{xy}, then:

$$\Delta E_\zeta = (E_{0x} - E_{nx}' - \Delta E_{xy}) + (E_{0y} - E_{ny}' + \Delta E_{xy}) = 0.5k[(A_{0x}^2 + A_{0y}^2) - (A_{nx}'^2 + A_{ny}'^2)]$$
$$= 0.5k[(A_{0r}^2 - A_{nr}'^2) \tag{7.10}$$

Compare (7.9) and (7.10), and obtain:

$$A_{nr} = A_{nr}' \tag{7.11}$$

When there is no energy transfer,

$$A_{nx} = A_{0x} e^{-2n\pi\zeta} \tag{7.12}$$

$$A_{ny} = A_{0y}e^{-2n\pi\zeta} \tag{7.13}$$

Therefore

$$A_{nr} = (A_{nx}^2 + A_{ny}^2)^{1/2} = (A_{0x}^2 + A_{0y}^2)^{1/2}e^{-2n\pi\zeta} = A_{0r}e^{-2n\pi\zeta} \tag{7.14}$$

Then, when there is vibration energy transfer and the vibration characteristics of the structure in the two directions are the same or close, the damping ratio of the structure can be obtained by the following formula:

$$\xi = \frac{\ln(A_{0r}/A_{nr})}{2n\pi} \tag{7.15}$$

The time history curves in two directions are combined (square root of sum of squares) to obtain a synthetic displacement, as shown in Fig. 7.15. The damping ratio of the structure can be calculated from the formula deduced above to be 0.3%.

The summary information of the model's natural vibration characteristics is shown in Table 7.14, where f is the frequency and ξ is the damping ratio.

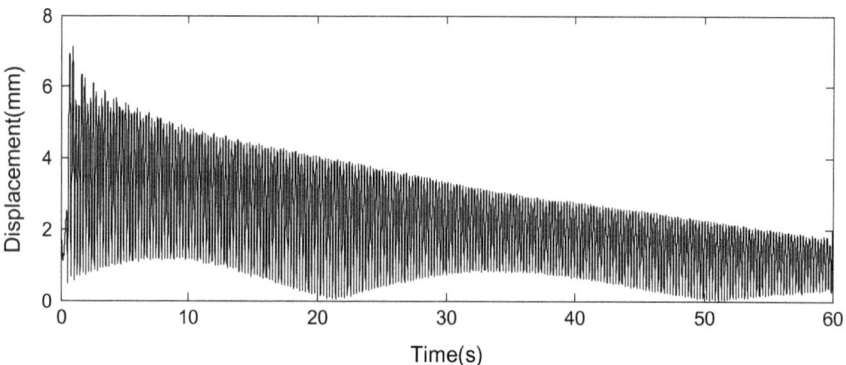

Fig. 7.15 Synthetic displacement time history curve

Table 7.14 Summary of model self-vibration characteristics

Self-vibration characteristics	Along-wind direction	Across-wind direction
f/Hz	2.16	2.17
ξ/%	0.3	0.3

7.2.2 Parametric Study

In China's national regulations on high-rise building design, acceleration and displacement peaks and R.M.S. response are important measures of structural vibration control. When assessing structural damage, it is necessary to study the effects of structural parameters on structural properties. The peak is often used to measure the maximum response of the structure. In random vibration, the R.M.S. response is usually used to represent the energy level of the random variable. In this test, the acceleration peak, the displacement peak, the acceleration R.M.S. and the displacement are used. The damper's damping performance is comprehensively investigated by multiple parameters listed above.

The acceleration and displacement time history response at the top of the model can be obtained by setting the accelerometer and laser displacement meter. Firstly, the free vibration method is used to compare and analyse the vibration properties of the main structures (controlled and uncontrolled). Secondly, the wind tunnel test is carried out, intercepting 80 s time-history data after smooth wind speed and calculating peak and R.M.S responses of displacement and acceleration. The corresponding damping effect is calculated by comparing the peak and R.M.S. responses of the controlled and uncontrolled structures.

The damping rate is defined as Eq. (7.16).

$$\frac{(\text{peak response of uncontrolled model}) - (\text{peak response of controlled model})}{(\text{peak response of uncontrolled model})} \times 100\% \quad (7.16)$$

The test studies the vibration control effect of PTMD on wind-induced vibration of highrise building from the relevant parameters, including particle filling ratio, mass ratio of the damper to structure, particle material, particle diameter, mass ratio of container to particles, tuning frequency, cavity type, wind speed and inclination angle.

The time history comparison curves of acceleration and displacement of typical uncontrolled and controlled structures are shown in Figs. 7.16 and 7.17. It can be seen from Fig. 7.16 that the peak acceleration of the controlled structure is significantly smaller than that of the uncontrolled structure. The values are attenuated to a greater extent over the entire time range. Similarly, the displacement time-history response of the controlled structure shown in Fig. 7.17, which is much smaller than that of the uncontrolled structure. These time history comparison graphs show that PTMD can effectively reduce the vibration of the structure.

Figure 7.18 compares the high-resolution plot of the five-cycle period of the controlled and uncontrolled structure time history curve. The comparison shows that under the controlled condition, the acceleration time history curve evenly distributes the high-frequency vibration curve, while under the uncontrolled condition, the curve is more 'smooth'. This is because the collisions between the particles in the PTMD and the collisions between the particles and the cavity wall is more likely to excite the higher-order modes of the structure, and the corresponding acceleration time-history

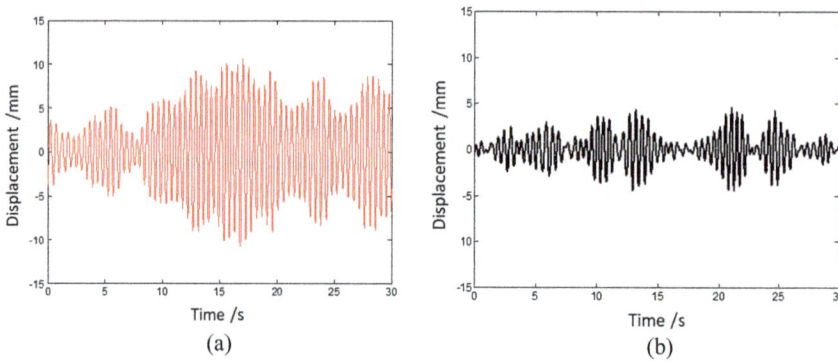

Fig. 7.16 Comparison of along-wind displacement: **a** Uncontrolled; **b** Controlled

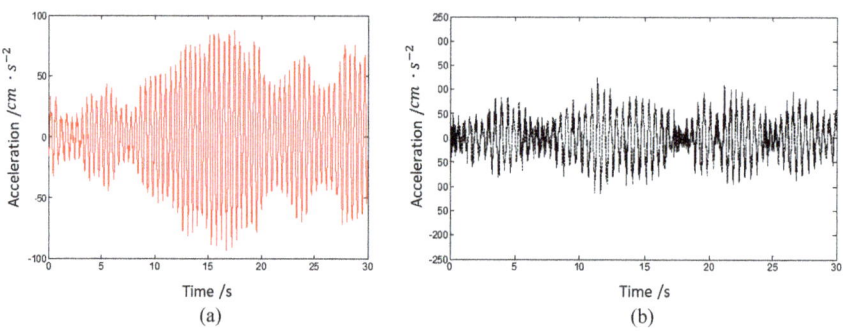

Fig. 7.17 Comparison of across-wind acceleration: **a** Uncontrolled; **b** Controlled

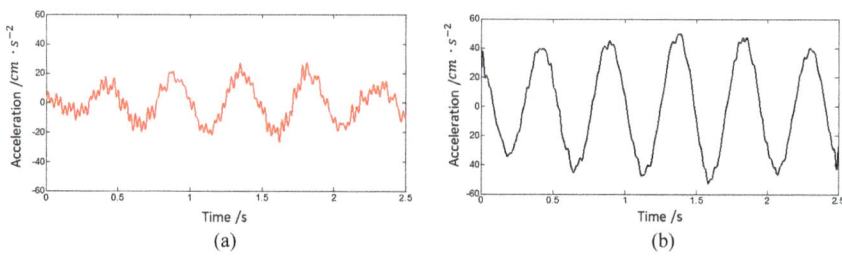

Fig. 7.18 Comparison of across-wind acceleration (5 cycles): **a** Controlled; **b** Uncontrolled

response uniformly distributes the high-order components. Furthermore, compared to impact damper, the amplitude of the high-order components is not large because the collision force between the particles is not so large that the vibration form of the structure is not substantially changed. It can be concluded that the damping effect of PTMD under wind load excitation is better than that of traditional impact dampers.

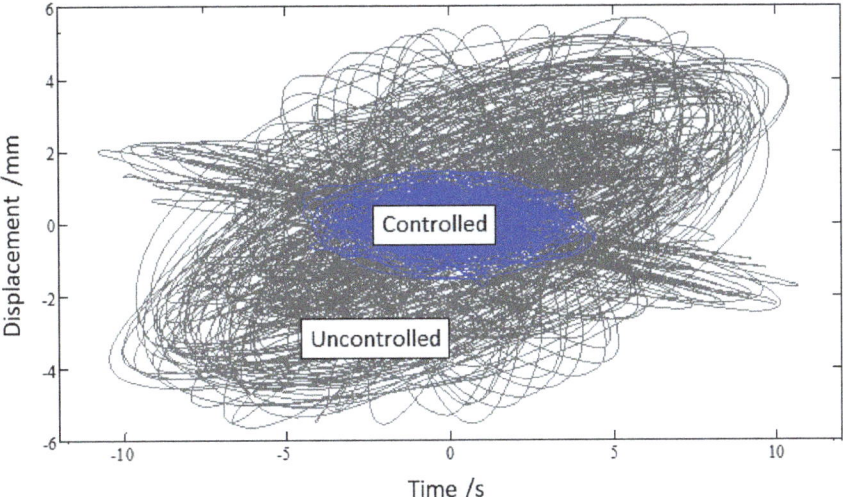

Fig. 7.19 Comparison of the trajectory of the along-wind direction and the across-wind direction of the uncontrolled structure and the controlled structure

Figure 7.19 compares the top motion trajectory of the uncontrolled structure and the controlled structure (eliminating the influence of the average wind response on the motion trajectory, only considering the influence of the pulsating wind response). From the figure, it can be seen intuitively that the response of the controlled structure is much smaller than the uncontrolled structure. Only from the aspect of amplitude, whether it is the across-wind direction or the along-wind direction, the amplitude of the uncontrolled structure is about three times that of the controlled structure, indicating that the PTMD has a good control effect on the vibration of the main structure under the wind load. In addition, the displacement of the across-wind direction under the two working conditions is about twice the displacement in the along-wind direction. This also verifies that the response of the high-rise building under the across-wind load is greater than the response of the along-wind load.

1. *Filling ratio*

The vibration control mechanism of PTMD on the structure is to reduce the vibration of the basic structure by tuning like TMD, and to dissipate the energy of the main structure by collision between particles and the cavity wall. The collisions between the particles are closely related to the number of particles and the distance between the particles and the damper cavity. An important parameter to measure these two indicators is the filling ratio of the particles.

The filling ratio is now defined as follows:

$$\text{Filling ratio} = \frac{\text{Particle amount}}{\text{Maximum amount can be reached in one layer}} \times 100\% \quad (7.17)$$

First, the tungsten carbide particles with a diameter of 6 mm are used. The mass ratio of the damper to the main structure is 1.5%. Secondly, the steel particle with a diameter of 6 mm is used. The mass ratio of the damper to the main structure is 1%. Finally, the lead particle working condition with a diameter of 2 mm is carried out, and the mass ratio of the damper to the main structure is 1%. During the test, the particle material, the suspension length, the mass ratio of the damper to the main structure, the wind tunnel wind speed (4 m/s) and the wind field class (C class) are kept unchanged, and the size and division of the damper cavity are changed. The number of layers achieves the same mass ratio and different filling ratios of the particles at the cavity size. A summary of the filling rate, damper cavity size and layer number in the test conditions is shown in Table 7.15.

(1) *Working condition of tungsten carbide particles*

When the Particle material is tungsten carbide with a diameter of 6 mm, the relationship between the PTMD damping effect and the filling ratio is shown in Fig. 7.20. It can be seen from the curve in the figure that the relationship between damping effect of the damper and the filling ratio shows an obvious "double peak" phenomenon, that is, when the filling ratio is about 50% and 90%, whether it is R.M.S. acceleration, displacement or peak response, the particle damper is better than other filling ratio conditions. This is because the change in filling ratio changes the distance between the particle and the cavity wall. As the filling ratio decreases, the distance between the particle and the cavity wall becomes longer, meaning that the particles take a longer distance to move effectively against the damper cavity wall under the same excitation. There are one or more optimal particle motion distances, so that the time it takes for the particles to move one cycle is in multiple relationship with the vibration period of the structure, which tends to have better vibration control effects.

(2) *Working conditions of steel particle*

When the particle material is a 6 mm steel particle, the relationship between the PTMD damping effect and the filling ratio is shown in Fig. 7.21. The curve in the figure also shows the 'double peak' damping law. When the filling ratio of the particles is 30 and 50% the damping effect is optimal. This agrees with the test results of the tungsten carbide particles.

(3) *Lead particle working conditions*

When the Particle material is 2 mm lead particles, the relationship between the PTMD damping effect and the filling ratio is shown in Fig. 7.22.

It can be seen from the figure that when the filling ratio is less than 100%, the optimal damping effect corresponds to a working condition with a filling ratio of 90%, which is consistent with the damping law in the tungsten carbide working condition. When the filling ratio exceeds 100%, the damping effect of the damper shows a downward trend because the stacking inside the particle damper cavity is generated, and the relative motion between the lower particles is small, resulting in small relative motion. There is no effective collision between the particles as well as

Table 7.15 Damper cavity size and filling ratio comparison table

Particle	Diameter (mm)	Density (g/cm^3)	Number of particles	Cavity size	Number of layers	Filling ratio (%)
Particle	6	14.7	167	9 cm × 9 cm	2	39.04
				8.5 cm × 8.5 cm	2	44.46
				7.5 cm × 7.5 cm	2	59.96
				9 cm × 9 cm	1	78.08
				8.5 cm × 8.5 cm	1	88.91
Particle	6	7.92	202	9 cm × 9 cm	3	31.48
				8.5 cm × 8.5 cm	3	35.85
				9 cm × 9 cm	2	47.22
				8.5 cm × 8.5 cm	2	53.77
				7.5 cm × 7.5 cm	2	72.53
				9 cm × 9 cm	1	94.44
Particle	2	11.47	3755	9 cm × 9 cm	3	64.24
				8.5 cm × 8.5 cm	3	70.71
				9 cm × 9 cm	2	96.36
				8.5 cm × 8.5 cm	2	106.06
				7.5 cm × 7.5 cm	2	145.50
				8.5 cm × 8.5 cm	1	212.14

particles and damper cavities, and only a small collision occurs between some of the particles in the upper layer, and the damping effect is limited.

However, when the filling ratio is more than 200%, the damping effect shows an upward trend. The reason is that the increase of the number of particles in the cavity causes the particle motion to appear as a whole flow state of the particle group, and the lower layer particles participate in the overall flow state, and the collision between

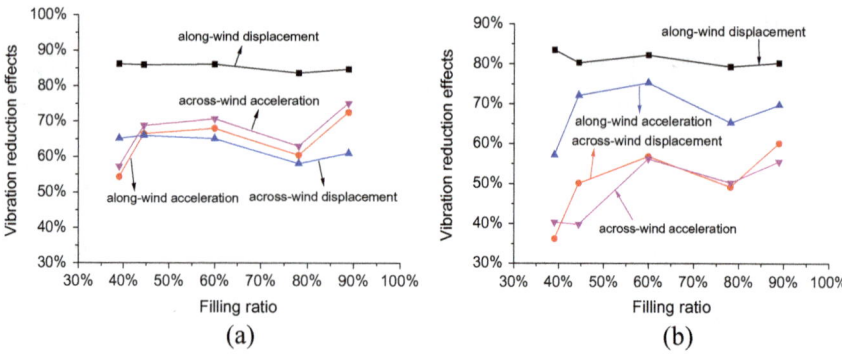

Fig. 7.20 Vibration reduction effect diagram (different filling ratio): **a** R.M.S. damping effect; **b** Peak damping effect

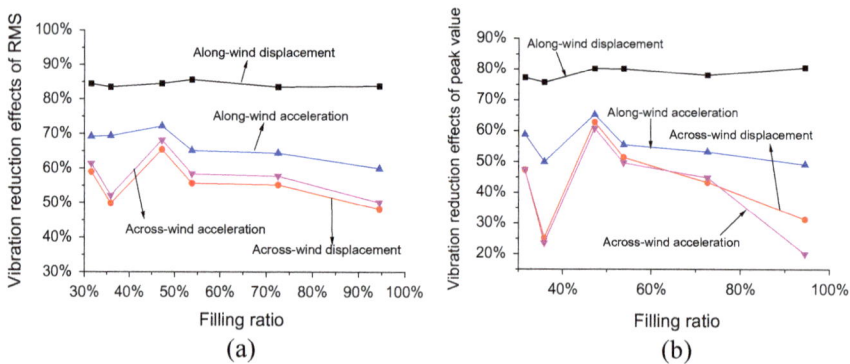

Fig. 7.21 The relationship between the damping effect and the filling ratio: **a** R.M.S. damping effect; **b** Peak damping effect

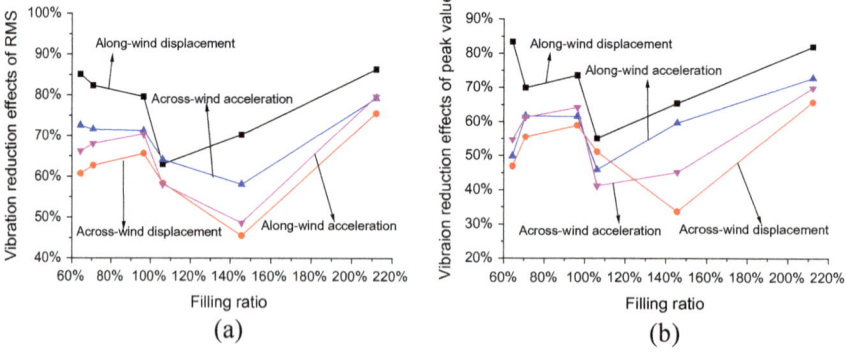

Fig. 7.22 The relationship between the damping effect and the filling ratio: **a** R.M.S. damping effect; **b** Peak damping effect

the particles and the damper cavity is more sufficient, so that a better damping effect is also produced.

Figures 7.23 and 7.24 show the comparison of the acceleration response and displacement response in the top across-wind direction with different particle filling ratios (96 and 146%). It can be seen from the figure that the damper can greatly reduce the acceleration and displacement peaks at the top of the main structure at a suitable filling ratio, and improve the damping effect of the damper.

At the same time, it can be found that the difference of the filling ratio can change the distribution and value of the structural vibration energy in different frequencies by the spectrum corresponding to the time history curve in Fig. 7.25. Under the condition of filling ratio of 146%, the peak corresponding to the spectrum is much larger than that of the filling ratio of 96%, and the spectrum of the latter is distributed

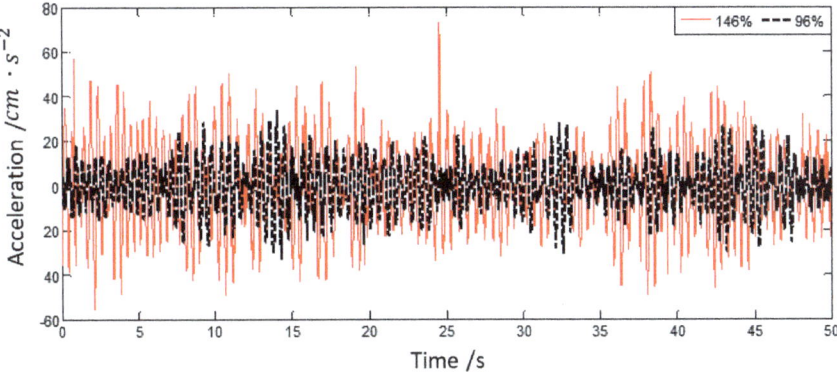

Fig. 7.23 Comparison of acceleration time history curves of damper in along-wind direction with different filling ratios

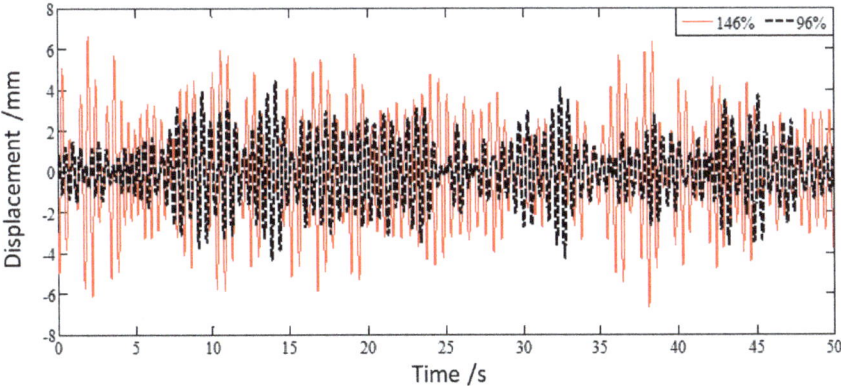

Fig. 7.24 Comparison of displacement time history curves of damper in along-wind direction with different filling ratios

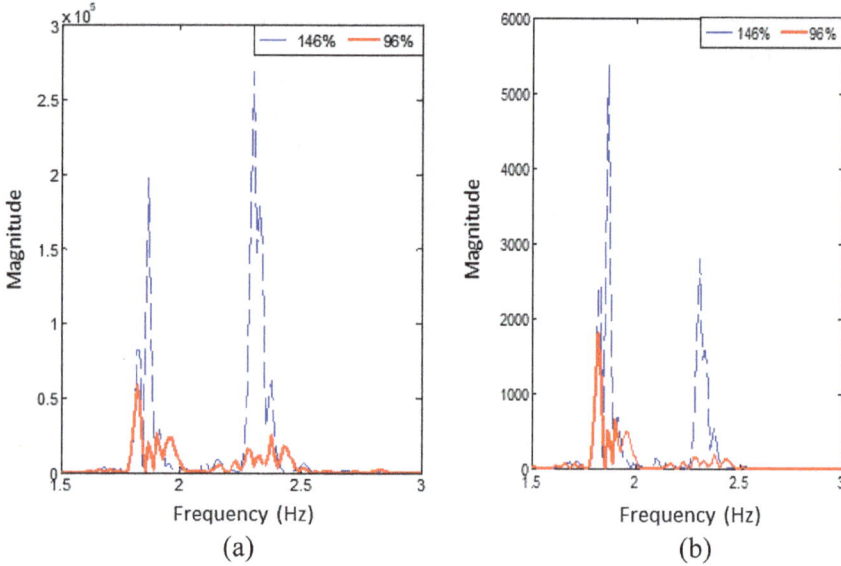

Fig. 7.25 Spectrum comparison chart at different filling ratio: **a** Acceleration response; **b** Displacement response

over a wider frequency band. It can be seen that a suitable filling ratio can significantly expand the damping band of the damper and enhance the robustness of the structure.

Through the comparative test analysis of different particle sizes and particle materials under different filling ratios, it can be concluded that when the particle filling ratio is less than 100%, there is a "multi-peak" damping between the filling ratio and the damping effect. That is, in the vicinity of the filling ratio of 30, 50, and 90%, the damping effect of the particle damper is superior to other filling ratio conditions. When the filling ratio is greater than 100%, the damping effect of the damper is worse than that of the filling ratio less than 100%, but when the filling ratio is greater than 200%, the particles in the particle damper are caused by the change of the overall movement pattern of the particles. The collision is more sufficient under the excitation of the wind load, which improves the damping effect of the damper.

It can be seen that the filling ratio has a great influence on the damping effect of the damper. The optimal particle size and filling ratio should be designed according to different working conditions, so as to achieve the effective damping effect of the particle damper.

2. *Mass ratio*

In order to investigate the effect of the mass ratio on the PTMD damping effect, the free vibration test and wind tunnel test of the main structure of PTMD with different mass ratios are carried out. Among them, the particles are made of lead with a diameter of 2 mm, and the damper cavity is a square three-layer wooden box with a side length of 9 cm, which keeps the mass of each layer of particles the same.

Fig. 7.26 Three-layer damper with a side length of 9 cm on each layer containing 2 mm lead particles of equal mass

The damper cavity and particle are arranged as shown in Fig. 7.26. Wind tunnel wind speed is 4 m/s. The mass ratio of the damper to the main structure is set to 0.5%, 0.75%, 1%, and 1.25%, respectively. The vibration damping characteristics of the damper under free vibration and the influence of wind load on the damping effect of the damper are examined.

(1) *Free vibration test*

Under the different mass ratios of the damper, the initial displacement of 1.5 cm is applied in both the along-wind direction and the across-wind direction at the top of the model. The free attenuation curves of the structure top under different working conditions is measured by the laser displacement meter, and the Gaussian formula is used to fit their envelope. The envelope diagram is shown in Fig. 7.27 (along-wind free attenuation curve) and Fig. 7.28 (across-wind free attenuation curve). From the free attenuation curve, it can be concluded that the free attenuation curve of the structure after adding damper in the across-wind direction and the along-wind direction exhibits the same characteristics with the change of the mass ratio of the damper to the main structure: that is, as the additional PTMD mass ratio increases,

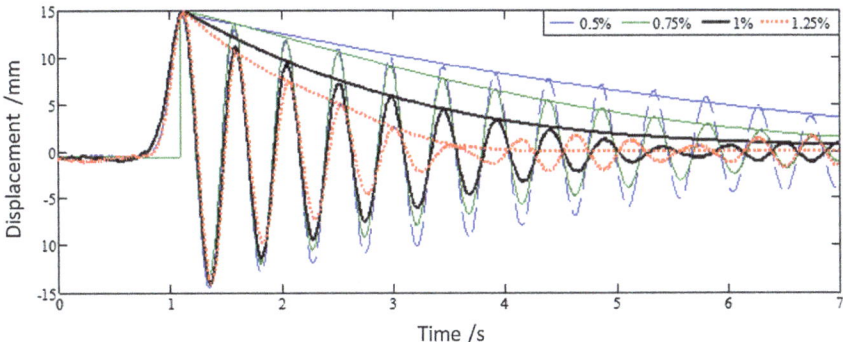

Fig. 7.27 Along-wind attenuation curve and envelope diagram of the controlled structure at different mass ratios

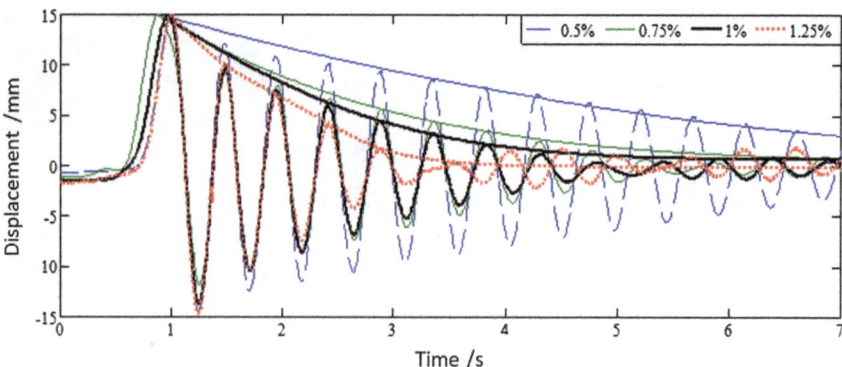

Fig. 7.28 Cross-wind attenuation curve and envelope diagram of controlled structure under different mass ratios

the displacement response of the structure can be attenuated faster in a certain period of time.

The logarithmic decay rate method is used to calculate the damping ratio of the main structure with different mass ratio dampers. Considering the additional attenuation of the structure after the adding damper does not strictly follow the logarithmic decay rate, the different segmentation starting points and end points will lead to different calculated damping ratio. In order to simplify the analysis, the logarithmic decay rate is still calculated. The first peak in the free attenuation curve is used as the starting point, and the second point is selected according to the peak size. The damping ratios of different numerical values can be calculated. Finally, the maximum value is chosen as its final damping ratio, as summarized in Table 7.16.

It can be seen from the table that as the mass ratio increases, the damping ratio also tends to increase gradually. When the mass ratio is 1.25%, the damping ratio of the structure can reach more than 10%. In the previous text, the damping ratio of the uncontrolled structure is 0.3%, and in the case of the additional mass ratio of 0.5%, the damping ratio of the main structure can reach nearly 3%, the damping ratio is increased by nearly 10 times. In the case of 1.25%, the damping ratio is increased by 30 times. It can be seen that the addition of a small mass ratio of PTMD can greatly improve the damping of the structure.

(2)   *Wind tunnel test*

The wind tunnel test of the main structure with different mass ratio is carried out. The R.M.S. and peak damping effect of each index changed with the mass ratio as shown

Table 7.16 Summary of damping ratios for PTMD systems with different mass ratios

Mass ratio (%)	0.50	0.75	1.00	1.25
Along-wind (%)	2.40	3.65	5.62	9.30
Across-wind (%)	2.78	5.83	5.84	11.01

in Fig. 7.29. The experimental results show that the acceleration and displacement response of the across-wind direction and the along-wind direction decrease with the increase of the mass ratio of the damper. The reason is that the increase of the mass ratio leads to an increase in the number of particles, and the momentum exchange between the particles and the friction dissipation increases. Therefore, the damping effect becomes better (as can be seen more clearly from the comparison of the time history curves in Figs. 7.30 and 7.31). However, the increase of the damping effect and the increase of the mass ratio are not purely linear. It can be seen that when the mass ratio increases to 1%, as the mass ratio increases, the increase of the damper damping effect is not obvious, and even decreases. This is due to the fact that when the size of the damper cavity is determined, the increase in the number of particles increases the filling ratio of the particles in the damper. The filling ratio is calculated

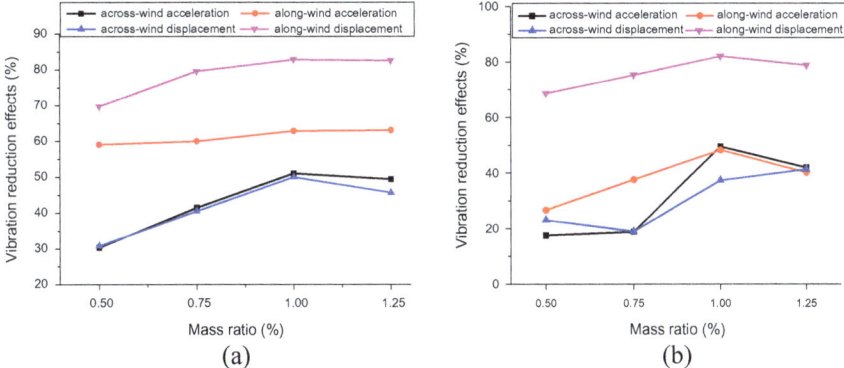

Fig. 7.29 The relationship between damping effect and mass ratio: **a** R.M.S.; **b** Peak

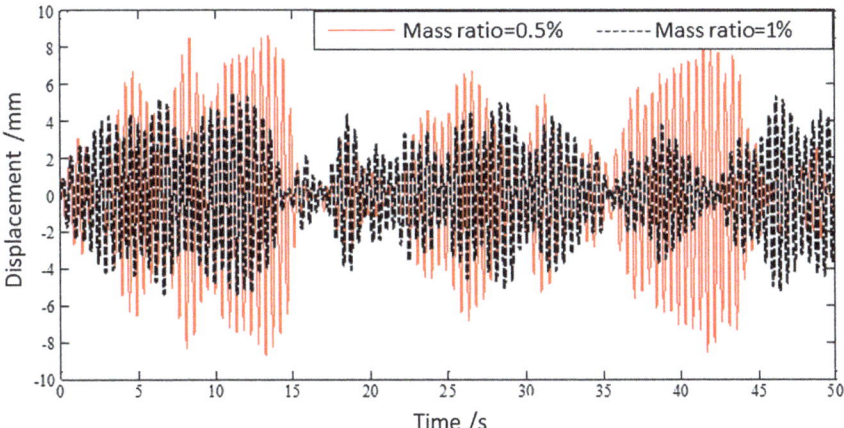

Fig. 7.30 Comparison of across-wind displacement time history curves at different mass ratios (0.5 and 1%)

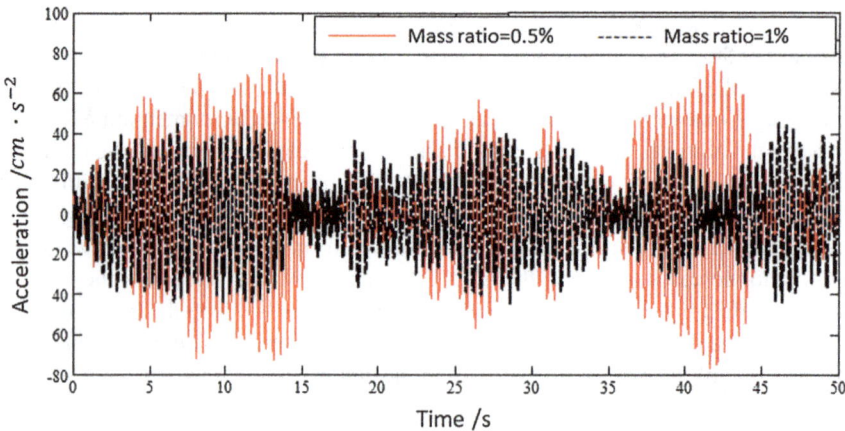

Fig. 7.31 Comparison of across-wind acceleration time history curves at different mass ratios (0.5 and 1%)

Table 7.17 Damper cavity size and filling ratio comparison table

Particle	Diameter (mm)	Density (g/cm³)	Cavity	Mass ratio (%)	Number of particles	Filling ratio (%)
Lead	2	11.47	9 cm × 9 cm, 3 layers	0.5	1783	30.34
				0.75	2779	47.29
				1	3775	64.24
				1.25	4771	81.20

as shown in Table 7.17. When the mass ratio is 1.25%, the filling ratio is 81.20%, and according to the conclusion in Sect. 7.2.2, the corresponding filling ratio is not the optimal filling ratio of the damper, but the corresponding filling ratio is relatively good when the mass ratio is 1%.

Another interesting phenomenon is that the increase in damping performance of the damper is limited after the particle has increased to a certain extent. This phenomenon can be explained by the principle of conservation of momentum between the particle and the system. When the mass of the particles increases, the absolute velocity and relative velocity of the particles after collision with the container cavity become smaller, and the time for completing a collision in the container becomes longer. When the mass of the particles reaches a certain value, the collision occurs, the speed cannot overcome friction and causes it to move to the other end to create the next collision. At the same time, the particles will go in the opposite under the action of friction, resulting in no collision between the particles and the container [12].

It can be seen from the spectrum in Fig. 7.32 that the vibration energy is concentrated at the same frequency at different mass ratios, but the amplitude is different. Whether it is the amplitude in the acceleration or displacement spectrum, the energy

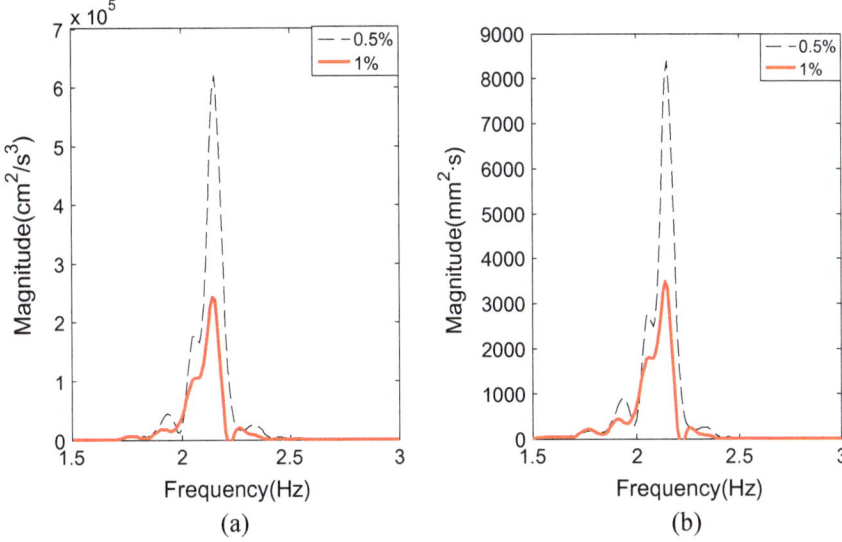

Fig. 7.32 Power spectral density under different mass ratios (0.5 and 1%): **a** Acceleration response; **b** Displacement response

of damper with larger mass ratio (1%) is lower than smaller mass ratio (0.5%). This means that an increase in the mass ratio can transfer more vibration energy to the damper without changing the width of the damping band.

Therefore, when designing PTMD, on the one hand, it is necessary to select the appropriate mass ratio so that the vibration damping effect is optimal under the same conditions. On the other hand, it is necessary to balance the requirements of the building space, and design a reasonable damper cavity size to make the damper of a specific mass ratio has the optimal filling ratio.

3. *Particle materials*

In order to investigate the effect of particle material in PTMD on its damping effect, the free vibration test and wind tunnel test of PTMD under different particle materials are carried out. The diameter of the control particle material is 6 mm, and the damper cavity is a two-layer square wooden box of 9 cm × 9 cm, and the wind speed of the wind tunnel is 4 m/s. The effect of particulate materials on PTMD damping performance is achieved by varying the different materials (steel, copper, lead and tungsten carbide) at the same mass ratio (1%).

(1) *Free vibration test*

Figure 7.33 shows the top free-attenuation response curve and the fitted envelope of the damper with two different density particle materials (copper particles and steel particles), respectively. It can be seen that under the same initial displacement, steel particle condition can make the peak of the structure attenuate faster than the copper particle condition, which indicates that the PTMD of the additional steel particles

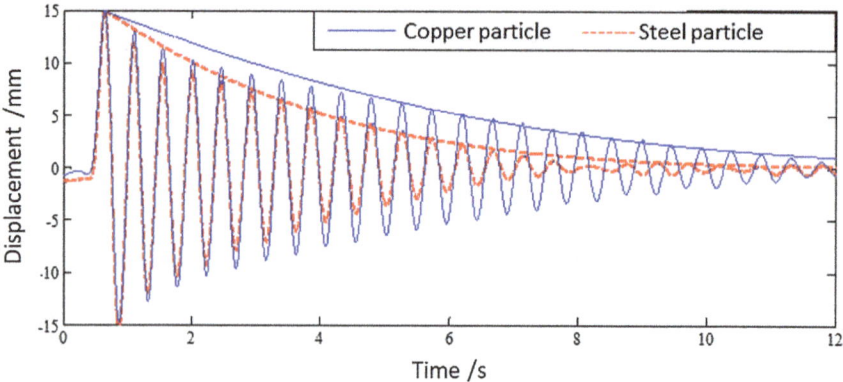

Fig. 7.33 Free attenuation curve and envelope diagram of dampers with different particle materials (steel and copper)

has better damping effect than the equal mass of the copper particle with higher density. For the structural free attenuation curve under the two working conditions, the damping ratio is calculated by the logarithmic attenuation rate in Sect. 7.2.2. When the particle material is steel, the damping ratio is 0.0425, and the damping ratio in the copper particle condition is 0.0236. It can be seen that in the case of the same mass ratio, different particulate materials have a great influence on the damping ratio of the structure.

(2) *Wind tunnel test*

The relationship between the R.M.S. and peak damping effect of the across-wind and the along-wind direction and the mass ratio is shown in Fig. 7.34. The experimental results show that the damping effect of the peak response is more affected by the particle material than the R.M.S. response. (The time history response curves in Figs. 7.35 and 7.36 can better interpretate this phenomenon) PTMD has good damp-

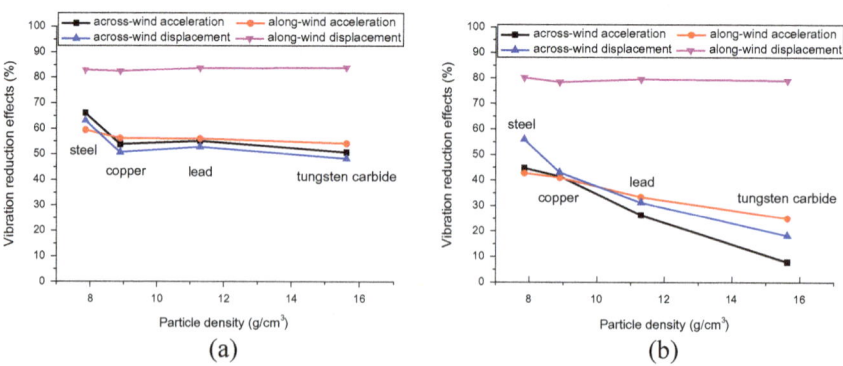

Fig. 7.34 Relationship between Vibration reduction effect and particle density: **a** R.M.S.; **b** Peak

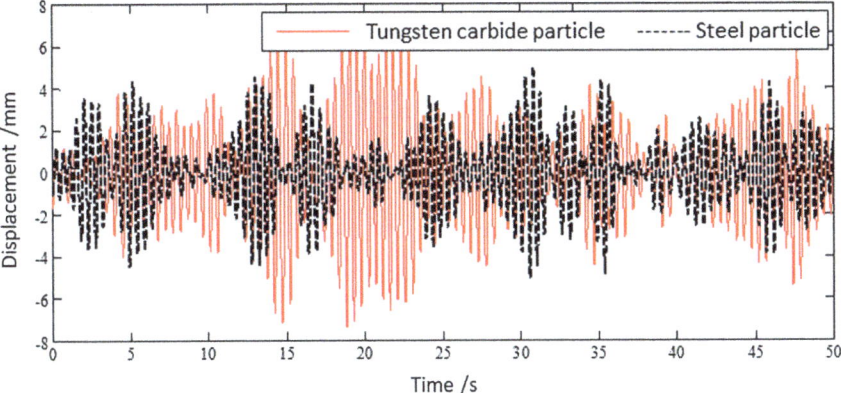

Fig. 7.35 Across-wind displacement time history curve with different Particle materials (tungsten carbide and steel particles)

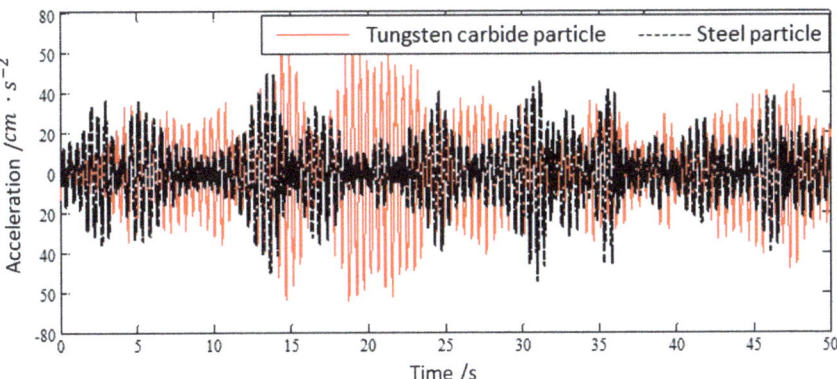

Fig. 7.36 Across-wind acceleration time history curve with different Particle materials (tungsten carbide and steel particles)

ing effect (48–84%), and the damping effect increases with the decrease of particle density, which correspond to researches on non-obstructive particle damper (NOPD) free vibration test of Zhao Ling, etc. [13]. The reason is that within a reasonable filling ratio range, as the particle density decreases, the number of particles increases. The increase in particle density causes the filling ratio of particles inside the damper (as shown in Table 7.18) to increase, and the number of particle collisions increases accordingly. The momentum exchange is increased, and the energy dissipated by the collision between the particles and the friction is increased, so that the vibration damping effect is improved. It can be concluded from Table 7.18 that the corresponding filling ratio in the four cases is within the "multi-peak" optimal damping frequency range in Sect. 7.2.2, which eliminates the effect of different filling ratio. However, the damping effect of lead particles is slightly better than that of copper

Table 7.18 Comparison of damper cavity dimensions and filling ratio

Diameter (mm)	Cavity	Particle material	Density (g/cm³)	Number of particles	Filling ratio (%)
6	9 cm × 9 cm, single layer	Steel	7.92	202	94.44
		Copper	8.40	191	89.30
		Lead	11.47	140	65.46
		Tungsten carbide	14.70	109	50.96

particles with lower density, because the particles of the four materials have oscillated and fully collided under external excitation, while the lead particles have rougher surface than other three materials. It effectively upgrades the frictional energy consumption when dissipating energy through collision. This test phenomenon is also shown in the literature [14].

Figures 7.35 and 7.36 show the time history curves of the top displacement and acceleration of the model when the wind speed is stable under the conditions of steel and tungsten carbide. It can be seen from the comparison that the displacement and acceleration response of the structure under steel conditions are significantly smaller than those of tungsten carbide particles, and the corresponding self-power spectrum (Fig. 7.37) also shows the peak value of conditions with the tungsten carbide particles is much larger than the condition where the particle material is steel. At the same time, the vibration of tungsten carbide particles is mainly concentrated

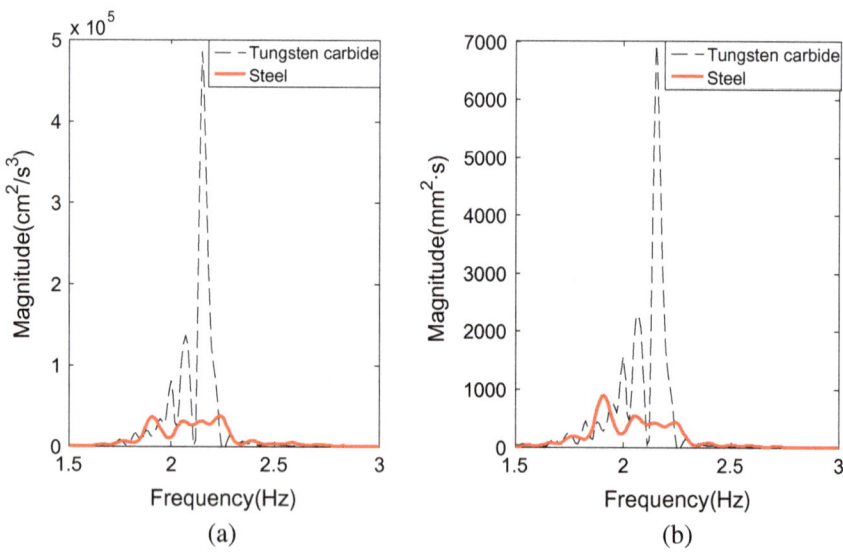

Fig. 7.37 Spectrum comparison of different Particle materials (tungsten carbide and steel): **a** Acceleration response; **b** Displacement response

in the fundamental frequency of the structure, while the structural vibration energy is distributed in a wider frequency band under steel particle conditions. According to the above analysis, appropriately reducing the density of particles in the PTMD system can improve the damping performance of the damper. At the same time, the particle damper of the small-density particles can distribute the vibration energy of the main structure over a wider frequency band, which has a strong vibration control robustness.

4. *Particle diameter*

In order to investigate the effect of particle diameter on the damping performance of PTMD, the same conditions are set. (The mass ratio is 1%. The particle material is lead. The damper cavity is 9 cm × 9 cm single-layer box. The same suspension length is 5 cm. The wind speed is the same. Six types of particles with different diameter are chosen, which are 2, 3, 4, 6, 8 and 10 mm.

The damping effect of the R.M.S. at different particle diameters is shown in Table 7.19. Observing the data in the table, it can be found that the damper has different damping effects at different diameters. For across-wind acceleration and displacement, along-wind acceleration, the damping effect of the damper ranges from 50% to 70%, while the along-wind displacement ranges from 75% to 85%. For a single indicator, the damping effect of the damper does not have a regularity with the change of the particle diameter. In this test, the across-wind acceleration and displacement have been controlled. It can be seen that the vibration damping effect is best when the diameter is 2 mm, and the vibration damping effect is the worst when 10 mm, which does not mean that the vibration is reduced with the increase of the particle size. Because the damping effect at 4 mm and 8 mm is better than the diameter of 3 mm and 6 mm respectively. At the same time, Figs. 7.38 and 7.39 respectively compare the along-wind direction displacement and acceleration time history curves with the particle materials of 3 and 8 mm. The comparison curves from the figure can not clearly show the advantages and disadvantages of the two damping effects. Moreover, the peaks are not much different.

Therefore, it can be concluded that the damper exhibits good damping effect at different particle sizes, but the damping effect is not directly related to the particle diameter, which correspond with relevant research of Marhadi [15].

Table 7.19 Summary of vibration reduction effects under different particle diameters

Particle diameter (mm)	2	3	4	6	8	10
Across-wind acceleration (%)	70.5	64.6	68.4	54.4	60.4	55.2
Along-wind acceleration (%)	63.5	70.8	69.8	63.0	64.5	59.6
Across-wind displacement (%)	66.2	60.9	65.7	51.9	57.1	51.8
Along-wind displacement (%)	75.2	83.4	84.2	82.7	81.9	82.7

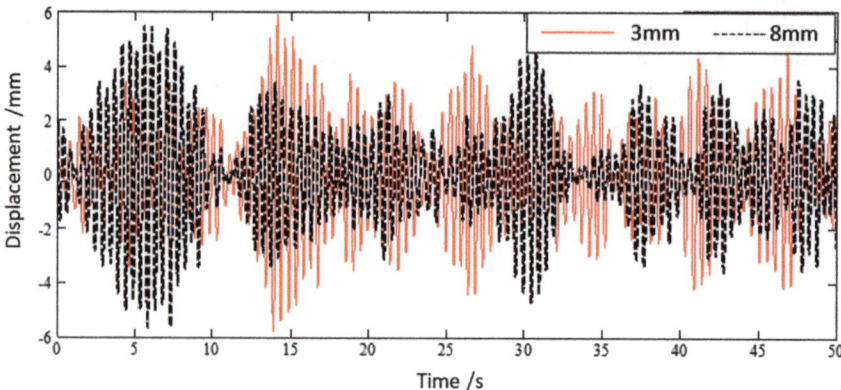

Fig. 7.38 Comparison of time history of along-wind direction displacement of structures with different particle diameters (3 and 8 mm)

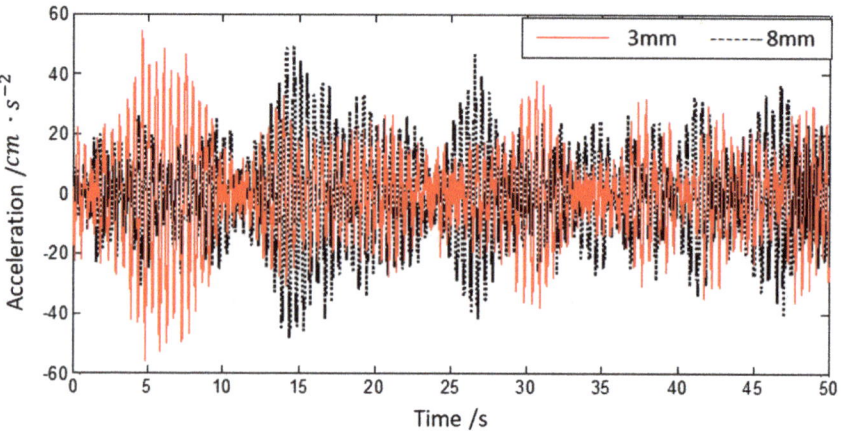

Fig. 7.39 Comparison of time history of across-wind acceleration with different particle diameters (3 and 8 mm)

5. *Mass ratio of container to particles*

In order to investigate the effect of mass ratio of container to particles on PTMD damping performance, by controlling the mass ratio of the damper to the main structure (1%), the mass ratio of container to particles is adjusted by reducing the mass of the particles and adding weight. The mass ratios are: 0.056, 0.25, 0.5, 1, 2, 4 and infinity. Of these, 0.056 is the original mass ratio of the cavity to the particles in the damper. Other parameters (lead particles of 2 mm, three-layer square wooden box with a damper cavity of 9 cm × 9 cm, a suspension length of 5 cm, and a wind tunnel wind speed of 4 m/s) remained unchanged during the test. The vibration control law of the main structure of PTMD under different mass ratios of container to particles is investigated by free vibration test and wind tunnel test.

(1) *Free vibration test*

From the across-wind direction and the along-wind free attenuation time history curve and its envelope diagram (Figs. 7.40 and 7.41), it can be seen that as the mass ratio of the cavity to particles in the damper decreases, the displacement response at the top of the main structure attenuate faster. At the same time, the along-wind direction and the across-wind direction have similar attenuation laws, and it is also proved that the structure has similar attenuation characteristics in two directions.

Figure 7.42 shows the frequency spectrum corresponding to the two directions of free attenuation curves. The frequency spectrum of the along-wind and across-wind directions show different attenuation characteristics. It can be seen that when

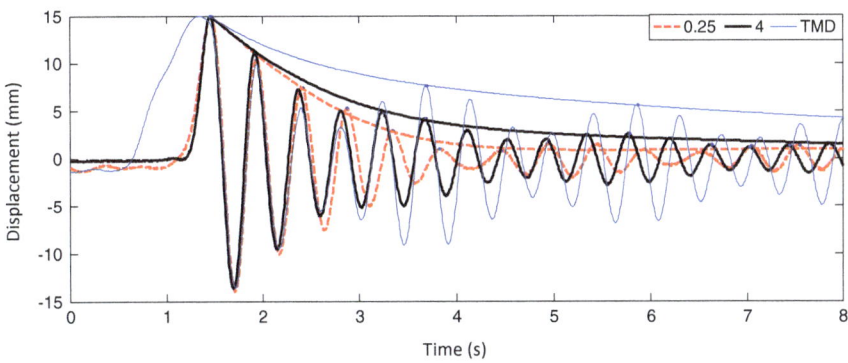

Fig. 7.40 Free attenuation curve and envelope diagram of the along-wind direction of the main structure under different mass ratio of container to particles

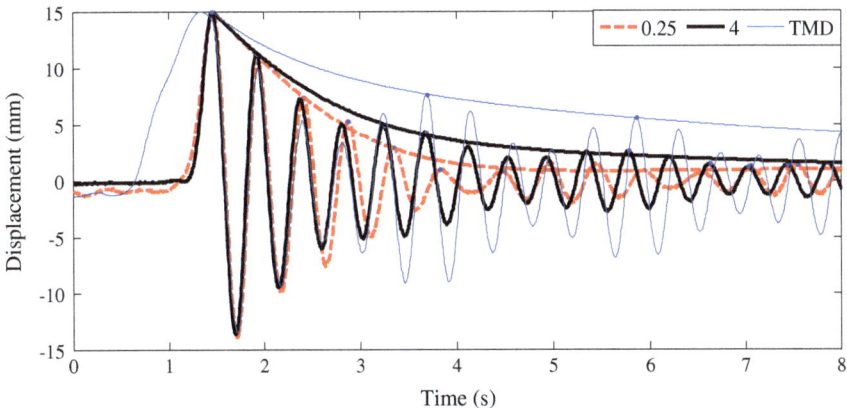

Fig. 7.41 Free wind attenuation curve and envelope diagram of the across-wind direction of the main structure under different mass ratio of container to particles

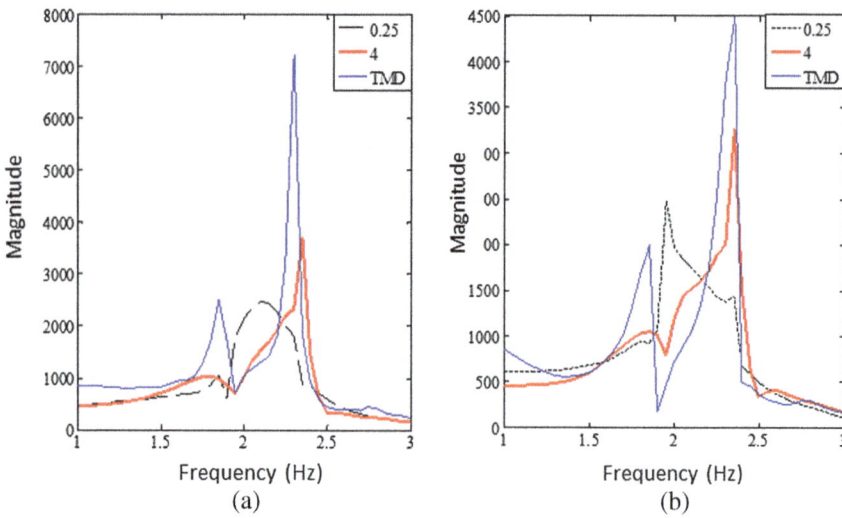

Fig. 7.42 Spectrum comparison of different mass ratio of container to particless (0.25, 4 and TMD):
a Along-wind direction; **b** Across-wind direction

the mass ratio of the damper cavity to the particle is infinite (TMD), the frequency
spectrum has two peaks, and the peak value of the main structure frequency is much
larger than the peak corresponding to the damper. With the decrease of mass ratio of
container to particles, the distribution width of the frequency spectrum is gradually
widened, and the amplitude of the peak is gradually decreased. When the mass ratio
is 0.25, the amplitude is the minimum value. It can be seen that in the case of free
vibration, as the mass ratio of container to particles is reduced, the damping band of
the damper can be effectively widened, and the vibration energy of the main structure
can be effectively reduced.

Table 7.20 lists the damping ratios of the PTMD structures in the along-wind
and across-wind free attenuation curves for different conditions. The results show
that under different mass ratio of container to particles, the damping ratio of the
particles existing in the damper cavity is greater than the damping ratio under the
TMD condition. As the ratio increases, the damping ratio of the structure tends to
decrease overall, but in the case of ratios of 2 and 4, the damping ratio increases. The
reason is that the damping performance of PTMD is not only affected by the mass of
the cavity, but also closely related to the filling ratio of the particles inside the damper.

Table 7.20 Summary of free attenuation calculated damping ratio of PTMD systems at different
mass ratios

Mass ratio of container to particles	0.056	0.25	0.5	1	2	4	TMD
Along-wind (%)	7.88	6.50	6.15	5.89	11.27	6.04	2.07
Across-wind (%)	7.94	8.30	6.23	5.98	7.31	5.96	1.75

Table 7.21 Damper cavity size and filling ratio

Particle	Diameter (mm)	Density (g/cm³)	Cavity	Total mass ratio (%)	Mass ratio of container to particles	Number of particles	Filling ratio (%)
Lead	2	11.47	9 cm × 9 cm 3 layers	1	0.056	3775	96.36
					0.25	3187	81.35
					0.5	2656	67.80
					1	1992	50.85
					2	1328	33.90
					4	797	20.35

When the mass ratio of container to particles is 2, there exists the optimum filling ratio that allows the additional PTMD structure to reach the maximum damping ratio.

The relationship between damper cavity size and filling ratio is shown in Table 7.21. It can be seen that in the design of PTMD system, it is not enough to consider the influence of single parameter. It is also necessary to integrate the influence of various parameters on the damping effect of the damper, clarify the relationship between the parameters, and distinguish the primary and secondary, which helps the designed PTMD system to achieve optimal damping control.

(2) *Wind tunnel test*

Figure 7.43 shows the variation trend of the damper wind vibration control effect as the mass ratio of container to particles changes. It can be seen from the figure that as the ratio increases, both the R.M.S. damping effect and the peak value are generally decreasing, and the damping effect is the worst when there is no particle. The reason is that PTMD not only reduces the vibration of the basic structure by means of tuning, but also through the collisions and friction between the particles and the damper to

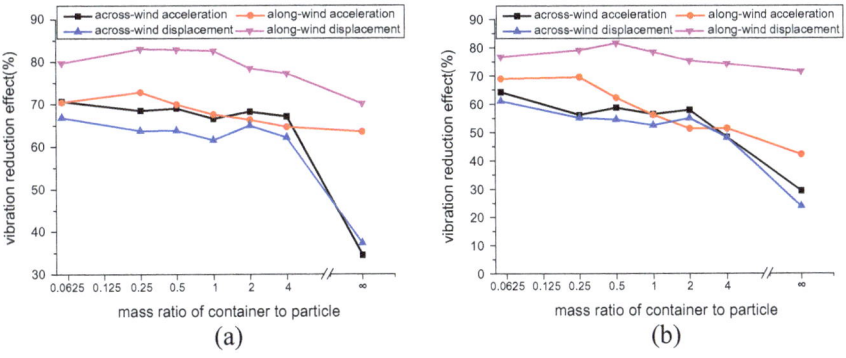

Fig. 7.43 Relationship between damping effect and mass ratio of container to particles: **a** R.M.S.; **b** Peak

consume the vibrational energy of the structure. When the mass ratio of container to particles is reduced, it means that the number of particles increases, and the increased particles increase the energy consumption of the damper by collision and friction. However, under the overall downward trend, the vibration damping effect of some working conditions is better than the working condition with smaller mass ratio, which is the same as the trend of the damping ratio calculated by free attentuation. The time history comparison curves of different damper cavities and particle mass ratios are shown in Figs. 7.44 and 7.45. The damping effect of the damper (both the acceleration and the displacement) after the increase of the particles can be greatly improved.

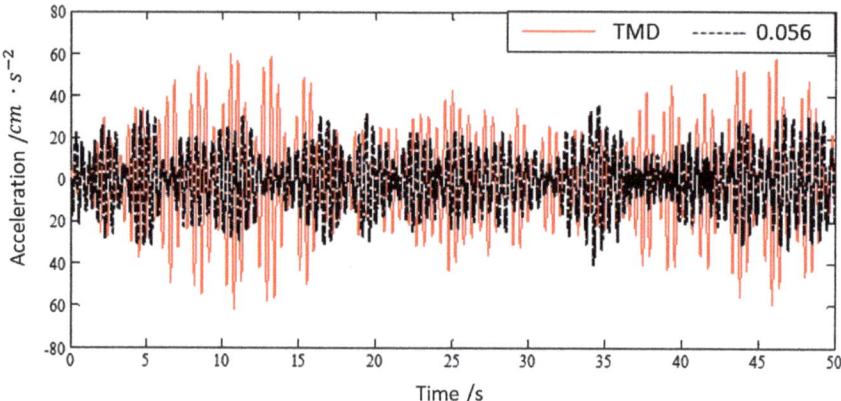

Fig. 7.44 Comparison of across-wind acceleration time history of different cavity mass ratios (0.056 and TMD)

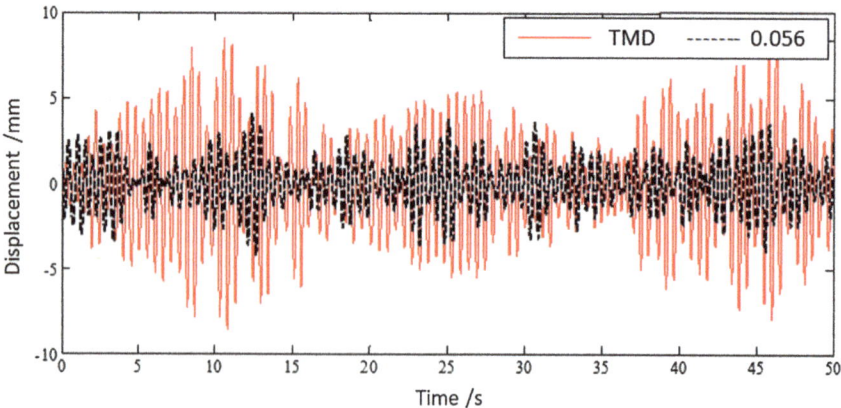

Fig. 7.45 Comparison of time along-wind direction displacement history of different mass ratio of container to particless (0.056 and TMD)

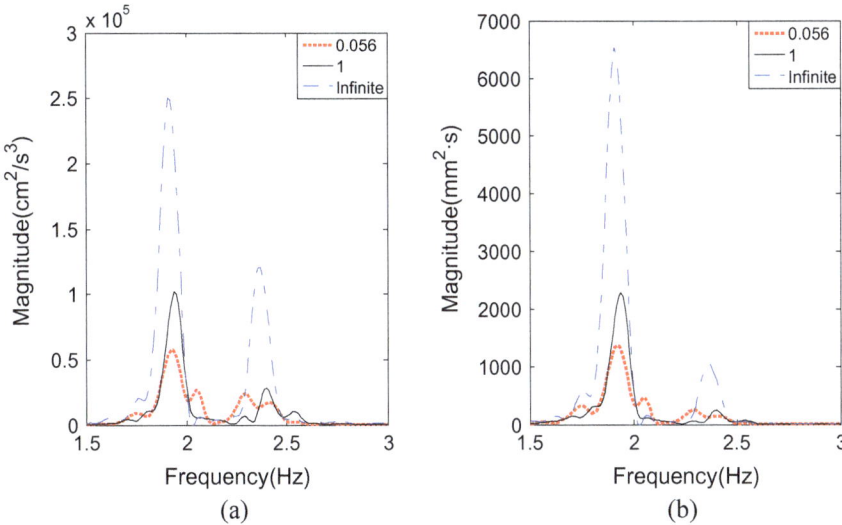

Fig. 7.46 Spectrum comparison of different particle mass ratios (0.25, 4 and TMD) in different wind vibration: **a** Along-wind direction; **b** Across-wind direction

Figure 7.46 shows the self-power spectrum of the structural vibration response (acceleration and displacement) for different mass ratio of container to particles (0.056, 1 and infinity). When the ratio is infinite, the corresponding amplitude is the largest. As the corresponding peak decreases, the bandwidth increases correspondingly, indicating that the increase of the particles can transfer more vibration energy of main structures to the damper, and reduce the vibration energy of the structure over a wider frequency band. It is consistent with the relevant conclusions of the free vibration curve of the additional PTMD structure. The above analysis shows that increasing the mass of the particles can effectively reduce the vibration of the structure without changing the overall mass of the damper.

In fact, when there is no particle, the mass of PTMD is concentrated on the mass of the damper cavity. At this time, it degenerates into the traditional TMD. The factors affecting the damping effect of TMD mainly include the tuning frequency and the optimal damping ratio. Since the optimal damping ratio cannot be accurately controlled during the test, only the tuning frequency of the damper is strictly controlled to be the same as the first-order natural frequency of the structure. In this test, PTMD is better than TMD to some extent. It is foreseeable that it has broad application prospects in the field of civil engineering vibration control. In engineering design, the additional mass should be concentrated on the moving particles as much as possible on the basis of tuning.

Combined with test results of the main structure of PTMD with different damper cavity and particle mass ratio in the free attenuation test and wind tunnel, it can be concluded that there exsits an optimal condition when the mass ratio of the damper to the main structure is constant that the control of the damper reach the maximum value.

Through reasonable calculation and optimization of the parameters, the damping effect of the PTMD can be maximized under the same mass ratio of damper to the structure

6. *Tuning frequency*

Under normal use conditions, the natural frequency of the building tends to change greatly with changes in the surrounding environment (such as temperature, humidity, heavy rain and strong wind). The traditional TMD can have a good damping effect in a specific frequency range, but when the natural frequency of the main structure changes, the damping law becomes unstable. On the other hand, PTMD can not only reduce the vibration of the structure by tuning, but also consume the vibration energy of the structure by collision between particles, friction and collision and friction between the particles and the cavity wall of the damper. Therefore, it is of great practical significance to study the damping effect of PTMD without tuning.

The different tuning frequencies of the damper can be achieved by adjusting the suspension length of the damper. When the suspension length is 4, 5, 6 and 7 cm, the natural frequencies of the corresponding dampers are 2.49, 2.29, 2.03 and 1.88 Hz. The damper particle material (tungsten carbide), mass ratio (1%), particle diameter (6 mm), damper cavity size (9 cm × 9 cm, 2 layers) and wind speed (4 m/s) remain unchanged.

(1) *Free vibration test*

Through the attenuation curve in the free vibration test and the fitted envelope diagram (Figs. 7.47 and 7.48), it can be clearly seen that the damping effect of PTMD is optimal under tuning, when the damper self-vibration frequency deviates from the tuning frequency about 20%, the damping effect is worse than that of tuning, but still has good damping effect. The damping ratio of the main structure with PTMD system under free attenuation conditions is calculated. The damping ratio is 7.94%

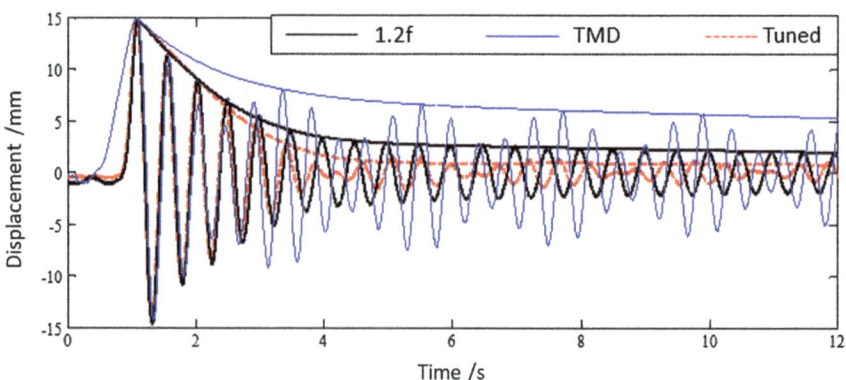

Fig. 7.47 Free attenuation curve and its envelope diagram under different tuning frequencies and TMD conditions

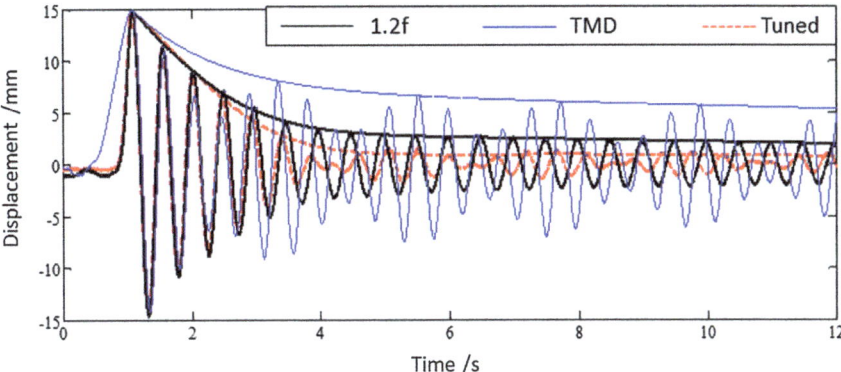

Fig. 7.48 Across-wind direction free attenuation curve and its envelope diagram under different tuning frequencies and TMD conditions

under tuned conditions, 4.43% under 1.2*f* untuned conditions, and 1.75% under TMD conditions. PTMD has a good damping effect under the condition of imbalance.

At the same time, the control effect of TMD on free vibration damping is also compared, shown in Fig. 7.49. It can be found that PTMD is still better than the tuned TMD under untuned conditions. It can be seen that PTMD has good robustness and vibration damping stability. The high-rise building will deviate from the designed natural vibration frequency in actual application, which brings certain limitations

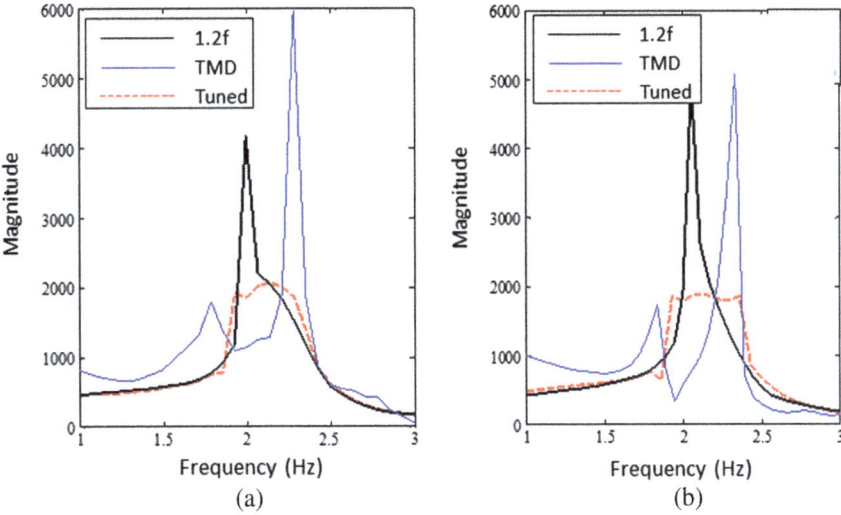

Fig. 7.49 Spectrum comparison chart for different tuning frequencies and TMD conditions under free vibration: **a** Along-wind direction; **b** Across-wind direction

Table 7.22 R.M.S. damping effect under different suspension lengths

Suspension length	4 cm (2.49 Hz)	5 cm (2.29 Hz)	6 cm (2.03 Hz)	7 cm (1.88 Hz)
Across-wind acceleration (%)	45.00	48.83	67.72	61.73
Along-wind acceleration (%)	45.03	54.08	60.90	60.19
Across-wind displacement (%)	40.89	46.96	65.97	64.70
Along-wind displacement (%)	77.49	83.76	86.06	76.30

to the application of TMD which can only have good damping effect in a specific frequency range. The good damping effect of PTMD under untuned conditions shows that it has a good application prospect in civil engineering.

Comparing the spectrum corresponding to the free vibration curve, it can be found that the amplitude under the TMD condition is larger, followed by the PTMD which deviates from the tuning frequency by 20%, and finally the tuned PTMD. It can be seen that PTMD can effectively reduce the energy of the main structure, and the vibration damping effect is better than TMD under the condition of untuned. At the same time, the tuned PTMD condition has the widest damping band, followed by the untuned PTMD, and finally TMD. It can be seen that PTMD can effectively broaden the damping band of the damper and enhance the robustness.

(2) *Wind tunnel test*

The experimental results are shown in Table 7.22. The optimal damping effect occurs under the condition of a suspension length of 6 cm (61–86%) because the additional damper reduces the natural vibration frequency while increasing the mass of the main structure. More importantly, under other suspension lengths (untuned conditions), the damper still maintains good damping control, indicating that PTMD has good robustness. The damping effect of the damper under all working conditions varies from 41% to 86%.

Figure 7.50 shows the relationship between the damping effect of the across-wind response and the frequency ratio of the damper to the main structure. It can be found that when the ratio is around 1, the damping effect is optimal. For the deviation from the tuning frequency, the damping effect is relatively poor under the optimal tuning frequency, but still has good vibration control effect. The reason is that tuning is only part of the damper damping control approach. Other vibrational energy can also be dissipated by nonlinear collisions between the particles inside the damper or between the particles and the damper. These energy dissipation paths can improve the damping stability.

Because the amplitude of the uncontrolled structure is much larger than the amplitude of the controlled structure, only the spectrum response of some uncontrolled structures is intercepted in the frequency spectrum curve. Through the comparative

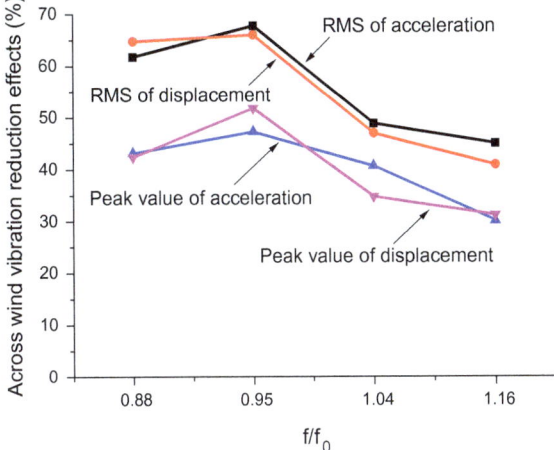

Fig. 7.50 The relationship between the damping effect of the across-wind response and the frequency ratio of the damper to the main structure

analysis of the spectrum in Fig. 7.51, the lowest amplitude is required for the 6 cm condition, and the widest frequency band occurs under the conditions of the suspension length of 5 and 6 cm. As the difference between the damper frequency and

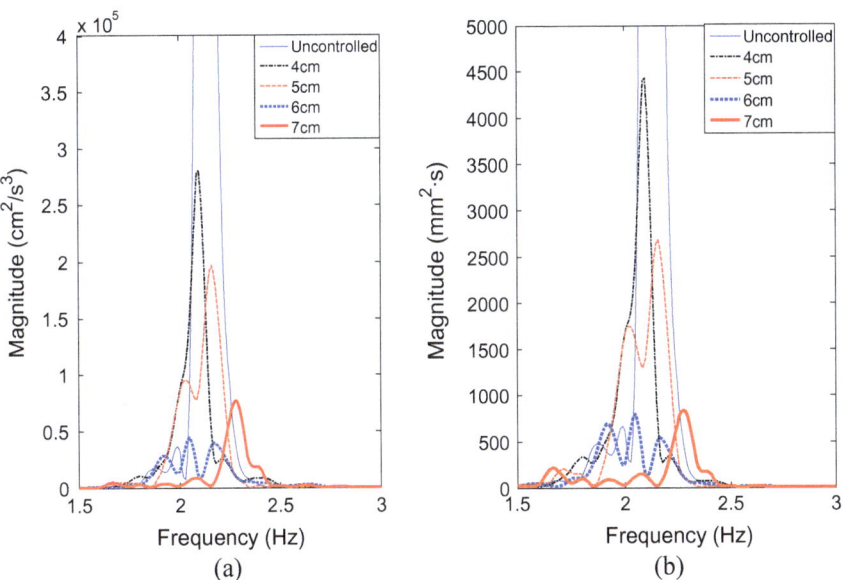

Fig. 7.51 Across-wind direction response frequency spectrum under different suspension lengths: **a** Acceleration response; **b** Displacement response

the frequency of the main structure is reduced, the corresponding damping band is gradually widened. The above analysis shows that PTMD still has good damping effect even under untuned conditions, which verifies its good robustness and good engineering application prospects in civil engineering.

7. *Cavity type*

PTMD can reduce the vibration of the structure by tuning like TMD, and can reduce the vibration energy of the main structure by friction and collision between the particles inside the cavity. The collisions between the particles and between the particles and the cavity wall are closely related to the division of the damper cavity. If the damper has multiple divisions, there is a large collision probability between the particles and the cavity wall, otherwise, the probability of collision is correspondingly reduced. In order to investigate the influence of the different division modes of the damper cavity on the damping effect of the damper energy dissipation, the cavity with the same bottom area is divided into different ways: no division, 4-square division ('+' shaped) and 9-square division ('#' shaped), and the size of each square in cavities are the same. A free vibration test and a corresponding wind tunnel test are performed.

(1) *Free vibration test*

Considering the influence of different division modes of damper cavity on the damping performance of PTMD, the same conditions are set (The mass ratio is 1%. The particle materials are all lead with a diameter of 2 mm. The damper cavity is 9 cm × 9 cm double layer. The particles are evenly distributed in the two layers. The suspension length is 5 cm. The wind speed is 4 m/s). Three division modes are tested.

For the model with PTMD, the same initial displacement is applied in different directions. Under different working conditions, the free attenuation curve of the along-wind displacement and its envelope comparison are shown in Fig. 7.52.

By comparison, it can be found that the PTMD with different division modes of the damper cavity will have different effects on the free attenuation of the main

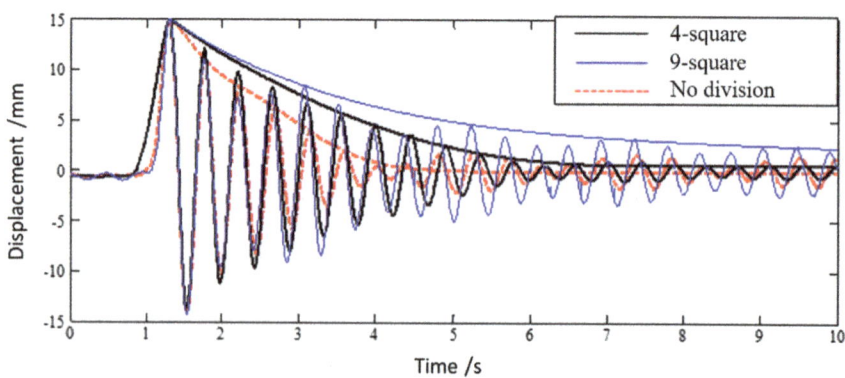

Fig. 7.52 Comparison of along-wind direction free attenuation curves

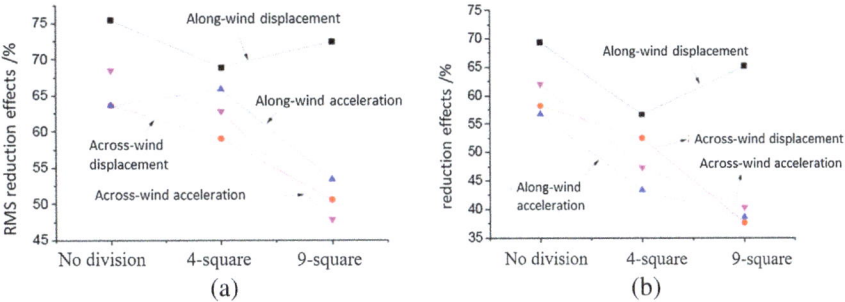

Fig. 7.53 Variation of wind-induced vibration control effect with different damper cavity division mode: **a** R.M.S.; **b** Peak value

structure. When the damper cavity type is no division, the suppression of the free attenuation of the main structure is the best. The worst case of vibration occurs when the damper cavity is in 9-square division. The damping effect of the 4-square division is somewhere in between. The damping ratio calculation results show that damper of no division is 7.94%, damper with the 4-square division is 4.34%, and the damper with 9-square division is 2.98%. It can be seen that different division modes of the damper cavity can change the collision mechanism between particles and between the particles and the cavity wall. Therefore, in the PTMD design process, a reasonable division mode should be adopted to maximize the damping effect of the damper.

(2)   *Wind tunnel* test—the *particle material is lead with a diameter of 2* mm

Under the action of wind load, the variation of the vibration control law of the main structure with the PTMD system of different division modes is shown in Fig. 7.53. It can be seen from the figure that the model of PTMD with different division modes exhibits the same damping effect as the free attenuated vibration under wind load. It can be found that the across-wind displacement and acceleration of the structure, the along-wind acceleration decreases with the increase of divisions, and the worst case of the damping effect for the along-wind displacement occurs at the 9-square division. The reason is that when the divisions are increased, the internal filling ratio is correspondingly increased. At the same time, different filling ratios have different effects on the damping effect of the damper. When there is no division, the filling ratio of the particles inside the damper cavity is 96.36%. When the 4-square-divided cavity is used, the increase of the damper cavity wall will increase the filling ratio of the damper cavity, and the filling ratio of the particles is 98.63%. When the 9-square-divided cavity is used, the filling ratio of the damper cavity is 100.84%. As the divisions increase, more particles collide with the walls of the damper cavity and the collisions between the particles are relatively reduced. At the same time, the reduction in the distance between the walls of the cavity reduces the time it takes for the particles to move back and forth for one cycle, and the strength of the collision between the particles and the walls of the cavity is reduced. By comparison, it can be found that for the 2 mm lead particles, when filling ratio is large in no division conditions, the increase of the divisions will lead to an increase in the filling ratio, and even there is no relative displacement between the particles, which reduces the damping effect of the PTMD system.

It can be seen from the along-wind direction displacement and acceleration response of structure with PTMD of different damper cavity divisions, the displacement and acceleration peaks in the no division condition are significantly smaller than the 9-square division conditions (Figs. 7.54 and 7.55). At the same time, the corresponding self-power spectrum is given in Fig. 7.56. It can be seen that the amplitude of the 9-square division condition is much larger than that of the no division condition, and the frequency band distribution is wider under no division conditions. It shows that the effect of vibration control of no division condition is better than that of 9-square division condition.

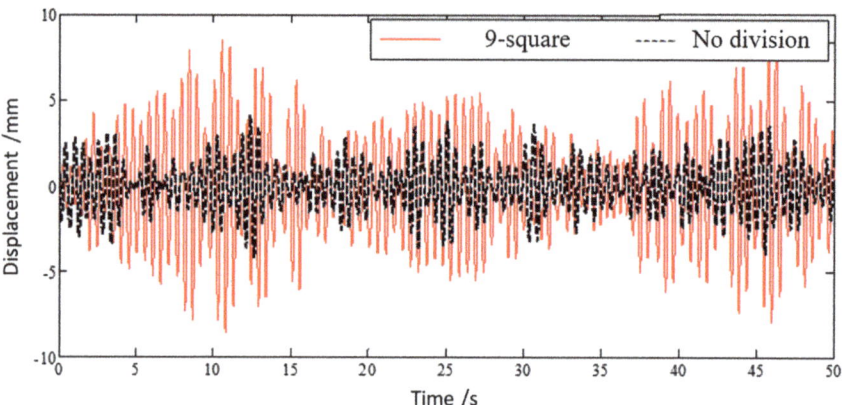

Fig. 7.54 Comparison of along-wind direction displacement time history of different cavity division modes

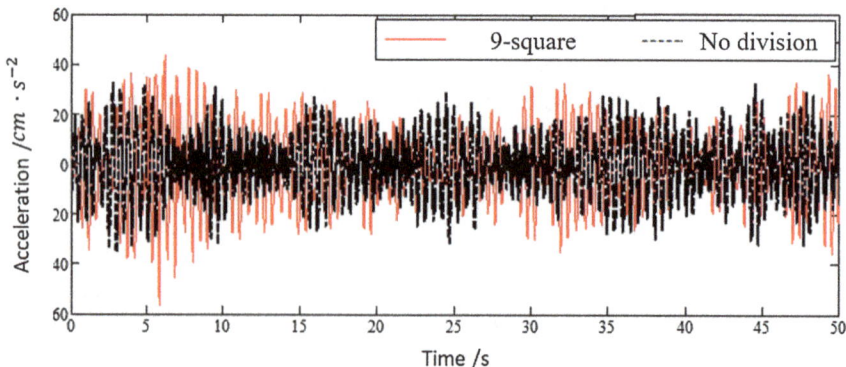

Fig. 7.55 Comparison of along-wind acceleration time history of different cavity division modes

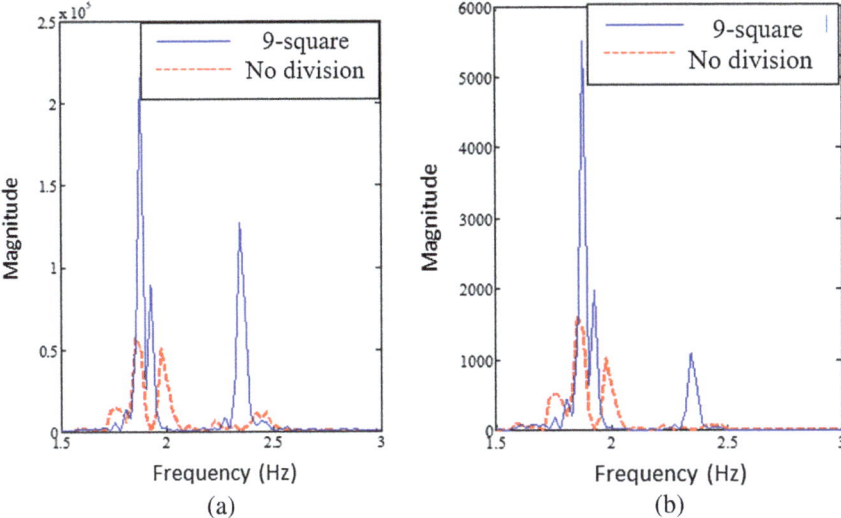

Fig. 7.56 Across-wind direction frequency response spectrum for different cavity division modes: **a** Acceleration response; **b** Displacement response

(3) *Wind tunnel test—the particle material is tungsten carbide with a diameter of 6 mm*

In order to further investigate the effect of different damper division modes on its damping effect, in this test, the working condition of additional tungsten carbide particles with a diameter of 6 mm is carried out. The mass ratio is 1%, and the damper cavity is 9 cm × 9 cm double-layer box. The particles are evenly distributed between the two layers, the suspension length is 5 cm, the wind speed is 4 m/s, and there are three different division modes.

Under the wind load, the variation of the vibration control law of the main structure with the PTMD system of different division modes is shown in Fig. 7.57. When the particles are 6 mm tungsten carbide, the vibration damping effect is optimal under the 9-square division condition, and the vibration damping effect is the worst under the 4-square division.

It can be seen from the displacement and the acceleration response of the structure top with PTMD of different division modes (Figs. 7.58 and 7.59) that the displacement and acceleration peaks of 9-square division are significantly smaller than those of no division. At the same time, the corresponding self-power spectrum is given in Fig. 7.60. It can be seen that the amplitude of the no division condition is much larger than the 9-square division condition. It shows that under such working conditions, the effect of vibration control of no division is worse than that of 9-square division.

It can be seen that adding appropriate divisions to the damper cavity can increase the damping effect of the damper. But it is essential to select a reasonable particle type and size, otherwise it will have the opposite effect.

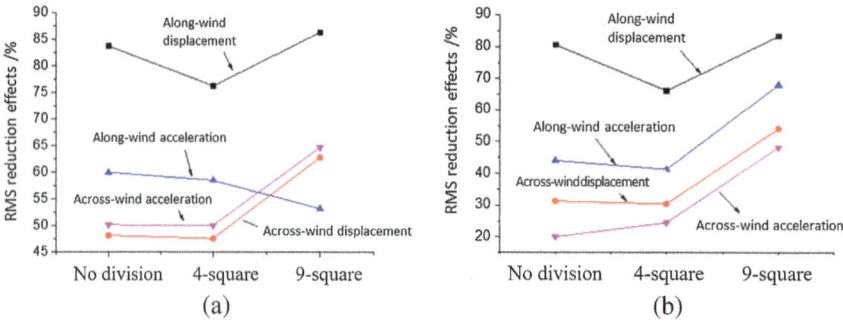

Fig. 7.57 Variation of wind-induced vibration control effect with different damper cavity division mode: **a** R.M.S.; **b** Peak

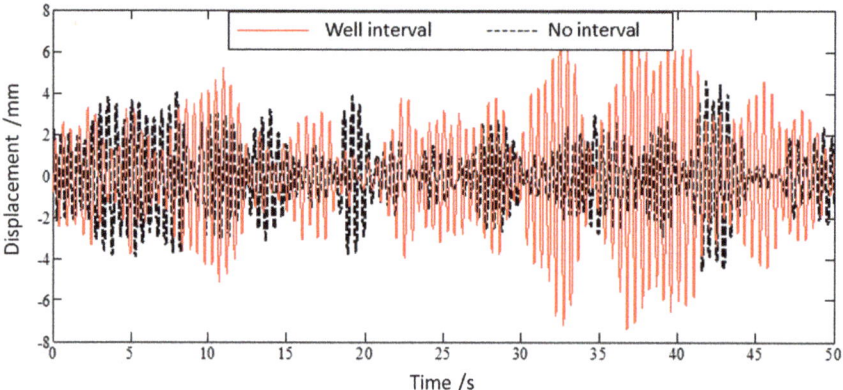

Fig. 7.58 Comparison of displacement time history curves of the along-wind direction of the top structure when the particles are tungsten carbide with a diameter of 6 mm

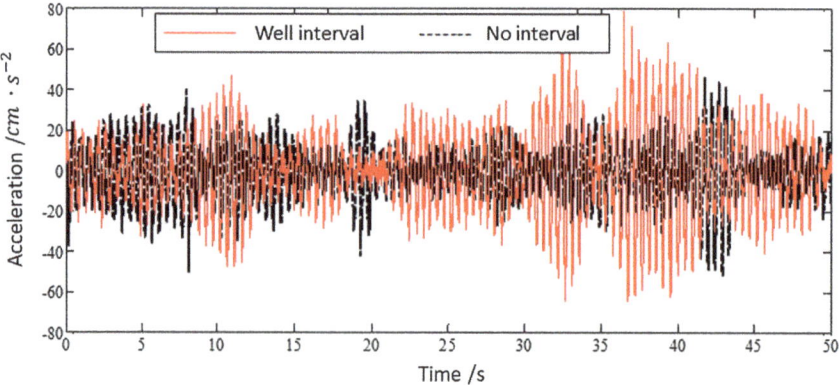

Fig. 7.59 Comparison of the acceleration time history curve of the along-wind direction of the top structure when the particles are 6 mm tungsten carbide

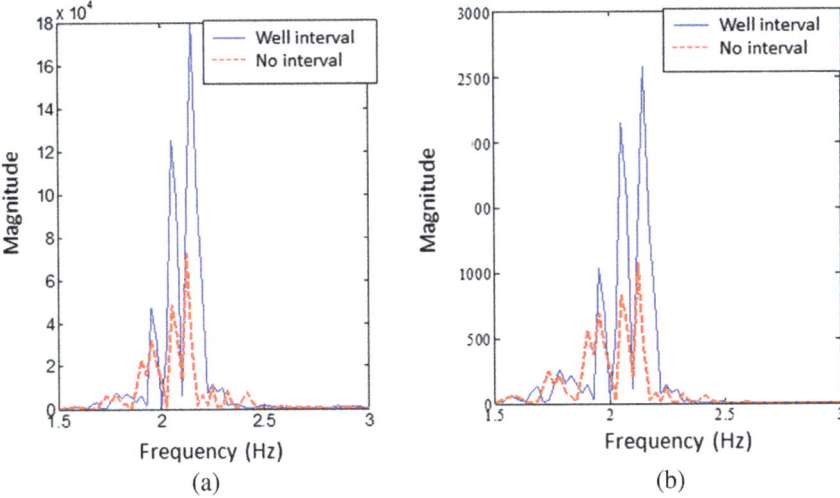

Fig. 7.60 The frequency spectrum of along-wind direction response of different cavity division modes when the particles are 6 mm tungsten carbide: **a** Acceleration response; **b** Displacement response

7.2.3   Effects of Wind Field

1. *Wind speed*

During the actual use of the building, especially for tall buildings, the ultimate wind load experienced by the building is often greater than the design wind load. Therefore, in the damper performance test, the influence of the most unfavorable wind load and the damping effect of the damper under the ultimate wind load should be considered.

Lead particles with a diameter of 2 mm are selected. The damper cavity is a square double-layer wooden box with a side length of 7.5 cm, which keeps the quality of each layer of particles the same. The mass ratio of the damper to the main structure is 1%, which is set in the wind tunnel. The wind speeds are 3 m/s, 3.5 m/s, 4 m/s, 4.5 m/s, and 5 m/s, respectively. The variation of the damping effect of the damper with the wind speed is examined.

The variation of R.M.S. and peak damping effect of each index with wind speed are shown in Fig. 7.61. The results show that the damping effect of PTMD increases with the increase of wind speed, because the increase of wind speed is intensified the vibration of the model, which increases the effective momentum exchange between the particles inside the damper and the energy dissipated by the friction, and the damping effect becomes better. It can be seen that the damping effect of the damper varies nonlinearly with the wind speed, and the degree of improvement of the damping effect decreases with the increase of the wind speed, which is due to the more and more sufficient collisions between the particles with the increase of the wind speed.

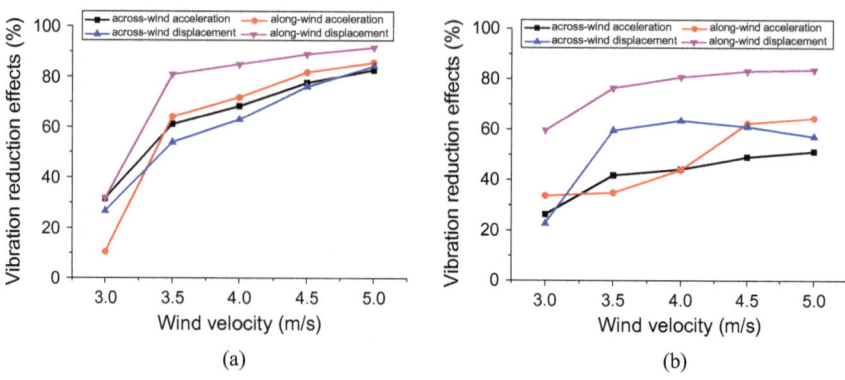

Fig. 7.61 Vibration control effects at different wind speeds: **a** R.M.S.; **b** Peak

When the collision is sufficient, the rate at which the energy is dissipated by the particle collision is gradually reduced, resulting in a gentle curve.

2. *Inclination angle*

Lead particles with a diameter of 2 mm is selected. The damper cavity is a square double-layer wooden box with a side length of 9 cm. The mass of each layer is the same, the mass ratio of the damper to the main structure is 1%, and the wind speed is 4 m/s, setting the inclination angle gradient as 15°. Since the model is a central symmetrical structure, the inclination angle varies from 0 to 90° to simulate all wind load conditions that the model may encounter.

In order to investigate the vibration characteristics of the model under different inclination angles, the vibration test of the uncontrolled structure under different inclination angles is first carried out. The variation of the R.M.S. response with the inclination angle is shown in Fig. 7.62. The results show that: The model has

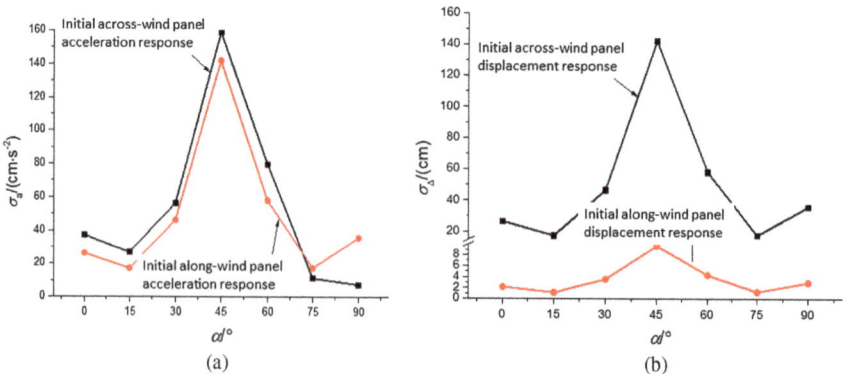

Fig. 7.62 Uncontrolled structural response with different inclination angle: **a** R.M.S. acceleration; **b** R.M.S. displacement

the largest vibration response at 45 degrees of inclination angle, and the vibration response at other angles is roughly symmetrically distributed at 45°. When the inclination angle is 15° and 75°, the vibration response of the model is minimal. The reason can be explained as: under the inclination angle of 45°, the wind receiving area of the model reaches the maximum, therefore the vibration response is the largest. When inclination angle is 15° or 75°, although the wind area of the model is not much different from 0° and 90°, the vibration of the model is also related to the shape of the wind, and the wind turbulence will lead to the vortex generated on the side of the model. The size of the vortex and the frequency of vortex shedding with time are directly related to the shape of the model. Different shedding frequencies will have different effects on the vibration of the model. When the shedding frequency is close to or equal to the natural vibration frequency of the model, resonance occurs, causing the model to generate large vibrations. When the two frequencies are far apart, the vibration response of the model is relatively small. At 0 and 90° of inclination angle, the vortex shedding frequency of the model is [16]:

$$v = n_s D/S_t \tag{7.18}$$

Among them, n_s is the vortex shedding frequency (Hz), D is the projected feature size of the object perpendicular to the average velocity plane, D = 21 cm in this experiment. S_t is the Storaja number, which is approximately 0.12. The vortex shedding frequency is 2.17 Hz (the natural frequency of the model), and the incoming wind speed is 3.8 m/s. The resonance lock zone is 1–1.3 v. Therefore, the vortex-induced resonance wind speed ranges from 3.8 m/s to 4.9 m/s. It can be seen that vortex-induced resonance occurs at this angle.

The PTMD's damping effect with the inclination angle is shown in Fig. 7.63b. It can be seen that the damper has different damping effects at different inclination angles. At 45° inclination angle, the structural vibration is the most severe. At this time, the damping effect of the damper is optimal, and the effect of wind speed on

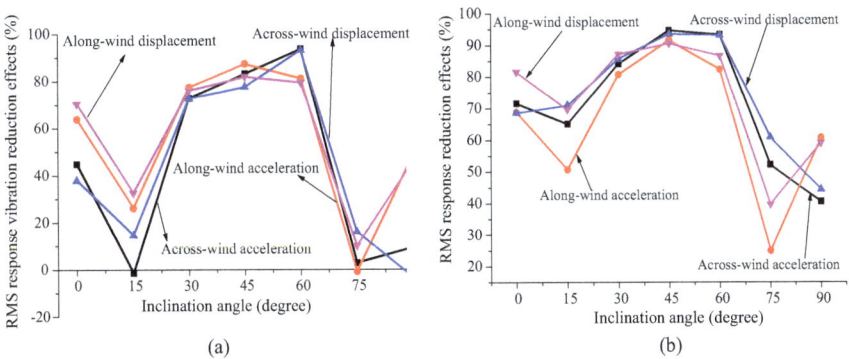

Fig. 7.63 Relationship between R.M.S. damping effect and inclination angle: **a** Additional TMD; **b** Additional PTMD

the damping effect of the damper is the same, that is, the more severe the structural vibrate, the better the damping effect of the damper is. Although the damping effect is different under different inclination angles, the damper has a good damping effect (20–90%) in the whole range of angle variation.

The damping effect of the two dampers with the inclination angle is shown in Fig. 7.63. It can be seen that the damper has different damping effects under different inclination angles, and the trend of the damping effect of the two dampers is basically consistent with the structural response with the inclination angle in Fig. 7.62. At 45° inclination angle, the structural vibration is the most severe. At this time, the damping effect of the damper is optimal, indicating that the severer structural vibrates, the better damping effect of the damper has. From the perspective of R.M.S. vibration reduction effect, the particle damper has a certain damping effect under different inclination angles, and the vibration response amplification phenomenon sometimes occurs in TMD.

Table 7.23 lists the damping advantages of particle dampers over TMD under the same conditions. The damping advantage is defined as:

$$\text{Damping advantage} = \frac{\text{TMD working condition} - \text{PTMD working condition}}{\text{TMD working condition}} \times 100\% \tag{7.19}$$

It can be seen that in most cases, the damping effect of the particle damper is better than TMD.

7.3 Numerical Simulation of Wind Tunnel Test

7.3.1 Brief Description of Simplified Simulation Method

1. *TMD simulation*

Based on the equivalent simplification principle and model in Chap. 3, the relevant equivalent theory can be derived to find the expression of d on particle density and mass and filling ratio:

The total volume of the particles is:

$$V_{pd} = m/\rho_{pd} \tag{7.20}$$

Among them, ρ_{pd} is the density of the particles;
The remaining volume in the damper cavity is:

$$V_{epd} = V_a - V_{pd} = d_x d_y d_z - m/\rho_{pd} \tag{7.21}$$

Table 7.23 Summary of the vibration response of the two-damper structure and the damping advantage of the particle damper compared to TMD

Inclination angle		0°	15°	30°	45°	60°	75°	90°
Across-wind direction acceleration	Additional tuned mass damper (cm/s²)	10.8	9.79	8.75	8.15	5.68	5.19	4.72
	Additional particle damper (cm/s²)	21.6	29.96	15.11	27.65	5.48	11.31	7.11
	Damping advantage (%)	50	67	42	71	−4	54	34
Along-wind acceleration	Additional tuned mass damper (cm/s²)	8.13	8.47	9.06	11.92	11.29	12.93	14.47
	Additional particle damper (cm/s²)	16.57	13.1	10.86	18.58	11.55	17.69	17.99
	Damping advantage (%)	51	35	17	36	2	27	20
Across-wind direction displacement	Additional tuned mass damper (mm)	1.44	0.95	0.94	1.15	0.70	0.49	0.50
	Additional particle damper (mm)	3.07	2.86	1.77	4.18	0.71	1.13	0.9
	Damping advantage (%)	53	67	47	72	1	57	44
Along-wind displacement	Additional tuned mass damper (mm)	0.42	0.36	0.49	0.91	0.65	0.79	1.27
	Additional particle damper (mm)	0.68	0.83	0.91	1.81	0.92	1.19	1.64
	Damping advantage (%)	38	57	46	50	29	34	23

Among them, V_a is the total volume of the damper cavity; d_x, d_y, d_z is the length, width and height of the damper cavity, respectively, and assumed $d_z = D_p$, where D_p is the diameter of the particles; $V_a = d_x d_y d_z = 1/4\pi D^2(D+d) = m/(\rho_{pd}\rho_v)$. The diameter of the equivalent single particle is:

$$D = (6m/\pi\rho_{pd})^{1/3} \tag{7.22}$$

The parameter d in the equivalent damper cavity length is:

$$d = 4(\pi\rho_{pd}/6m)^{2/3}(m/\rho_{pd}\rho_v - 3m/2\rho_{pd}) \tag{7.23}$$

Therefore, the relevant simplified parameters in the particle damping system can be solved according to the above calculation method.

2. *Main structure simulation*

The main structural model of the wind tunnel test with additional PTMD is shown in Fig. 7.64. Because of the symmetry, small damping ratio and uniform distribution quality, the main structure can be simplified to cantilever beams with n degrees of freedom. PTMD is attached to the top of the model and can be simplified to the $(n+1)$th degree of freedom of the main structure. Its mass, damping and stiffness are m_c, c_c and k_c, respectively. Therefore, the whole system can be regarded as a multi-degree-of-freedom system with $(n+1)$ degrees of freedom, as shown in Fig. 7.64. The simplification of the collision force between the particle and the damper cavity wall can be applied as an external impact load to the $n+1$th degree of freedom of the simplified structure.

Based on the above simplified analysis, the motion expression of the structure with PTMD can be expressed as Eq. (7.24):

$$\begin{cases} \mathbf{M}\ddot{x} + \mathbf{C}\dot{x} + \mathbf{K}x = \mathbf{W}(t) + \varphi(c_c\dot{z}_1 + k_c z_1) \\ m_c\ddot{x}_c + c_c\dot{z}_1 + k_c z_1 - c_p H(z_2, \dot{z}_2) - k_p G(z_2) = 0 \\ m_p\ddot{x}_p + c_p H(z_2, \dot{z}_2) + k_p G(z_2) = 0 \end{cases} \tag{7.24}$$

$$\mathbf{M} = \text{diag}[\, m_1 \; m_2 \; \ldots \; m_n \,] \tag{7.25}$$

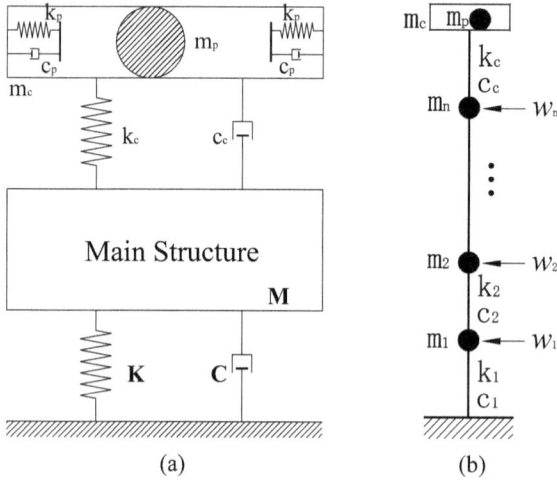

Fig. 7.64 Test modeland simplified schematic diagram with PTMD: **a** Main structure; **b** Wind load and main structure simplified model

$$\mathbf{C} = \begin{bmatrix} c_1 + c_2 & -c_2 & & & \\ -c_2 & c_2 + c_3 & -c_3 & & \\ & -c_3 & \cdots & & \\ & & & \cdots & -c_n \\ & & & -c_n & c_n \end{bmatrix} \tag{7.26}$$

$$\mathbf{K} = \begin{bmatrix} k_1 + k_2 & -k_2 & & & \\ -k_2 & k_2 + k_3 & -k_3 & & \\ & -k_3 & \cdots & & \\ & & & \cdots & -k_n \\ & & & -k_n & k_n \end{bmatrix} \tag{7.27}$$

$$\mathbf{W}(t) = [\, w_1(t) \;\; w_2(t) \;\; \cdots \;\; w_n(t) \,]^T \tag{7.28}$$

$$\varphi = [\, 0 \; 0 \ldots 0 \,]^T \tag{7.29}$$

$$z_1 = x_n - x_c \tag{7.30}$$

$$z_2 = x_c - x_p \tag{7.31}$$

where $\mathbf{M}$, $\mathbf{C}$ and $\mathbf{K}$ are the mass matrix, damping matrix and stiffness matrix of the main structure respectively, x is the N-dimensional displacement vector of the main structure. $\mathbf{W}(t)$ is the N-dimensional wind load excitation vector, and φ is the N-dimensional control force position vector. The nth element is 1 and the other elements are 0. x_c is the displacement of the damper cavity. x_p is the displacement of the simplified particle. z_1 is the displacement of the damper cavity relative to the main structure, and z_2 is the equivalent particle displacement relative to the cavity.

At the same time, other parameters of the damper cavity are defined as follows: $k_c = m_c \omega_c^2$, $c_c = 2m_c \zeta_c \omega_c$, $\omega_c = 2\pi f_c$, $f_c = 1/(2\pi)(g/l)^{0.5}$, where l is the suspension length from the top of the model to the top of the damper cavity.

Similarly, the parameters of the particles in the equivalent single particle damper can be as follows: $k_p = m_p \omega_p^2$, $cp = 2m_p \zeta_p \omega_p$, ζ_p is the damper equivalent shock damper ratio, ω_p is the damper equivalent natural vibration circular frequency.

Nonlinear functions $G(z_2)$ and $H(z_2, \dot{z}_2)$ as shown in Fig. 7.65 represents the nonlinear characteristics of the damper. Through reasonable selection of ω_p, based on previous research results, the nonlinear spring function $G(z_2)$ can simulate rigid walls of arbitrary precision, and the parameters ζ_p and $H(z_2)$ provide the path of simulations from inelastic collision to elastic collision. Therefore the coefficient of restitution e of any value can be used to set reasonable ζ_p [17, 18]. The relationship between the damping ratio and the coefficient of restitution is shown in Fig. 7.66 [19].

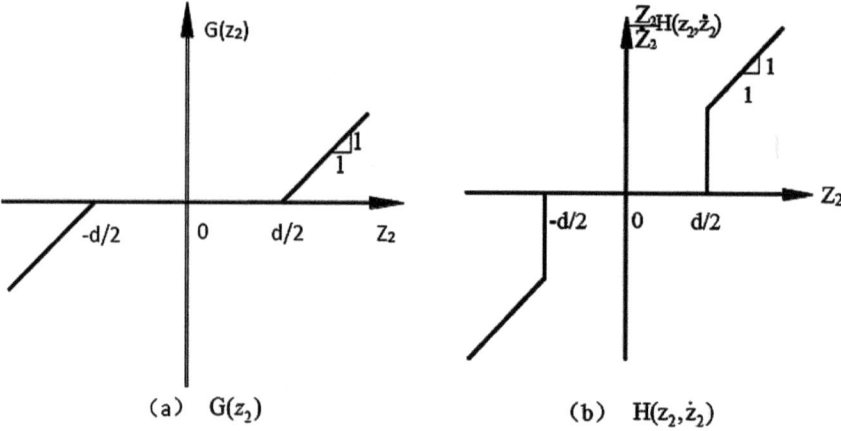

Fig. 7.65 Nonlinear function in particle damper

Fig. 7.66 Relationship between particle damping ratio and coefficient of restitution

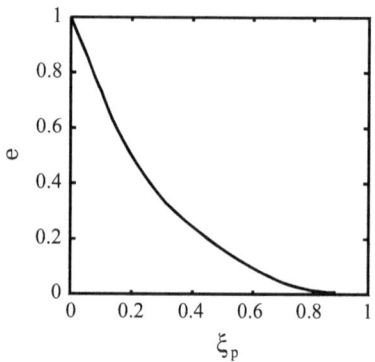

3. *Wind load simulation*

In the wind tunnel test, the main structure is subjected to the motion of the approx-imate elliptical trajectory under the wind load excitation. The accurate simulation of the motion form is often complicated and difficult to implement. For the sake of simplicity, only the wind load in a specific direction is often simulated. The purpose of this simulation is to investigate the effect of PTMD on the vibration control of the top of the model under wind-induced excitation, and lay the foundation for more complicated research in the future. Considering the limitations of existing research, there will be more accurate calculations and analysis in the future [3].

In the actual high-rise building structure design, the largest vibration response of the structure often occurs in the across-wind direction, therefore the across-wind load plays a controlling role in the design. It can also be understood by the measurement of the uncontrolled structural vibration response that the response of the structural across-wind direction is greater than that of the along-wind direction. However, the across-wind load is mainly caused by the vortex shedding in the direction perpendicular to the wind speed. The complex formation mechanism and the high correlation with the shape of the building make it difficult to accurately simulate. The wind load used in this test simulation is based on the results of the wind tunnel pressure test conducted by the Benchmark model at the University of Sydney by Samali [20]. The test data and detailed data can be downloaded from the relevant website.

The test environment for the pressure-measuring wind tunnel test by Samali et al. is as follows. The length of the model is 1:400. The wind speed scale is 1:3. The model is divided into 16 segments of equal length along the height direction. The time scale is 1:133. And 27 s wind speed time history data is obtained. It is actual 3600 s time history data after the scale conversion. Only the wind load in the across-wind direction is measured, as shown in Fig. 7.67. The pressure measurement data of the 32 panels is given by a simple pressure coefficient at the center of the panel. The pressure coefficient is not equal to 0, and the value varies with height.

In order to convert the across-wind load pressure coefficient on the model obtained by the pressure test to the pressure in the actual structure, the conversion is performed by the following formula:

$$F(t) = 0.5 \cdot \rho \cdot \bar{U}^2 \cdot C_p(t) \cdot A \tag{7.32}$$

where ρ is the air density, generally taking 1.29 kg/m^3; $\bar{U}$ is the wind speed at the top of the model $\left(\text{ms}^{-1}\right)$; A is the area of a single panel (m^2); C_p is the combined pressure coefficient.

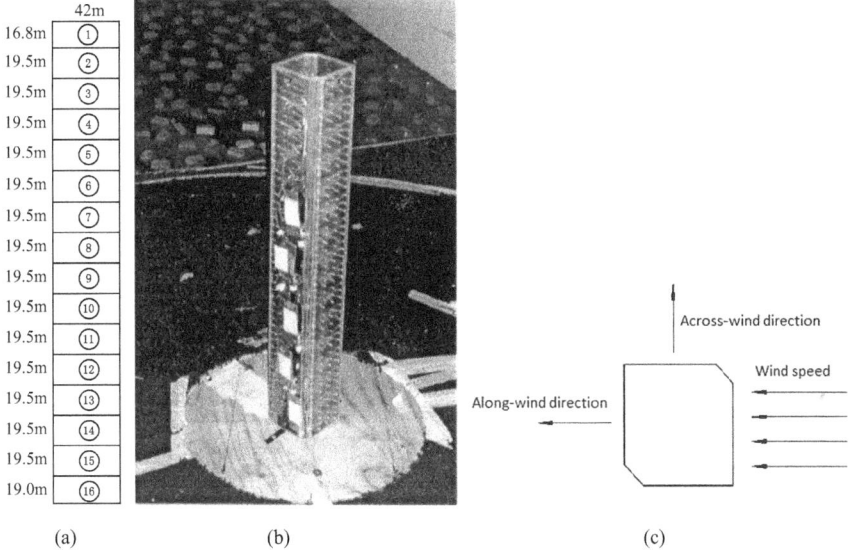

Fig. 7.67 Benchmark model pressure test: **a** Height direction segmentation; **b** Pressure measurement wind tunnel model; **c** Wind speed flow direction

In the wind tunnel test, the wind speed at the top of the model is based on the wind speed specified in the design wind load at 10 m height in Australia. $\bar{U}_r = 13.5$ m/s. After conversion, the wind speed at the top of the specified actual structure is 47.25 ms^{-1} [21], The wind speed profile index is 0.365, D class wind field.

In this test, the wind speed at the top of the model is $\bar{U} = 4$ m/s. In this test, the wind speed at the top of the model is $\bar{U} = 4$ m/s. The length of the model is 1/200, therefore the wind load time history data required by the model can be calculated according to the wind load of the actual building. $F(t)_m$ can be calculated through formula (7.33):

$$F(t)_m = F(t) \cdot \left(\frac{\bar{U}_m}{\bar{U}}\right)^2 \cdot \left(\frac{A_m}{A}\right)^2 = F(t) \cdot \left(\frac{4}{47.25}\right)^2 \cdot \left(\frac{1}{200}\right)^2 \qquad (7.33)$$

The actual time series is the time history data of 3600 s column vector. Figures 7.68, 7.69 and 7.70 list the time history curves of the along-wind direction at the 76, 75 and 74th degrees of freedom. In the actual model, the time history data of the across-wind loads on the 3rd, 7 and 10th degrees of freedom are shown in Figs. 7.71, 7.72, 7.73.

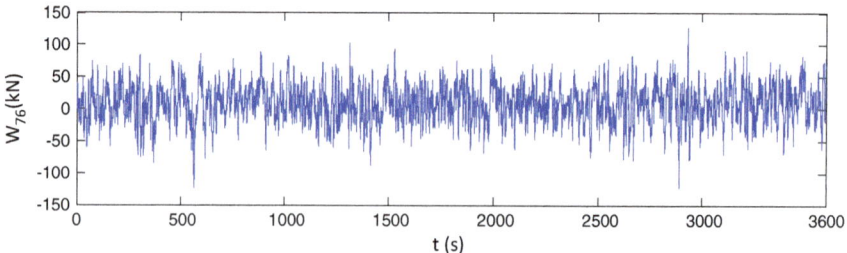

Fig. 7.68 The actual across-wind wind load time history curve at the 76th degree of freedom

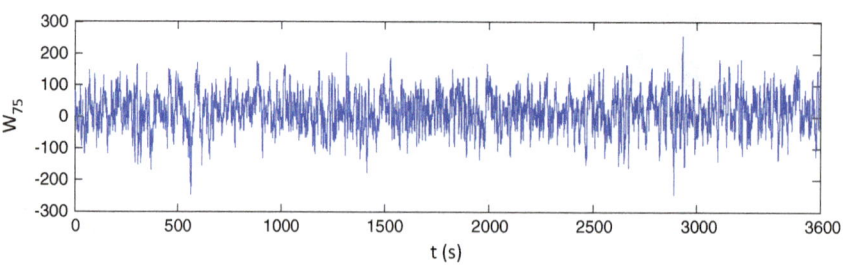

Fig. 7.69 The actual across-wind wind load time history curve at the 75th degree of freedom

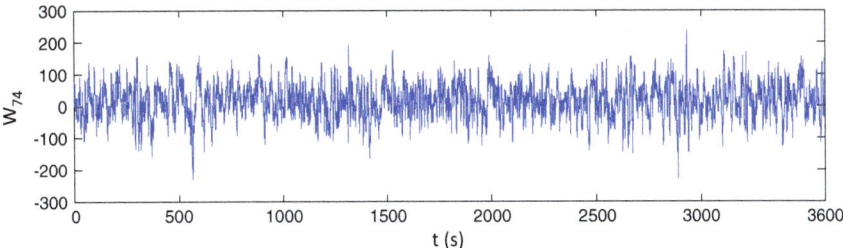

Fig. 7.70 The actual across-wind wind load time history curve at the 74th degree of freedom

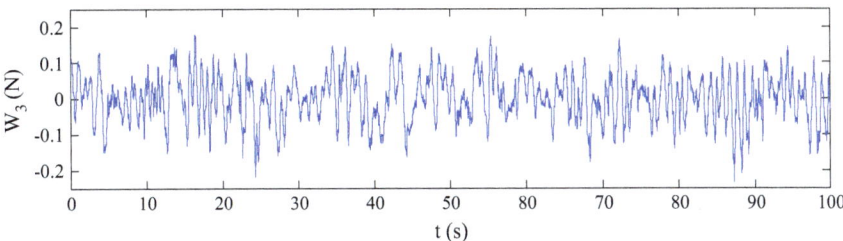

Fig. 7.71 The actual across-wind wind load time history curve at the 3rd degree of freedom

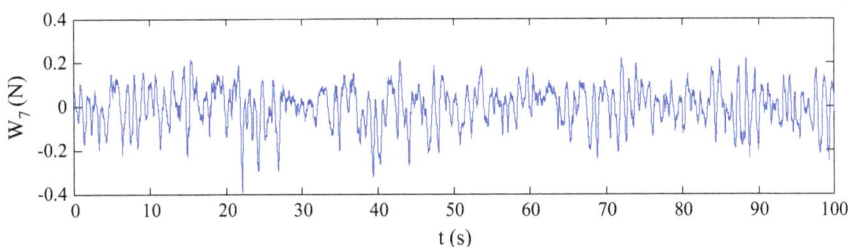

Fig. 7.72 The actual across-wind wind load time history curve at the 7th degree of freedom

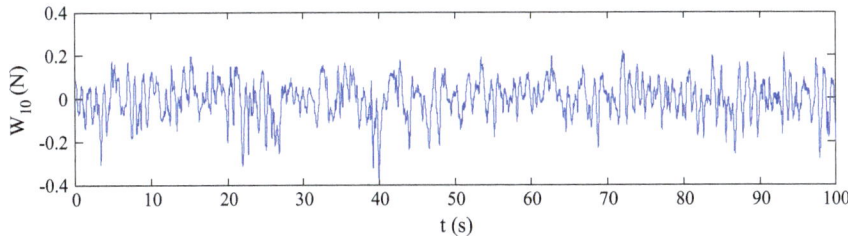

Fig. 7.73 The actual across-wind wind load time history curve at the 10th degree of freedom

7.3.2   Simulation Parameter Determination

The numerical simulation is used to simulate the test results. The particle diameter is 6 mm, the mass ratio is 1%, and the damper cavity size is 8.5 cm × 8.5 cm. The summary of the simulation parameters is shown in Table 7.24.

Because the damping ratio of the wooden damper cavity is difficult to determine, it is obtained by trial and error calculation in the actual simulation. Figure 7.74 shows the variation of the vibration response at the top of the main structure at different cavity damping ratios. The data shows that when the damping ratio is less than 0.05, the vibration response of the main structure is maintained at a specific level. This phenomenon shows that when the damper cavity damping ratio is less than 0.05, the final result is less affected by the damper cavity damping ratio. Considering that the damper cavity and the main structure are connected by four equal length nylon wires, the damper cavity can move freely in the space of the top of the model, and

Table 7.24 Summary of system parameter values

Main structure	Total mass (kg)	19.2
	Circular frequency (rad/s)	13.63
	Damping ratio	0.003
Damper cavity	Quality (kg)	0.01
	Circular frequency (rad/s)	13.63
	Damping ratio	0.005
Particle	Total mass (kg)	0.182
	Equivalent circular frequency (rad/s)	13.63
	Recovery factor	0.5
	Equivalent damping ratio	0.2
	filling ratio (%)	27.5
	Particle density (kg/m^3)	7644
	Diameter (mm)	6
Main structure	Total mass (kg)	19.2

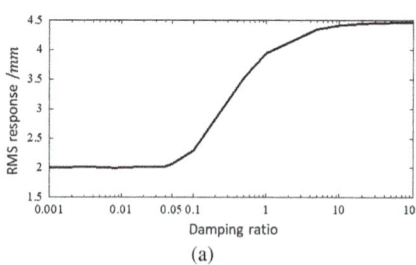 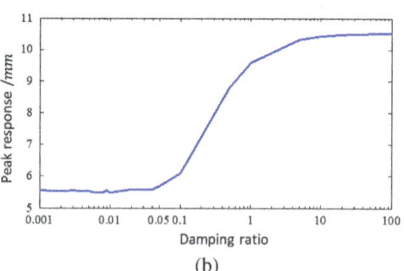

(a)   (b)

Fig. 7.74 Vibration response of the top of the main structure under different damper cavity damping ratios: **a** Displacement R.M.S. response; **b** Displacement peak response

the damping is small, therefore it can be inferred that the damping is relatively small, assuming 0.05.

7.3.3 Simulation Results

Through the established vibration model, the numerical simulation is carried out using the fourth-order Runge-Kutta method. The simulation results and test results are shown in Table 7.25, where σ_x and $\sigma_{\ddot{x}}$ are the displacement and acceleration R.M.S. response of the 10th degree of freedom of the main structure, x and $\ddot{x}$ are displacement and acceleration responses, respectively. x_{max} and $\ddot{x}_{max}$ are the displacement and acceleration peak responses, respectively.

It can be seen from Table 7.25 that under the uncontrolled condition, the numerical simulation results are in good agreement with the experimental results, which indicates that the simulation of the wind load and the dynamic characteristics of the main structure is resonable. The simplified simulation of the wind load and the main structure is the key and foundation of the simulation of control structure with PTMD. At the same time, the simulation of the controlled structure with PTMD is also compared in Table 7.25. The results show that the proposed simplified method can effectively simulate the test results within a resonable range.

Figures 7.75 and 7.76 show the numerical simulation of the 10 s high-resolution results, and compare the uncontrolled structure with the controlled structure. It can be seen that PTMD has good damping effect under wind load excitation, and the time history curves have good similarity with the test results in Chap. 3. The high frequency component of acceleration response in the controlled structure can be clearly shown in the simulation result.

It can be seen from the above simulation results that the particle damper is appropriately simplified, and the multi-particles are simplified into single particle by a reasonable equivalent principle, and a resonable simulation result under uncontrolled and controlled conditions can be obtained. The numerical simulation results show

Table 7.25 Summary of test results and simulation results

	σx (mm)			$\sigma\ddot{x}$ (cm/s^2)		
	Test	Simulation	Error (%)	Test	Simulation	Error (%)
Uncontrolled	4.05	4.16	2.66	77.73	76.31	−1.82
Controlled	2.00	2.00	0.00	37.64	37.90	0.70
	$x$max (mm)			$\ddot{x}$max (cm/s^2)		
	Test	Simulation	Error (%)	Test	Simulation	Error (%)
Uncontrolled	11.02	11.61	5.40	203.51	204.83	0.65
Controlled	5.56	5.55	−0.27	127.61	123.32	−3.36

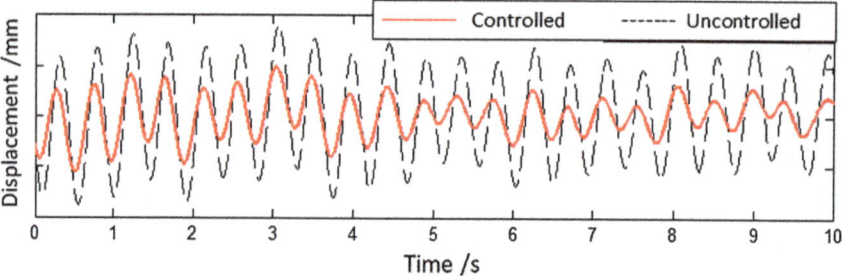

Fig. 7.75 Numerical simulation results of displacement response at the top of the model

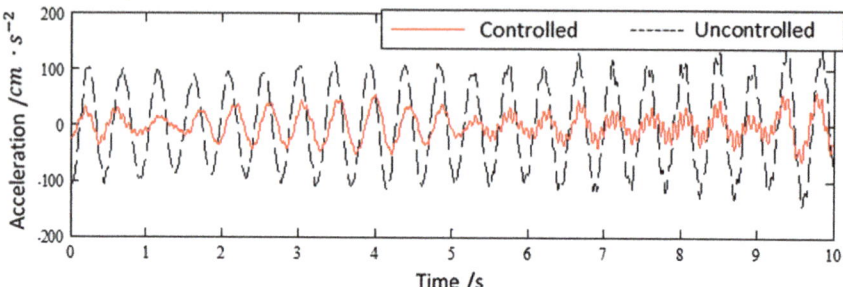

Fig. 7.76 Numerical simulation results of acceleration response at the top of the model

that the R.M.S. and peak response of both acceleration and displacement can be well matched with the experimental results.

References

1. Spencer Jr, B.F., S.J. Dyke, and H.S. Deoskar. 1998. Benchmark problems in structural control: Part I—Active Mass Driver system. *Earthquake Engineering and Structural Dynamics* 27(11): 1127–1139.
2. Spencer, B.F.J. and S.J. Dyke. 1999. *Next generation benchmark control problems for seismically excited buildings*.
3. Yang, J.N., et al. 2004. A benchmark problem for response control of wind-excited tall buildings (Second Generation Benchmark Problem). *Journal of Engineering Mechanics* 130 (4): 437–446.
4. Samali, B., et al. 2004. Wind tunnel tests for wind-excited benchmark building. *Journal of Engineering Mechanics* 130 (4): 447–450.
5. Zhang, J.S., Y. Wu, and D. WU. 2011. Sensitivity analysis of structural wind resistance design. *Journal of Vibration Engineering* 2011(06): 682–688.
6. Wu, J.C. and J.N. Yang. 1997. Continuous sliding mode control of a TV transmission tower under stochastic wind. In *American control conference*.
7. Samali, B., J.N. Yang, and C.T. Yeh. 1985. Control of lateral-torsional motion of wind-excited buildings. *Journal of Engineering Mechanics* 111 (6): 777–796.

8. Huang, P., Y. Quan, and M. GU. 1999. Study on passive simulation method of atmospheric boundary layer in Tj-2 wind tunnel. *Journal of Tongji University (Natural Science Edition)*, 1999(02): 11–15 + 19.

9. Yang, Y.M., and D.Z. Liu. 2014. *Civil engineering structure test*. Wuhan University Press.

10. Coulgh, J., and R. Penzien. 2006. *Structural dynamics,* 2nd edn *(Revised Edition)*. Higher Education Press.

11. Quan, Y. 2002. *Study on cross wind direction wind load and response of super high-rise buildings: Dissertation*. Tongji University.

12. Butt, A.S. 1995. *Dynamics of impact-damped continuous systems*, in *College of engineering*, 172. Louisiana Tech University: Louisiana.

13. Zhao, L., P. Liu, and Y. Lu. 2009. Experimental study on damping characteristics of non-obstructive ranular damping columns. *Journal of Vibration and Shock* 28 (8): 1–5.

14. Huang, Y.W. 2011. *Research on particle damper performance: Dissertation*. Beijing University of Technology.

15. Marhadi, K.S., and V.K. Kinra. 2005. Particle impact damping: effect of mass ratio, material, and shape. *Journal of Sound and Vibration* 283 (1–2): 433–448.

16. Huang, B.C. and C.J. Wang. 2008. *Principle and application of structural wind resistance analysis,* 2nd edn. Tongji University Press.

17. Masri, S.F., and A.M. Ibrahim. 1973. Response of impact damper to stationary random excitation. *Journal of the Acoustical Society of America* 53(1): 200–211.

18. Masri, S.F., and A.M. Ibrahim. 1972. Stochastic excitation of a simple system with impact damper. *Earthquake Engineering and Structural Dynamics* 1(4): 337–346.

19. Lu, Z. and X.L. Lu. 2013. Numerical simulation of vibration control of particle damper. *Journal of Tongji University (Natural Science Edition)* 2013(08): 1140–1144 + 1184.

20. Samali B, K.K., Wood G et al. 1999. *Wind tunnel tests for wind excited benchmark problem*. Technical Note.

21. Simiu, E., and R.H. Scanlan. 1996. Wind effects on structures. *Wiley* 185 (92): 301–317.

Chapter 8
Optimization Design of Impact Dampers and Particle Dampers

Regarding the design of particle impact dampers, there are many design parameters affecting their damping performance, such as the mass ratio of the PID to the primary structure, number, size, and material of the particles, restitution and friction coefficients of the particles, gap clearance, and filling ratio (volume fraction). Therefore, the solution of applying PIDs to dissipate vibration provides more possibilities for engineers, as there are plenty of design parameters available for a possible implementation. In this chapter, some innovative and effective simulation approaches have been carried out to determine the optimal parameters of impact dampers and particle dampers by which the maximum damping effectiveness can be acquired.

8.1 Optimization Design of Impact Dampers

A simple and efficient real-time simulation method is presented for investigating certain classes of vibration problems by employing a hybrid electro-mechanical analog computer. This approach combines the efficiency of electric-analog techniques with the advantages of using actual mechanical components to generate "genuine" functions that are not known or not well defined analytically. The response of a two-degree-of-freedom system equipped with an impact damper was investigated by using this approach and the effects of mode shape, frequency ratio, mass ratio, coefficient of restitution, and damper clearance ratio on the response of the system were determined.

© China Machine Press and Springer Nature Singapore Pte Ltd. 2020
Z. Lu et al., *Particle Damping Technology Based Structural Control*,
Springer Tracts in Civil Engineering,
https://doi.org/10.1007/978-981-15-3499-7_8

8.1.1   Hybrid System

The basic idea behind the use of a hybrid electromechanical analog computer is that such a system, employing electronic analog devices, as well as actual mechanical components, utilizes the advantages of both approaches while minimizing their disadvantages. The equation of motion of the system shown in Fig. 8.1 without an impact damper is

$$\ddot{x} = -2\zeta\omega\dot{x} - \omega^2 x + \frac{1}{M}F(t) \tag{8.1}$$

where $\omega = (k/m)^{1/2}$; $\zeta = c/(2\sqrt{kM})$, x is the displacement of primary system, M is the mass of primary system (single-degree-of-freedom), k is the spring stiffness, m is the mass of particle. If the primary mass, M, is provided with an impact damper (Fig. 8.1), the corresponding equation of motion becomes

$$\ddot{x} = -2\zeta\omega\dot{x} - \omega^2 x + \frac{1}{M}F(t) + \frac{1}{M}f(t) \tag{8.2}$$

where f(t) is the force due to the interaction of M with the auxiliary mass, m.

It has been established both analytically and experimentally that f(t) is directly proportional to the particle mass, m, so that f(t) can be expressed as ma(t), where a(t) is the absolute acceleration of m. Using this, Eq. (8.2) becomes

$$\ddot{x} = -2\zeta\omega\dot{x} - \omega^2 x + \frac{1}{M}F(t) + \mu a(t) \tag{8.3}$$

where $\mu = m/M$.

Fig. 8.1 Model of a single-degree-of-freedom system provided with an impact damper

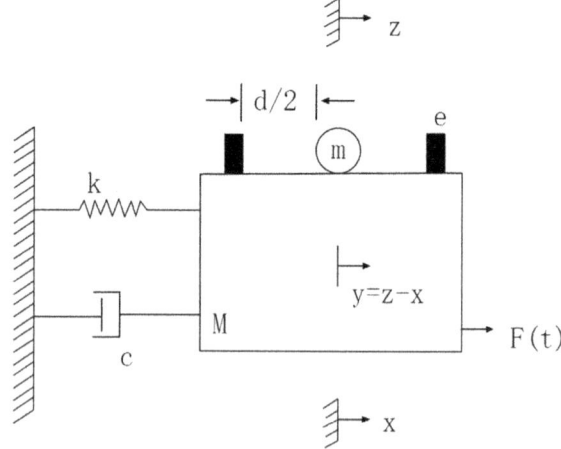

Suppose that Eq. (8.1) is wired in the conventional manner on an electronic analog computer. Let the signal of the amplifier, whose output corresponds to the displacement x, be used as the input signal to the power amplifier that drives an electrodynamic shaker (a "small" one is sufficient). The motion of the shaker would then correspond to the actual physical motion of the single degree-of-freedom system being simulated on the analog computer.

If a container enclosing a solid particle, m, is rigidly attached to the shaker armature, then the container (and, consequently, m) will undergo the same motion it would if it were attached to an actual mechanical model corresponding to that of Fig. 8.1. Attaching an accelerometer to the particle, m, and using its output as an additional feedback to the analog system, completes the loop and results in the correct solution of Eq. (8.3).

The aforementioned setup is illustrated schematically in Fig. 8.2. The mass ratio, μ, can be easily varied by changing the potentiometer that controls the magnitude of the amplified accelerometer signal which is fed to the analog system. By measuring the computer output at the time of an impact, and by using the momentum equation as expressed in the form

$$\mu = -(\dot{x}_+ - \dot{x}_-)/(\dot{z}_+ - \dot{z}_-) \tag{8.4}$$

(where $-$ and $+$ subscripts indicate conditions immediately preceeding and succeeding an impact, respectively. z is the displacement of particle); the value of μ corresponding to a given potentiometer setting may be calculated from Eq. (8.4)

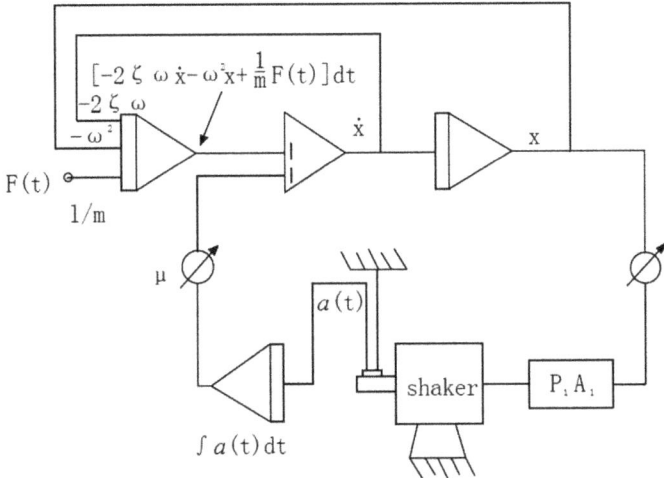

Fig. 8.2 Circuit diagram for electromechanical analog of a single-degree-of-freedom system provided with an impact damper

since the values of the quantities on the right-hand side of Eq. (8.4) can be determined from the computer output. The absolute velocity, $\dot{z}$, of the particle is the time integral of the accelerometer output.

The clearance, d, could be varied either by physically adjusting the length of the container, or by changing the scale factor between the analog system and the shaker power amplifier. The coefficient of restitution e could be varied by changing the material of the free particle and the ends of the container.

Figure 8.3 shows details of the mechanical portion of the system with the damper "particle" removed from the container. The free mass, together with its attached accelerometer, represents the auxiliary mass, m, which was suspended on a long string, resulting in a pendulum whose frequency is very low as compared to the natural frequency of the primary system.

Oscillograph traces illustrating the behavior of the analog model for a typical case are shown in Fig. 8.4
where

$$\omega = 200 \, \text{rad/s}; \quad \Omega/\omega = 1; \quad \zeta = 0.01$$

$$\mu = 0.10; \quad e = 0.75; \quad d/(F_0/k) \approx 25$$

Fig. 8.3 Photograph of mechanical portion of hybrid computer

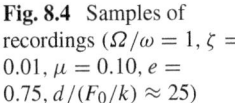

Fig. 8.4 Samples of recordings ($\Omega/\omega = 1$, $\zeta = 0.01$, $\mu = 0.10$, $e = 0.75$, $d/(F_0/k) \approx 25$)

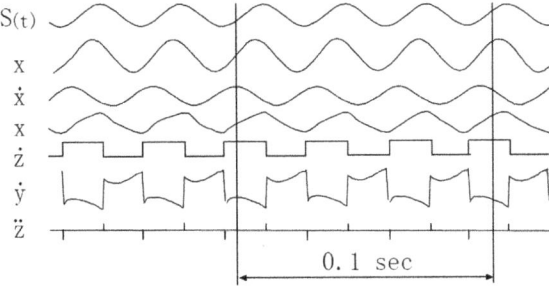

where Ω is excitation frequency. It should be noted that the rapid velocity change in $\dot{x}$ is as severe as that experienced by the actual structure. Since e is defined as the ratio of the relative velocity immediately after impact to that immediately preceeding impact, i.e., $e = -\dot{y}_+/\dot{y}_-$, its value, in this particular case, can be verified from the waveform shown in Fig. 8.4.

It is obvious from the accelerometer output $\ddot{z} = a(t)$, that the impulse resulting from the collision of the particle with the restraining end of the container is symmetric. The steady-state motion represented in Fig. 8.4 consists of two symmetric (i.e., equispaced in time and on opposite ends of the container) impacts per cycle of the excitation.

The slight deviation of the $\ddot{z}$ waveform from that of a rectangular waveform, corresponding to an ideal impact damper in which the particle moves frictionlessly, is due to the existance of a small amount of sliding friction between m and its container.

Typical analog computer measurements of the response levels for the system as a function of the clearance for different values of μ are compared to available theoretical results in Fig. 8.5. The slight difference in the results is primarily due to the presence of the aforementioned friction forces which are not accounted for in the theory.

8.1.2 Application

Theoretical studies of the response of multi-degree-of-freedom systems, equipped with impact dampers, to any type of excitation are unavailable in the literature. The technique described in this work was used to investigate the response of the impact damped two-degree-of-freedom system shown in Fig. 8.6 to a sinusoidal base excitation.

The equations of motion for the system shown in Fig. 8.6 without the impact damper are

$$\ddot{x}_1 = -\omega_1^2(x_1 - s) - 2\zeta_1\omega_1(\dot{x}_1 - \dot{s}) - \frac{m_2}{m_1}[\omega_2^2(x_1 - x_2) + 2\zeta_2\omega_2(\dot{x}_1 - \dot{x}_2)]$$
$$\ddot{x}_2 = \omega_2^2(x_1 - x_2) + 2\zeta_2\omega_2(\dot{x}_1 - \dot{x}_2) \tag{8.5}$$

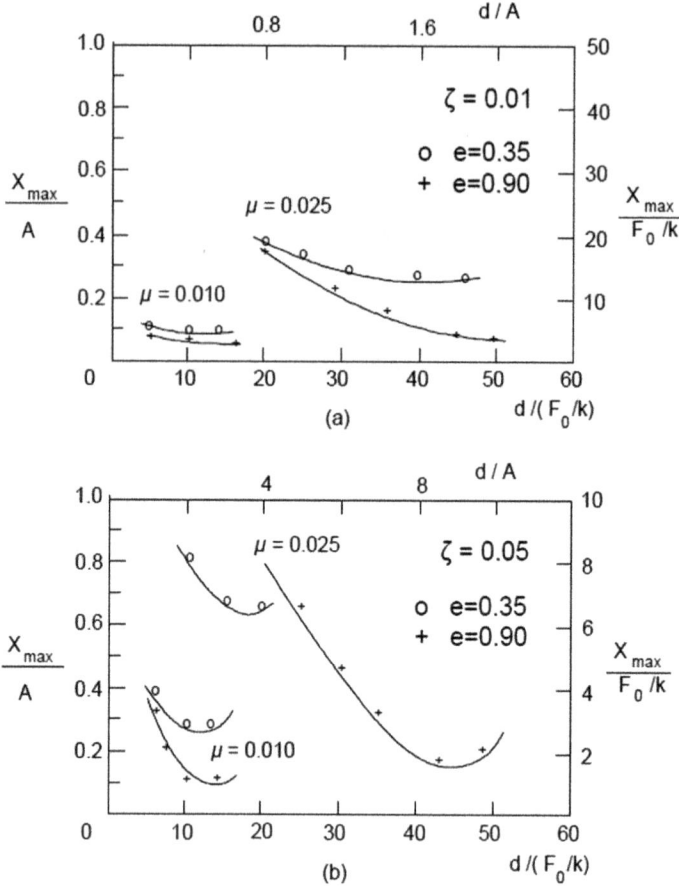

Fig. 8.5 Comparison of theoretical and experimental response levels for an impact-damped single-deg-of-freedom system with $\Omega/\omega = 1$; **a** $\zeta = 0.01$, **b** $\zeta = 0.05$

where m_i represents ith mass in multi-degree-of-freedom system.

$$\omega_1^2 \equiv k_1/m_1 \quad \omega_2^2 \equiv k_2/m_2$$

$$\zeta_1 \equiv c_1/(2\sqrt{k_1 m_1}) \quad \zeta_2 \equiv c_2/(2\sqrt{k_2 m_2})$$

With the introduction of the impact damper, Eq. (8.5) becomes

$$\ddot{x}_1 = -\omega_1^2(x_1 - s) - 2\zeta_1\omega_1(\dot{x}_1 - \dot{s}) - \frac{m_2}{m_1}[\omega_2^2(x_1 - x_2) + 2\zeta_2\omega_2(\dot{x}_1 - \dot{x}_2)]$$
$$\ddot{x}_2 = \omega_2^2(x_1 - x_2) + 2\zeta_2\omega_2(\dot{x}_1 - \dot{x}_2) + \mu_2 a(t) \tag{8.6}$$

for the configuration shown in Fig. 8.6a, and

Fig. 8.6 Model of a two-deg-of-freedom system provided with an impact damper: **a** damper attached to m_2, **b** damper attached to m_1

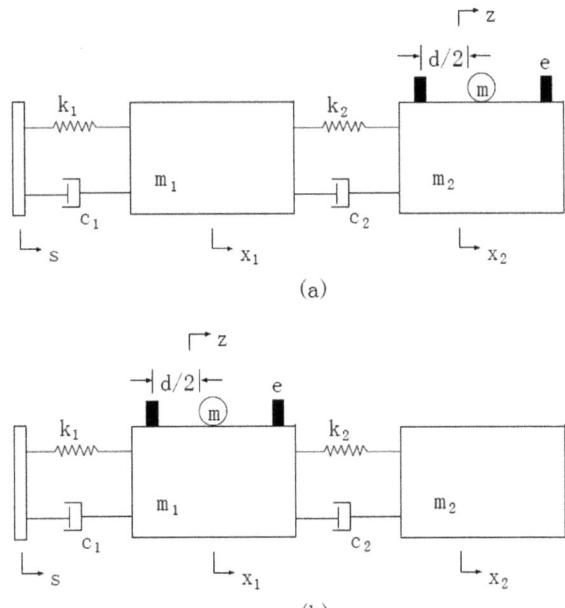

(a)

(b)

$$\ddot{x} = -\omega_1^2(x_1 - s) - 2\zeta_1\omega_1(\dot{x}_1 - \dot{s}) - \frac{m_2}{m_1}[\omega_2^2(x_1 - x_2) + 2\zeta_2\omega_2(\dot{x}_1 - \dot{x}_2)] + \mu_1 a(t)$$
$$\ddot{x}_2 = \omega_2^2(x_1 - x_2) + 2\zeta_2\omega_2(\dot{x}_1 - \dot{x}_2)$$

(8.7)

for the configuration shown in Fig. 8.6b, where $\mu_1 \equiv m/m_1$ and $\mu_2 \equiv m/m_2$.

The hybrid computer circuit diagram for Eq. (8.6) is shown in Fig. 8.7 sign changers have been omitted for clarity. It consists of wiring Eq. (8.5) in the conventional manner and using the signal, x_2, to drive the shaker-damper system in order to generate the impact force, $f(t)$. The loop is then closed by feeding $f(t)$ to the amplifier that governs the motion of m_2. The circuit diagram for Eq. (8.7) utilizes a similar arrangement, except that the shaker is driven by x_1, and $f(t)$ is fed back to m_1. Although several different primary systems were investigated, the system reported in this section had the following characteristics:

$$\omega_1 = \omega_2 = 100 \text{ rad/s}; \quad m_2/m_1 = 1; \quad \zeta_1 = \zeta_2 = 0.01$$

The resulting natural frequencies and mode shapes were
1st natural frequency $p_1 = 61.5$ rad/s;   1st mode: $A_2^{(1)}/A_1^{(1)} = 1.6$
2nd natural frequency $p_2 = 162$ rad/s;   2nd mode: $A_2^{(2)}/A_1^{(2)} = -0.63$
The system was subjected to a sinusoidal base excitation, $s(t) = S_0 \sin \Omega t$, and the effects of the following parameters were determined:

(1) clearance ratio, d/S_0
(2) coefficient of restitution, e

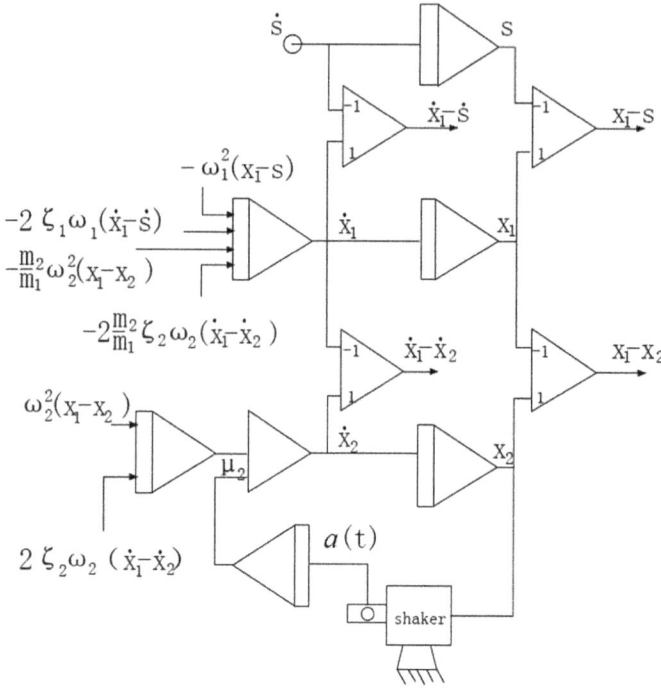

Fig. 8.7 Analog circuit for system shown in Fig. 8.6a

(3) mass ratio, $\mu_1 = \mu_2 = \mu$
(4) position (i.e., whether damper attached to m_1 or m_2)
(5) frequency ratio, Ω/p_1 (related to mode shape).

The excitation and displacement response of the two-degree-of-freedom system as a function of time are shown in Fig. 8.8 for a typical case, where the primary system with the damper attached to m_2 is started from rest. It is clear that when the

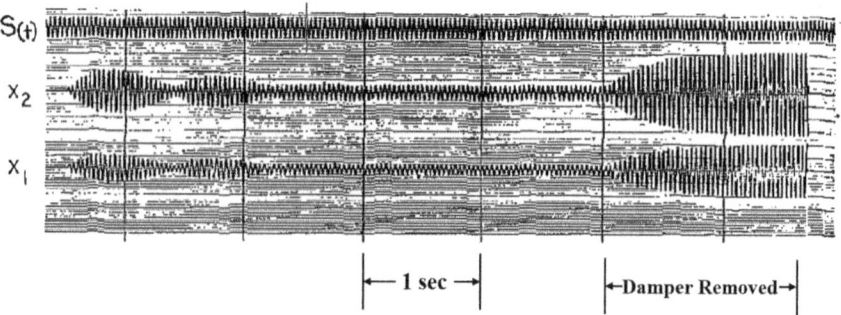

Fig. 8.8 Time histories of response

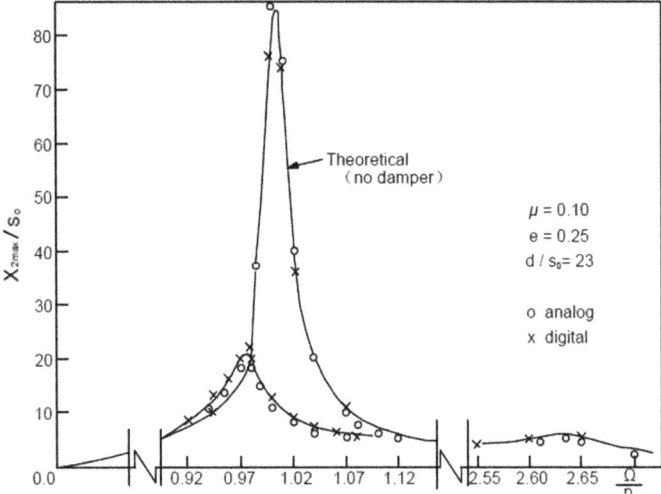

Fig. 8.9 Frequency response characteristics of an impact-damped two-degree-of-freedom system

damper is operating in a stable range, the motion of the primary system settles down to steady-state conditions fairly quickly. Removal of the damper (i.e., setting $\mu = 0$) while the system is vibrating results in an appreciable increase in amplitude, as shown in Fig. 8.8.

The response of the primary systems under discussion as a function of excitation frequency for a practical clearance ratio is given in Fig. 8.9.

The abscissa of the graph has been reduced to dimensionless form by dividing the excitation frequency, Ω, by p_1, the lower natural frequency for the two-degree-of-freedom system. The ordinate has similarly been reduced to dimensionless form by dividing the displacement response of mass m_2 by S_0, the amplitude of base displacement.

The digital computer results shown in Fig. 8.9 were obtained by simulating the motion of the system under discussion on a digital computer. This simulation method involves the step-by-step construction of the solutions of the equations that govern the motion of an idealized system whose motion changes discontinuously during impact in accordance with the momentum equation and the coefficient of restitution. It is clear that the experimental and digital computer results, although derived by using completely different techniques and models, are in relatively good agreement.

The effects of damper clearance, coefficient of restitution, damper mass, primary system mode shape, and damper location are presented in Figs. 8.10, 8.11, 8.12, 8.13. The lower abcissas of the graphs represent the dimensionless clearance ratio, while the left-hand side ordinates represent the dimensionless ratio of the amplitude of the ith mass with the damper operating to its amplitude in the absence of the damper. In this form, the response of the primary system without an impact damper would be represented as a straight horizontal line on the graph at $(x_{i\max}/A_i^{(i)})_{r-pj} = p_{r-pj} = 1$.

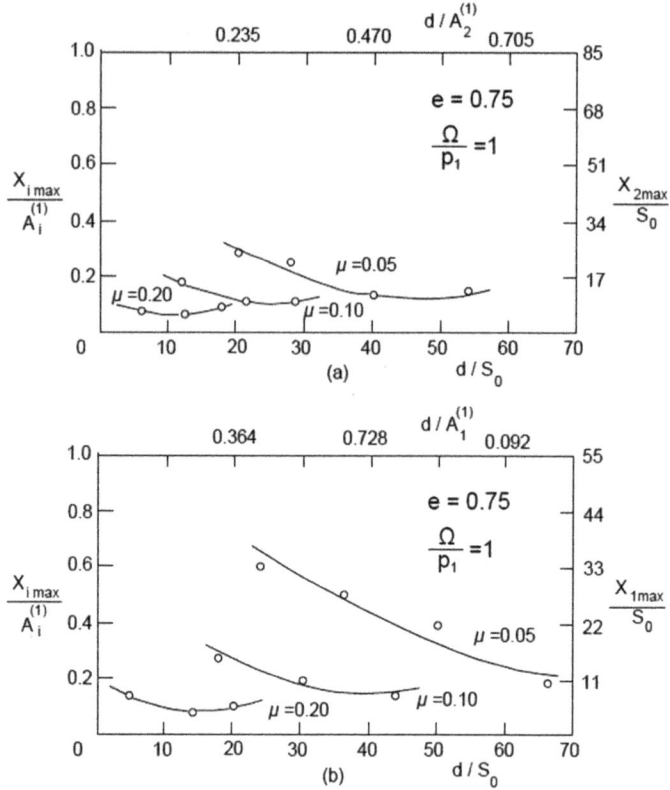

Fig. 8.10 First mode response levels with $e = 0.25$: **a** damper attached to m_2, **b** damper attached to m_1

It should be noted that the left-hand side ordinates apply to each of the two masses; i.e., for a given set of parameters, each mass undergoes the same percent reduction in amplitude.

The top abcissa scale of Figs. 8.10, 8.11, 8.12, 8.13, $d/A_i^{(j)}$ is the dimensionless clearance ratio referred to the amplitude of mass m_i (to which the damper is attached) in the jth mode (in which the damper is operating). Similarly, the right-hand side ordinate in Figs. 8.10, 8.11, 8.12, 8.13, $x_{i\max}/S_0$ is the dimensionless amplification ratio of the maximum displacement of the ith mass (to which the damper is attached) to the amplitude of base excitation. As can be verified from Fig. 8.9, the aforementioned parameters are related by:

$$A_1^{(1)}/S_0 = 55, A_1^{(2)}/S_0 = 10, A_2^{(1)}/S_0 = 85, A_2^{(2)}/S_0 = 6.25.$$

The optimum response curves (in regard to vibration attenuation) for the system under discussion are shown in Figs. 8.14 and 8.15 where the minimum amplification ratio (which can be attained with the optimum d/S_0) is plotted as a function of μ.

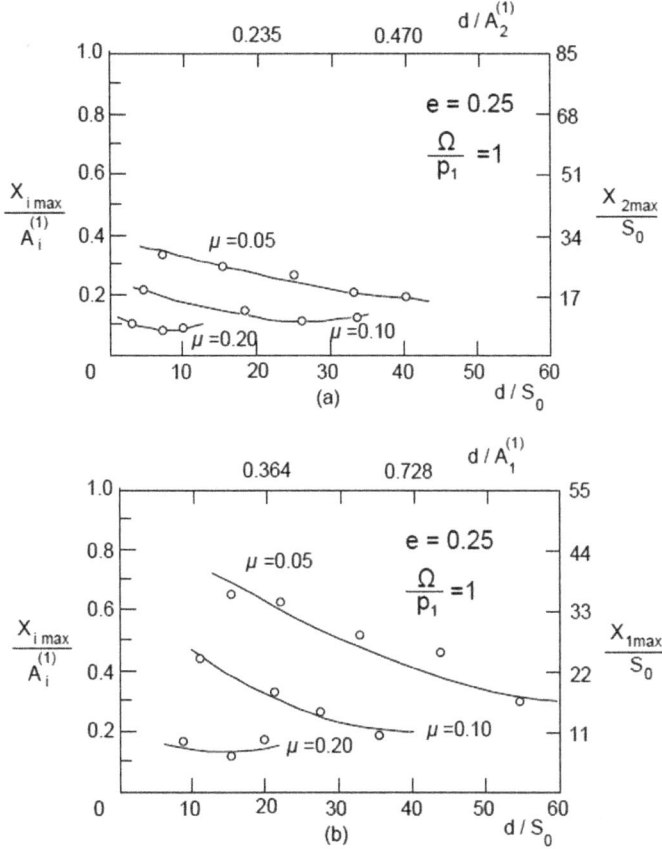

Fig. 8.11 First mode response levels with $e = 0.75$: **a** damper attached to m_2, **b** damper attached to m_1

8.1.3 Discussion

The influence of various system variables on the performance of impact dampers can be summarized as follows:

1. The displacement ratio, x_{imax}/S_0 for both masses exhibits a definite minimum for a certain clearance ratio, d/S_0, which is, in turn, a function of other system parameters. The optimum clearance ratio for the first mode is $\mu d/S_0 \approx 2.5$ and for the second mode is $\mu d/S_0 \approx 2$. It should be noted that the effective ratios of critical damping in the first and second modes are 0.007 and 0.02, respectively. These observations match the behavior of similarly damped equivalent single-degree-of-freedom systems.
2. If the primary system is to operate with a fixed excitation frequency, the use of a damper with high e will lead to the maximum attenuation of vibration amplitude.

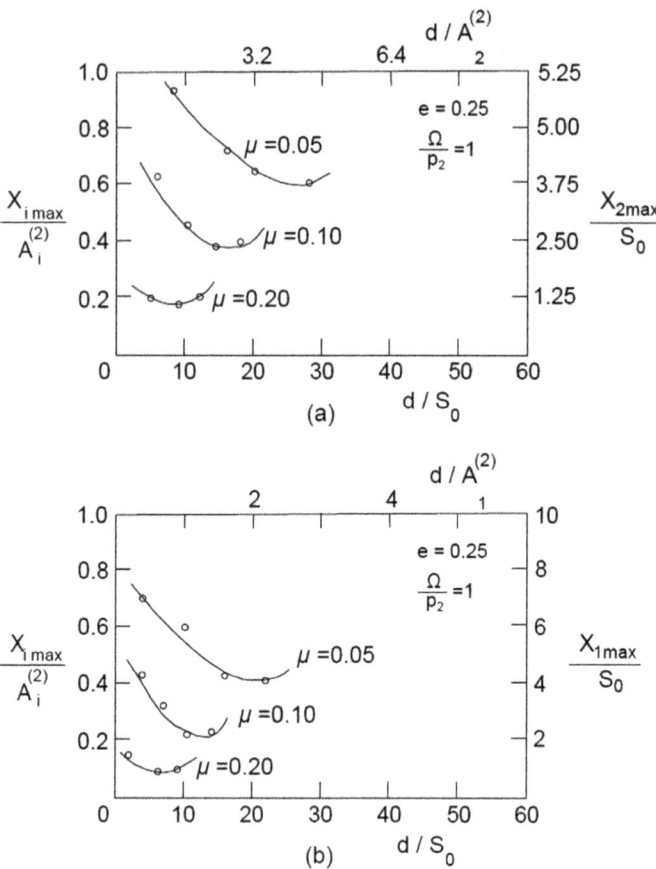

Fig. 8.12 Second mode response levels with $e = 0.25$: **a** damper attached to m_2, **b** damper attached to m_1

3. Even with a relatively small damper mass ($\mu \approx 0.05$), a significant reduction (up to 90%) in the amplitude of the primary system can be achieved. If the data shown in Figs. 8.10, 8.11, 8.12, 8.13 are plotted in terms of $\mu x_{i\,max}/A_i$ and $\mu d/S_0$, respectively, the results will be nearly independent of μ. However, the reduction in amplitude per unit μ decreases as μ increases.
4. In order to operate at maximum efficiency, the damper should be attached to the primary system at the point of maximum displacement. For the system under discussion, the damper should be attached to m_2 in the first mode since $x_2^{(1)}/x_1^{(1)} \approx 1.6$, and it should be attached to m_1 in the second mode since $x_2^{(2)}/x_1^{(2)} \approx 0.7$.
5. For primary systems operating over wide frequency ranges, energy dissipation, due to inherent damping in the primary system or due to the inelasticity of the particle impact, will extend the stable regions of operation.

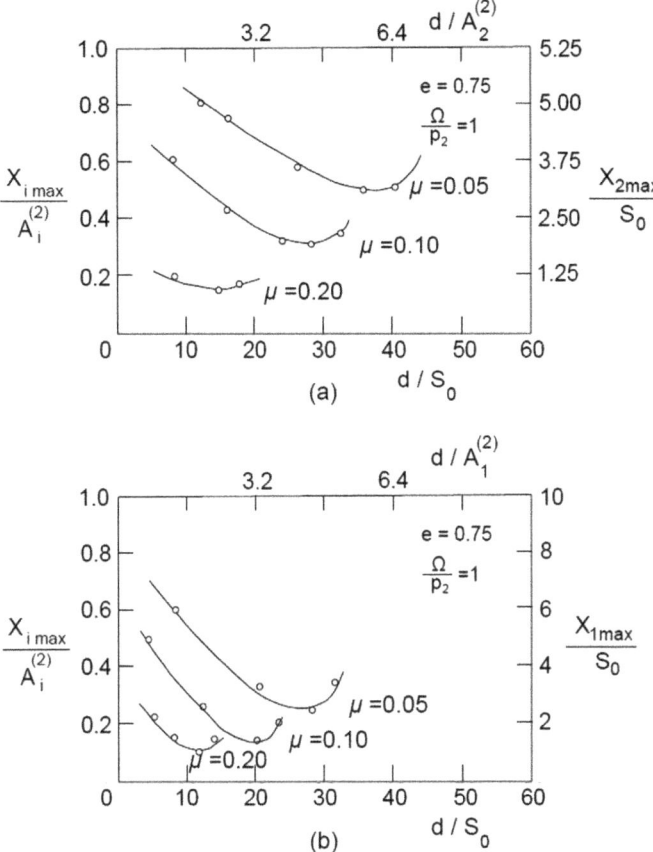

Fig. 8.13 Second mode response levels with $e = 0.75$: **a** damper attached to m_2, **b** damper attached to m_1

The hybrid computer technique under discussion can be incorporated in a computer setup where the parameters of the dynamical system can be varied automatically so as to minimize the criterion function (e.g., amplification ratio in the case of the impact damper). By using gradient methods, such as the method of steepest descent, the hybrid analog computer can be programmed to perform the optimization procedure automatically without any human interaction.

8.2 Optimization Design of Particle Dampers

This section proposes a performance-based optimal design method for a tuned impact damper system, which is composed of a solid mass in a container and is located at the top of a 20-story complex nonlinear benchmark building of the third generation.

Fig. 8.14 First mode optimum response levels **a** damper attached to m_2, **b** damper attached to m_1

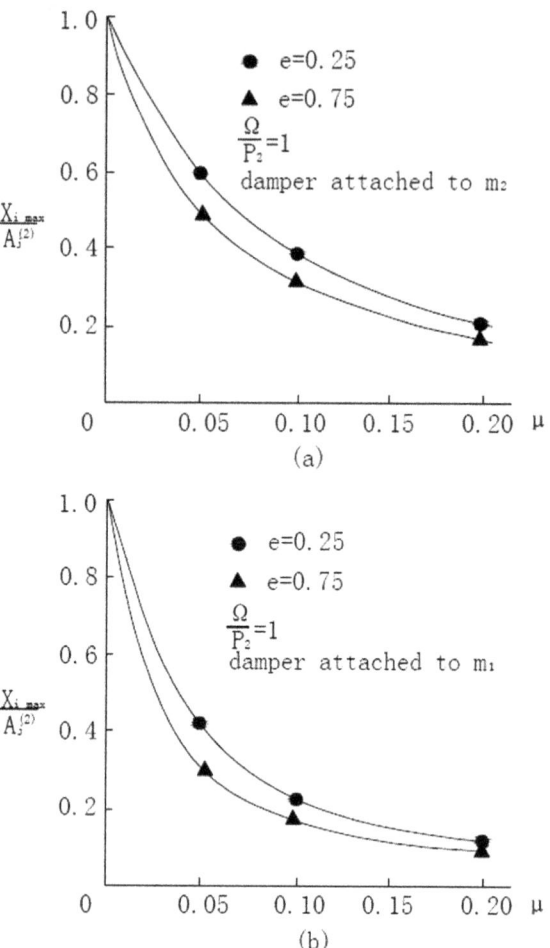

This benchmark building is a steel frame, accounting for nonlinear response via material non-linearity (bi-linear hysteresis) concentrated at the ends of moment-resisting beam-column joints, and is designed for the SAC Phase III Steel Project. In order to illustrate the real-world implementable optimal design approach for designing an optimal tuned impact damper system attached to practical complex structures, a reduced-order data-driven physical model is developed by using the response data specifically from finite element model and the optimization-based parameter identification algorithm. The reduced-order model has been iteratively constructed to be consistent favorably with the finite element model on the aspects of both modal frequencies and mode shapes. Based on the reduced-order model, the optimal parameters of the tuned impact damper system are subsequently designed by adopting proposed performance indices and the differential evolution algorithm conveniently. The optimal vibration control effects are evaluated by using the original finite element

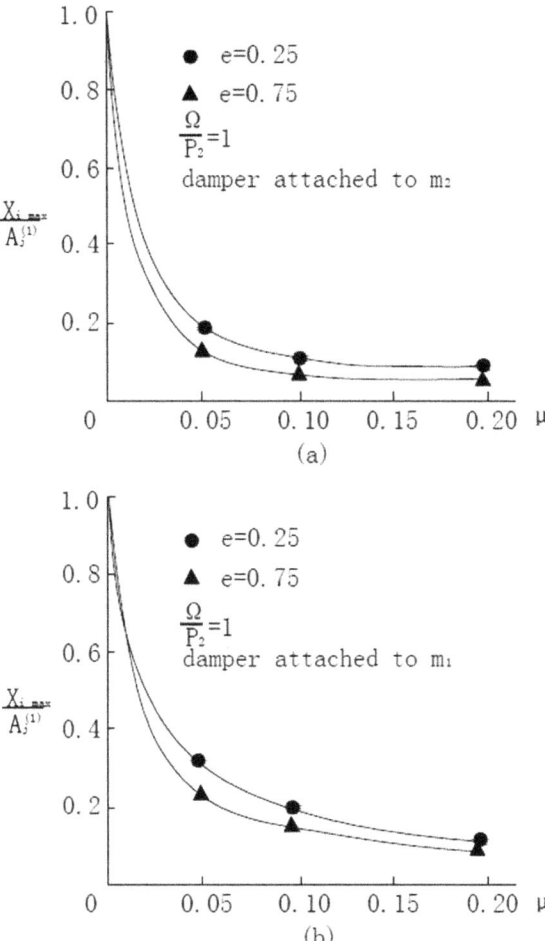

Fig. 8.15 Second mode optimum response levels **a** damper attached to m_2, **b** damper attached to m_1

model structure with the optimal tuned impact damper. Such optimal performance is also compared with that of the original finite element model with a conventionally designed tuned impact damper system.

8.2.1 Methodology and Finite Element Model

Performance-based optimal design method of tuned impact damper (TID) including three phases is proposed in this section, as shown in Fig. 8.16. At Phase 1, for a complex high-rise structure, such as Shanghai Center Tower, TaiPei 101, Carton Tower et al., there are often a large number of degrees of freedom and high frequency dynamic response in the original finite element (FE) model. In order to

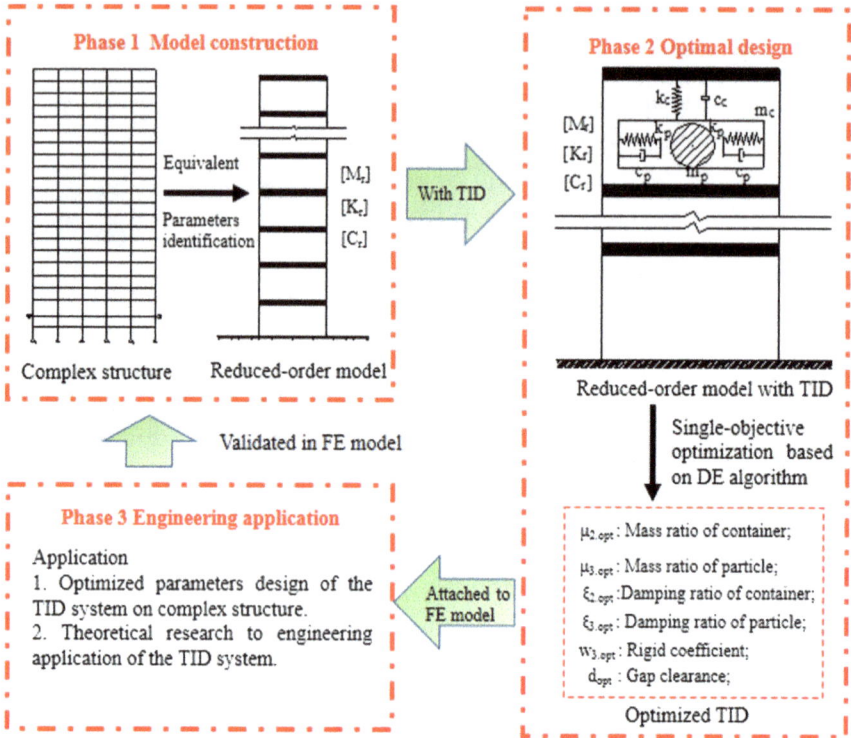

Fig. 8.16 Flowchart of the engineering application of the TID system

design a series of optimal parameters of the TID system efficiently, a 3-D equivalent reduced-order model, which accounts for displacement response of two orthogonal directions and torsional response along the elevation direction, has to develop specifically from original FE model. As the first step to apply TID system to realistic engineering structures, the 20-story nonlinear benchmark building selected in this section is condensed to a 2-D reduced-order model by differential evolution (DE) algorithm at Phase 1, which has been elaborately tuned to be consistent well with the FE model on the aspects of both modal frequencies and mode shapes for capturing the main dynamic response of the original FE model [1]. At Phase 2, a series of optimal parameters of TID system are designed based on the linear reduced-order model by DE algorithm similarly, including mass ratios of particle and container, damping ratios of particle and container, rigid coefficient and gap clearance; therefore, the computation can be simplified significantly. At Phase 3, a TID system with optimal parameters is attached to an original nonlinear complex structure to validate its vibration control effect based on the FE model according to some performance indexes, especially for some nonlinear performance indexes, such as number of plastic hinges, component energy consumption et al. Furthermore, the robustness of the

TID system based on performance optimal design is evaluated and some conclusions for practical engineering application of the TID system are presented.

The differential evolution algorithm has received much attention in solving the complicated optimization problem since it is first proposed by Storn and Price [2] with a gradient-free characteristic as a simple and efficient heuristic methodology. The differential evolution algorithm is a hybrid algorithm that combines the larger population concept of genetic algorithm, the adaptive mutation of evolutionary algorithm and adopts the greedy selection strategy, which makes the DE algorithm more robust and faster than evolutionary algorithms and genetic algorithms. The standard procedure of the DE-based Optimization algorithm can be briefly summarized for better reference as follows:

Step 1: Initialization
The first step of the DE algorithm is to initialize the parent population with a generation size of NP in the search space of S by:

$$z_j = (z_{j1}, z_{j2}, \dots z_{jm})^T \in S, j = 1, 2, \dots, NP$$
$$z_{jl} = z_{\min,l} + U_l(0, 1)(z_{\max,l} - z_{\min,l})$$

where z_j is m-dimension vector consisting of variables that can be optimized in the process of analysis; $z_{\min,l}$ and $z_{\max,l}$ are the lower and upper limits of the search range of the lth parameter, and $U_l(0, 1)$ is a random variable distributed uniformly in the interval range [0, 1]. Usually, engineering knowledge and experience confine the search interval range of the variables that can be optimized to $z_{\min,l}$ and $z_{\max,l}$.

Step 2: Mutation
After the initialization, a new offspring population with a parameter values is developed in the process of mutation, which can be calculated based on values inherited from the parent population $\mathbf{z}$. Both the search diversity in the parameter space and the suitable variation of the parameter values are taken into account during the process of mutation. The "random" combination is chosen as:

$$v_j^{(G+1)} = z_{r1}^{(G)} + F(z_{r2}^{(G)} - z_{r3}^{(G)}) + F(z_{r4}^{(G)} - z_{r5}^{(G)})$$

where F is a positive scale parameter named a mutation constant, which controls the scale factor of the difference between two individual vectors; G is the number of the current population generation; and $r1, r2, r3, r4$ and $r5$ are mutually different integers, which can be chosen from the interval range [1, NP] randomly.

Step 3: Crossover
Following the mutation phase, the crossover process is then introduced to the parent and mutant vectors to generate the trial vector $\mathbf{u}_j^{(G+1)}$

$$u_{jl}^{(G+1)} = \begin{cases} v_{jl}^{(G+1)} & if\ rand(l) \leq CR \\ z_{jl}^{(G)} & if\ rand(l) > CR \end{cases}$$

where $rand(l)$ is lth independent random number uniformly distributed in the interval range of [0, 1], and CR is the crossover constant prior defined within [0, 1] to control the diversity of the population.

Step 4: Selection

In the process of selection, the retained vectors in the population $z_j^{(G+1)}$ is determined by comparing the performance of the parent and offspring populations ($z_j^{(G)}$ and $u_j^{(G+1)}$). In terms of the single-objective $J_1(z)$, the next population is evolved by the vector with better fitness. In order to develop dual objective $J(z)$, the non-dominant sorting and distance crowding ranking can be utilized. The selection provides Pareto optimal models which simultaneously compromise the overall fit in the measured responses with that in the stiffness variations.

In this section, the DE algorithm is used to identify the stiffness and damping parameters of the nonlinear benchmark building and design the optimal parameters of the TID system based on reduced-order model.

The nonlinear seismic benchmark problem was proposed to provide a consistent standard to compare the seismic control effects of different vibration control systems or devices when applied to a practical engineering structure when subjected to various earthquake excitations. In this section, the 20-story steel frame structure that can resist the lateral force when subjected to external horizontal loads, designed for the SAC Phase III Steel Project, is selected for numerical simulations. The benchmark building is 30.48 m in width, 36.58 m in length, and 80.77 m in height. The bays are 6.10 m in all directions on the center, the building has five bays each in the north-south (N-S) and six bays in the east-west (E-W) direction. The benchmark building's lateral force-resisting system consists of steel perimeter moment-resisting steel frames (MRFs). The interior bays of the building consist of steel frame with composite floors. For a detailed introduction and model information of the nonlinear benchmark structure, see Ohtori et al. [3]. The seismic performance analysis of the structure is carried out using MATLAB (SIMULINK) with earthquake excitations imposed along the N-S direction served as a principal direction of a planar frame where the damping devices can be carried out within the perimeter frames. The nonlinear response was accounted for in the benchmark model by means of material nonlinearity served as bi-linear hysteresis characteristics occurred at beam-column and column-column joints to form the concentrated plastic hinge model. The Rayleigh damping formulation is considered in the benchmark model and the modal damping ratios of the first and five mode-order are 2%. The first ten natural frequencies of the 20-story nonlinear benchmark model are: 0.261, 0.753, 1.30, 1.83, 2.40, 2.44, 2.92, 3.01, 3.63 and 3.68 Hz, respectively.

The ground excitation used to evaluate the seismic response performance of the nonlinear benchmark model consists of two far-field earthquake waves including El Centro and Hachinohe and two near-field earthquake waves including Northridge and Kobe. In order to emerge different kinds of earthquake excitations, the following amplification factors were applied to the original earthquake waves including 0.5, 1.0 and 1.5 for the El Centro and Hachinohe waves, while 0.5, 1.0 for the Northridge and Kobe waves. In this case a total of ten earthquake excitations are accounted for in the nonlinear benchmark seismic analysis.

8.2.2 Optimal Design Procedure

The detailed flowchart about performance-based optimal design of the tuned impact damper (TID) system attached to a nonlinear benchmark building will be elaborated in this section.

Since the original finite element (FE) model includes a large number of degrees of freedom and high frequency dynamic response, the high-order modes have little effects on the structure dynamic response, it is necessary to generate an equivalent reduced-order model of the benchmark building for the optimal design of the TID system.

Data-driven parameter identification of nonlinear benchmark building is carried out to get the reduced-order model from original FE model by differential evolution (DE) algorithm [1], which can be elaborated as follows:

$$Minimize \ J(z) = [J_1(z), J_2(k)] \tag{8.8}$$

$$s.t., \ z \in S, \ S = \{z : z_{\min,J} \leq z_l \leq z_{\max,J}, \forall l = 1, 2, \ldots m\} \tag{8.9}$$

$$J_1(z) = \frac{1}{N} \sum_{i=1}^{N} \frac{R.M.S.(a_i(t) - a_i(z, t))}{R.M.S.(a_i(t))} \tag{8.10}$$

$$J_2(k) = \sqrt{\frac{1}{N} \sum_{i=1}^{N} (k_i - \mu_k)^2} \tag{8.11}$$

$$\mu_k = \frac{1}{N} \sum_{i=1}^{N} k_i \tag{8.12}$$

where S represents the search space; $z = (k_1, k_2, \ldots k_{20}, \xi_1, \xi_2)$ represents the identification parameters; $z_{\min,l}$ represents the lower limit of the lth identification parameter; $z_{\max,l}$ represents the upper limit of the lth identification parameter; m represents the number of identification parameters; μ_k represents the average of the identified stiffness; $J_1(z)$ represents the objective function of response error; $J_2(k)$ represents the structural stiffness uniformity index; R.M.S. represents root mean square response.

According to DE algorithm mentioned above, the 2-D FE model is simplified into a story shear-type model (including story stiffness and story mass at each floor) and the number of degrees of freedom in the FE model is reduced from 414 to 20 via neglecting all motions other than the horizontal movement at each floor conveniently.

It is conceivable that for the high-rise complex structure, although the 3-D reduced-order model is developed to account for displacement response of two orthogonal directions and torsional response along the elevation direction, the degrees of freedom can be reduced further more. The reduction of the FE model was deemed to be reasonable because the damping device can only reduce the horizontal motion of the nonlinear benchmark structure and the rotational inertia is ignored. The identified inter-story stiffness of the 20-story benchmark structure are shown in Table 8.1, where k_i represents the inter-story stiffness of the ith floor. The first and second mode damping parameters identified by DE algorithm are: 0.012 and 0.014, respectively.

As shown in Table 8.1, the stiffness of each floor of the reduced-order is evenly distributed, which is consistent with the actual situation.

The reduced-order model is simulated based on the story shear-type form using the identified stiffness parameters and original mass parameters, the Rayleigh damping formulation is used by the identified damping coefficients mentioned above to account for the structural damping. The first eight modal frequencies between reduced-order model and the original FE model is shown in Fig. 8.17. The comparison of the first four mode shapes between the reduced-order model and the original FE model is shown in Fig. 8.18. It can be seen that the modal properties of the complex FE model have been well captured by the condensed shear-type model, especially for the first four modes, where the discrepancies of the frequencies are 0.06, 0.73, 0.32 and 0.99%, and the modal assurance criteria (MAC) values of the mode shapes are 99.68, 97.42, 90.68, and 80.40%. From both figures, it can be seen that there is only a slight difference in terms of the modal frequencies and mode shapes of the first four order between the reduced-order model and the original FE model, with the errors within the reasonable range. It may be indicated that the global responses of the target model can be acceptably simulated by the condensed model, because the dynamic response is mainly controlled by the first four-order modal formation, and the high-order modal formation has little effect on the structural dynamic response [4–8].

In order to further investigate the effectiveness of the reduced-order model, the displacement and acceleration time history of the reduced-order model is compared with those of the original FE model under the action of El Centro and Hachinohe waves, which are shown in Fig. 8.19. The R.M.S. discrepancy of the displacement response for El Centro and Hachinohe wave is 1.64% and 1.68%, respectively. The peak discrepancy of the displacement response for El Centro and Hachinohe wave is 3.82% and 4.44%, respectively, which indicates that the displacement response of reduced-order model and the original FE model is particularly good under the action of El Centro and Hachinohe waves, in terms of the acceleration response, there is a slight difference, however still with an acceptable accuracy.

According to the illustrations above, from the engineering prospective, the reduced-order model of the nonlinear benchmark building is validated well to capture the main modal characteristics and dynamic responses of the original FE model [9]. Such well condensed models from the Phase 1 in Fig. 8.16, can be further utilized for the optimal design stage as the Phase 2, which ensures that we can design the

Table 8.1 Stiffness parameters recognition results by DE algorithm ($\times 10^8$ N/m)

k_1	k_2	k_3	k_4	k_5	k_6	k_7	k_8	k_9	k_{10}
2.55	2.40	2.38	2.50	2.67	2.82	2.80	2.69	2.72	2.66
k_{11}	k_{12}	k_{13}	k_{14}	k_{15}	k_{16}	k_{17}	k_{18}	k_{19}	k_{20}
2.54	2.45	2.30	2.17	2.16	2.26	2.42	2.66	2.67	2.53

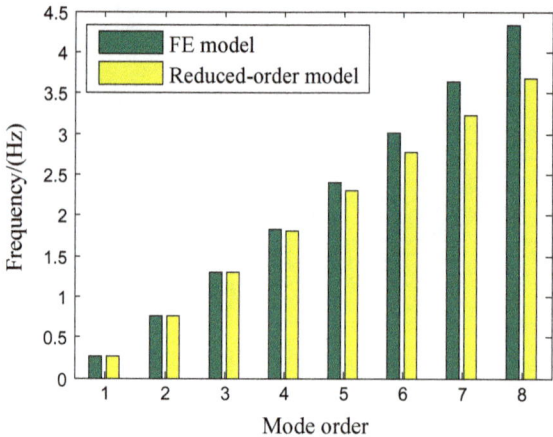

Fig. 8.17 Comparison of modal frequency between the original FE model and reduced-order model

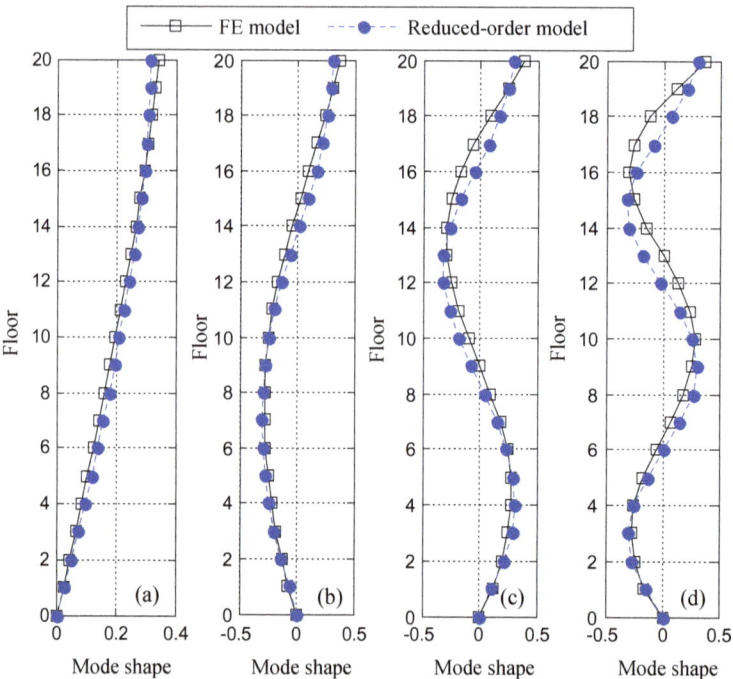

Fig. 8.18 Comparison of the mode shapes between FE model and reduced-order model. **a** Mode 1; **b** Mode 2; **c** Mode 3; **d** Mode 4

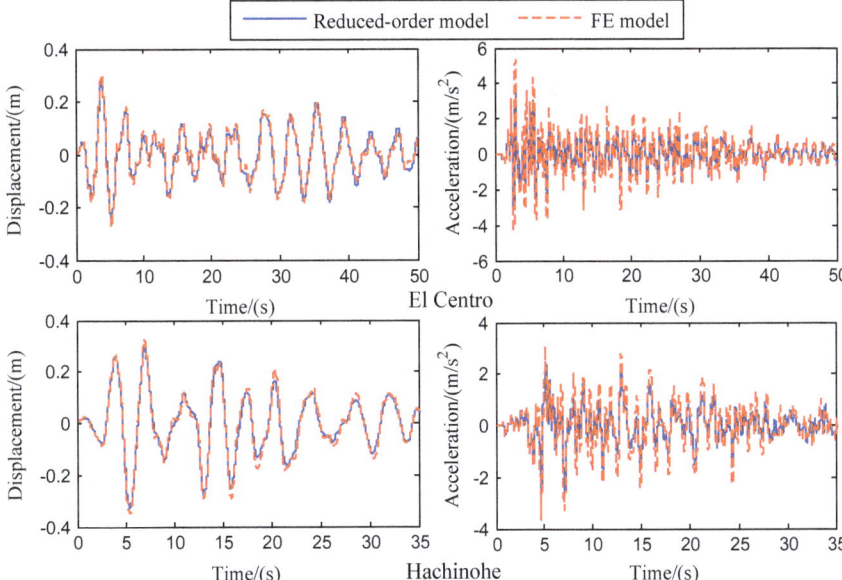

Fig. 8.19 Comparison of displacement and acceleration time history at the top of the building between reduced-order model and the original FE model under the action of El Centro and Hachinohe waves

optimal parameters of the TID system based on this reduced-order model instead of the original FE model.

Based on the reduced-order model mentioned above, a TID system is applied to this model with a particle and container, the simplified model of the reduced-order model with the TID system is shown in Fig. 8.20 and the governing equation of the whole system can be written as Eq. (8.13):

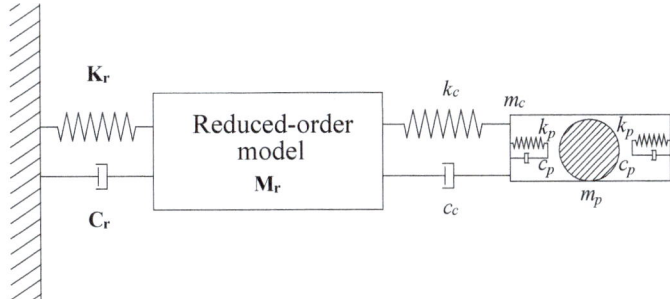

Fig. 8.20 Simplified model of the reduced-order model with the TID system

$$\begin{cases} \mathbf{M_r\ddot{U}} + \mathbf{C_r\dot{U}} + \mathbf{K_r U} = \mathbf{M_r E\ddot{x}_g} + \mathbf{\psi} F_p \\ m_c\ddot{u}_c + F_p - k_p G(y) - c_p H(y,\dot{y}) = 0 \\ m_p\ddot{u}_p + k_p G(y) + c_p H(y,\dot{y}) = 0 \\ F_p = k_c(u_c - u_{20}) + c_c(\dot{u}_c - \dot{u}_{20}) \end{cases} \tag{8.13}$$

where $\mathbf{U}$, $\mathbf{\dot{U}}$, $\mathbf{\ddot{U}}$ represent the 20-dimensional displacement, velocity and acceleration time history vectors of reduced-order model, respectively; $\mathbf{M_r}$, $\mathbf{K_r}$, $\mathbf{C_r}$ represent the mass, stiffness and damping matrices of reduced-order model, respectively; $\mathbf{E}$ denotes location vector of seismic force, $\mathbf{\psi}$ denotes the location vector of the TID system, $y = u_p - u_c$ represents the relative displacement of the particle with respect to the container, $G(y)$ and $H(y,\dot{y})$ are nonlinear functions, as shown in Fig. 8.21, respectively [10]. In addition, d represents the gap clearance between the particle and the container, k_c and c_c represent the interactions between the container and the primary structure; k_p and c_p represent collisions between the particle and the container; F_p represents the nonlinear restoring force between main structure and container.

The optimal parameters of TID system can be designed by DE algorithm through solving the equation above, including mass ratio of particle, mass ratio of container, damping ratio of particle, damping ratio of container, rigid coefficient and gap clearance, the objective function is the R.M.S. displacement at the top of the building, which is as follows:

$$\min J(z) = abs\left[J_1(z) - J_{obj}\right] \tag{8.14}$$

$$J_1(z) = \frac{abs[R.M.S.(d_{20}(t)) - R.M.S(d_{20}(z,t))]}{R.M.S.(d_{20}(t))} \tag{8.15}$$

$$z = (\lambda, \xi_1, \xi_2, \mu_1, \mu_2, d) \tag{8.16}$$

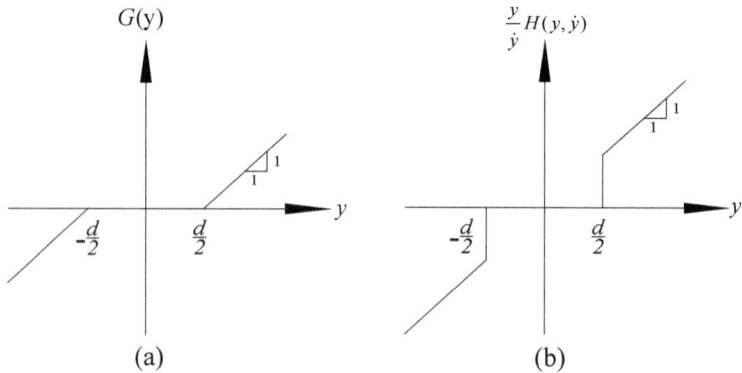

Fig. 8.21 The graph of nonlinear function: **a** function $G(y)$; **b** function $\frac{y}{\dot{y}}H(y,\dot{y})$

where $J_{obj} = 0.60$ represents objective vibration control effect of the TID system; $J_1(z)$ represents vibration control effect of the TID system with optimal parameters; $d_{20}(t)$ represents displacement at the top of the uncontrolled building; $d_{20}(\mathbf{z}, t)$ represents displacement at the top of the building with TID; $\mathbf{Z}$ represents parameters vector of the TID; λ represents rigid collision coefficient between particle and container; ξ_1 represents damping ratio of the container; ξ_2 represents damping ratio of the particle; μ_1 represents mass ratio of container; μ_2 represents mass ratio of particle; d represents gap clearance between the particle and the container.

Figure 8.22 shows the error convergence process $J(z)$ and Table 8.2 shows the optimal parameters of the TID system. It can be seen from Fig. 8.22 that the vibration control effect of the TID system with optimal parameters approximately maintains at the value of 30% when the iteration process is terminated. Table 8.2 shows that the optimal rigid collision coefficient is not 20 recommended by Masri [11], but 11.3624 by DE algorithm, thus demonstrating the conventional design parameters of the TID system is not the optimal solution for the main structure. Such optimal parameters

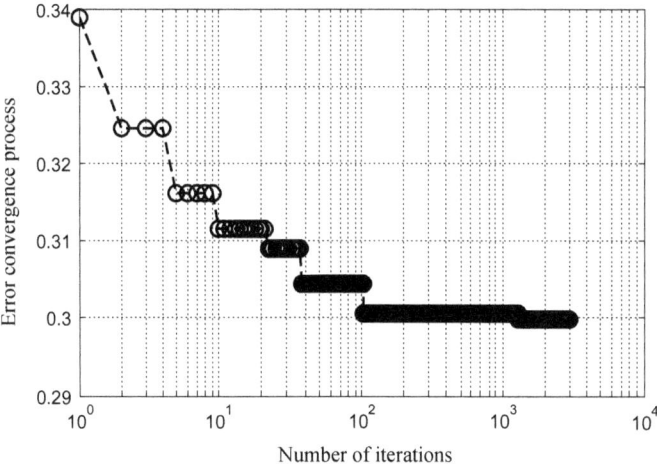

Fig. 8.22 Error convergence process of the parameters design by DE algorithm

Table 8.2 Parameters of the TID system by conventional design and optimal design

Parameters	Conventional design	Optimal design
λ	20	11.3624
ξ_1	0.10	0.1971
ξ_2	0.375	0.0603
μ_1	0.02	0.0297
μ_2	0.02	0.0058
d (m)	0.30	0.0433

of TID system from Phase 2 in Fig. 8.16, can be further utilized for the performance object validation in the original FE model as the Phase 3.

In order to elaborate the superiority of optimal design, two series of TID parameters are put into the original FE model to compare performance indexes of optimal design with that of the conventional experienced design. Some experienced design methods are introduced in paper [10]. Table 8.2 shows the parameters of the TID system by such conventional experienced design and the proposed optimal design, respectively.

The acceleration and displacement time history at the top of the building with TID system are shown in Fig. 8.23, the reduction control effect is not obvious at the beginning of the excitation, which is occurred because sufficient collisions between particle and the container take some time. As time progresses, the TID system can fully mitigate the response of main structure. The peak and R.M.S. acceleration and displacement response of each floor of the main structure subjected to El Centro with scale 1.5 are shown in Figs. 8.24 and 8.25, respectively. From Fig. 8.24, it can be seen that the maximum vibration control effect of peak acceleration for conventional design and optimal design is 3.8 and 5.4% occurred at floor 16, similarly for peak displacement response at floor 16, it can reach 8.1 and 16.2%. From Fig. 8.25, the maximum vibration control effect of R.M.S. acceleration for conventional design and optimal design is 8.3 and 15% occurred at floor 14, similarly for R.M.S. displacement

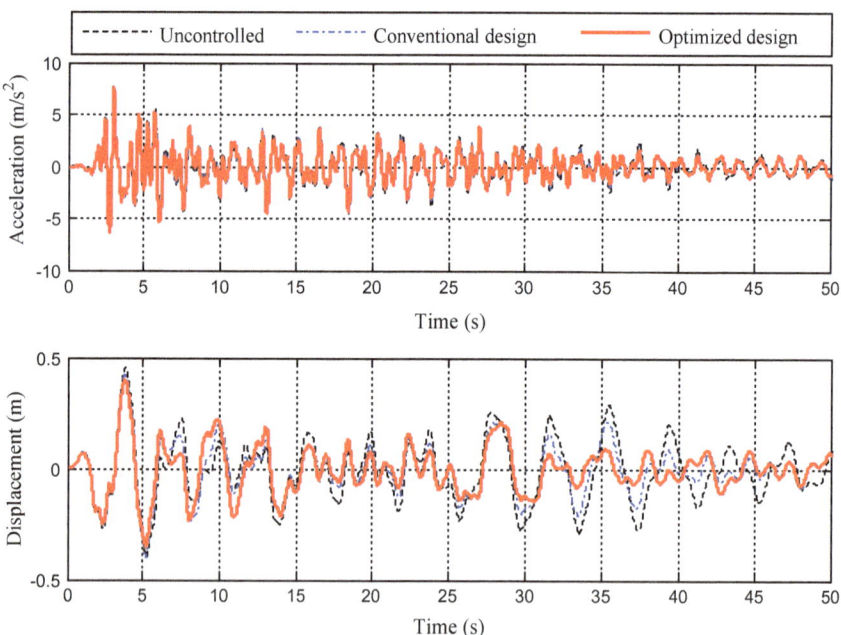

Fig. 8.23 Comparison of the acceleration and displacement response at the top of the building between conventional design and optimal design under the action of El Centro with scale 1.5

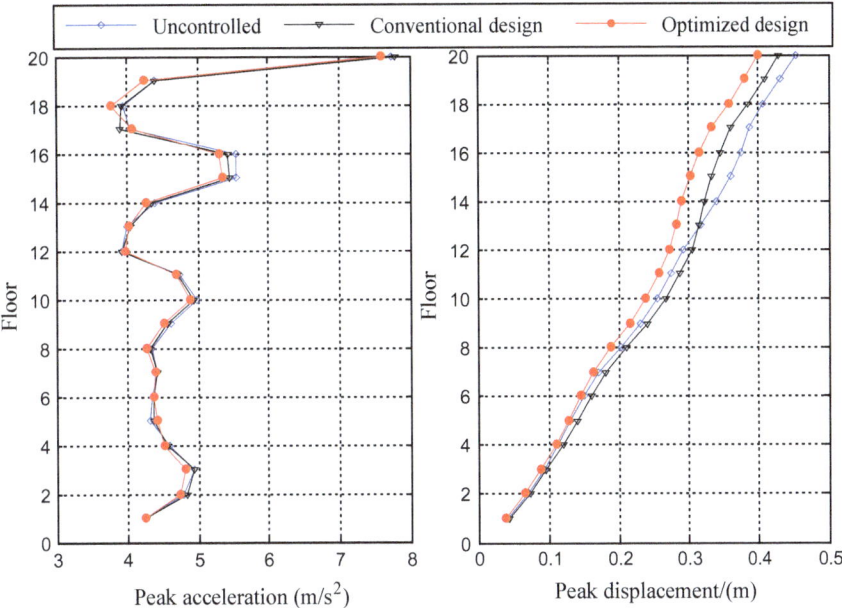

Fig. 8.24 Comparison of the peak acceleration and displacement response at top of the building between conventional design and optimal design under the action of El Centro with scale 1.5

response at floor 14, it can reach 17.3 and 26.6%, indicating that the TID system can effectively reduce the peak and R.M.S. displacement response at the top of the building; however, the TID system does not perform well in mitigating acceleration response, especially for peak acceleration response. This is because for this specific benchmark building, the acceleration response is mainly controlled by the second-order mode of vibration instead of the dominant first mode [4–8], while the damper is practically designed to be installed on the top of the main structure with the largest response under the first mode. On the opposite, the vibration control effects for the displacement are quite good for TID system. For example, the R.M.S. displacement at the top floor is reduced by 16.2% under the conventional design, and by 24.0% under the optimal design, the improvement rate of optimal design can reach 48%, which further validates the superiority of performance-based optimal design method, where the improvement rate is defined as follows: the improvement rate = (vibration control effect of optimal design – vibration control effect of conventional design)/vibration control effect of conventional design × 100%.

Figure 8.26 shows the inter-story drift ratio of each floor of the main structure subjected to El Centro with scale 1.5 and Fig. 8.27 shows the base shear of the main structure. It is noticed from Fig. 8.26 that the TID system can effectively reduce the inter-story drift ratio below floor 17, which occupies 85% of the total building floors; however, it has little effectiveness above floor 18, maybe it is the impact force between particle and container, as well as the tensile force between

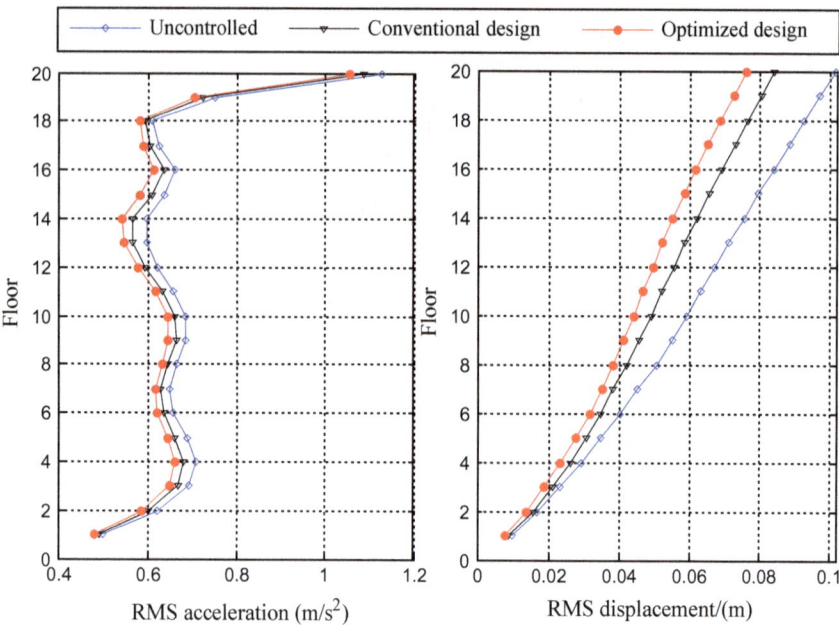

Fig. 8.25 Comparison of the R.M.S. acceleration and displacement response at the top of the building between conventional design and optimal design under the action of El Centro with scale 1.5

Fig. 8.26 Comparison of the inter-story drift ratio between conventional design and optimal design under the action of El Centro with scale 1.5

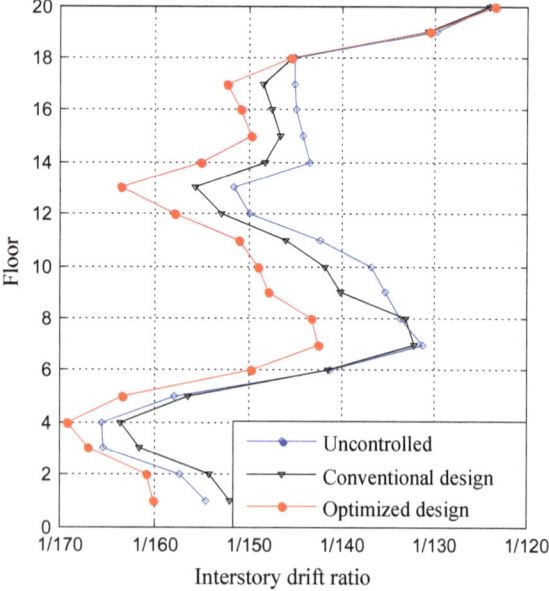

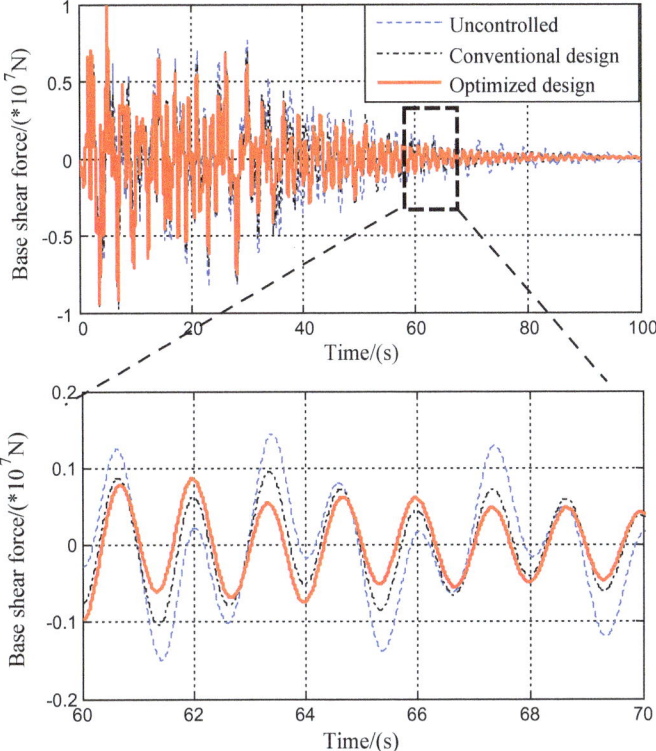

Fig. 8.27 Comparison of the base shear between conventional design and optimal design under the action of El Centro with scale 1.5

container and main structure that cause the larger inter-story drift ratio. As to the optimal design scheme, it can effectively improve the vibration control effects of the conventional designed damper system, for example, compared with the conventional design, the improvement rate of the optimal design at floor 13 is up to 185%. In addition, Fig. 8.27 shows the optimal TID system can significantly reduce the base shear of main structure, which is related to the fact that the damper system can mitigate the R.M.S. acceleration of the main structure. Furthermore, the vibration control effect of base shear for optimal design is better than that for conventional design, which further validates the superiority of optimal design method mentioned in Fig. 8.16.

Considering that the beam-column joints of the main structure can respond in a nonlinear state when subjected to large earthquake excitations, some nonlinear performance indexes are investigated between optimal design and conventional design, including number of plastic hinges and component energy consumption. In this section, the building model accounts for nonlinear response via material non-linearity (moment-curvature bi-linear hysteresis model for structure member bending), the members are assumed to be elastic and these plastic hinges only occur at the ends of the moment resisting beam-column and column-column joints. The number of plastic hinges and component energy consumption of each floor of main structure subjected to El Centro with scale 1.5 is shown in Fig. 8.28 and the distribution of plastic hinge of main structure is shown in Fig. 8.29, where the blue dots represent plastic hinges possessed by uncontrolled structure, with conventional TID system and with the optimized TID system, and the red dots represents plastic hinges reduced by the optimized TID system. Accounting for the damage mechanism of beam hinge, there are up to ten plastic hinges each floor occurred at the end of the beam.

It can be seen from Figs. 8.28 and 8.29 that the optimal designed TID can significantly decrease the number of plastic hinges from 86 to 62 of the main structure subjected to El Centro with scale 1.5, but for conventional designed TID, there is no decreasing, only the distribution of plastic hinges is different from uncontrolled structure. The component energy consumption of the main structure with TID system by optimal design is fewer than that with conventional designed TID, thus the damage and inelastic response of the main structure with optimal TID can be reduced to a minimum value. While the number of plastic hinges is important, the plastic hinge rotation in hinges is also important. In this section, the ratio of joint curvature and yield curvature is introduced to evaluate the development deepness of plastic hinges. Figure 8.30 shows the maximum ratio of joint curvature and yield curvature at each floor for three cases under the action of El Centro with scale 1.5. It can be seen that the optimal designed TID system can significantly reduce the development

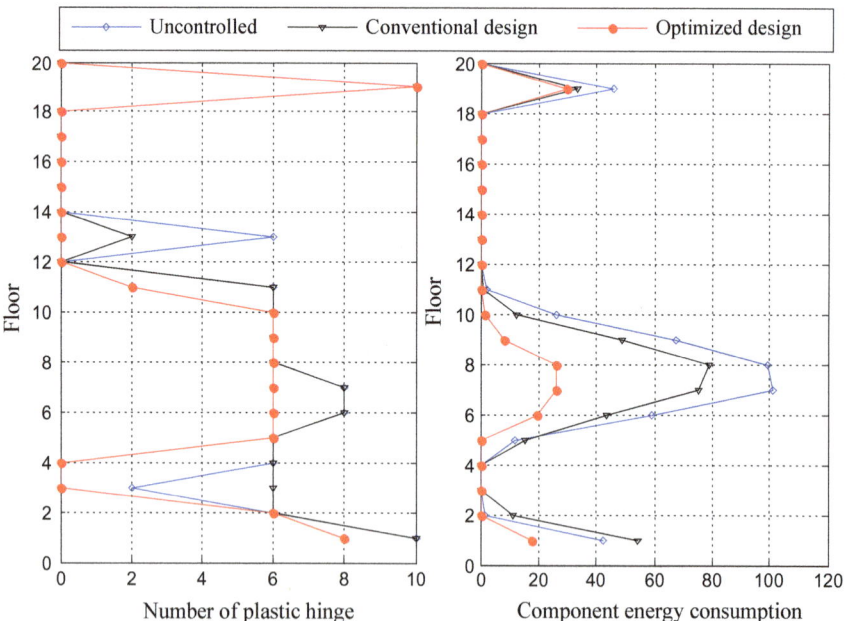

Fig. 8.28 Comparison of the number of plastic hinges and component energy consumption between conventional design and optimal design under the action of El Centro with scale 1.5

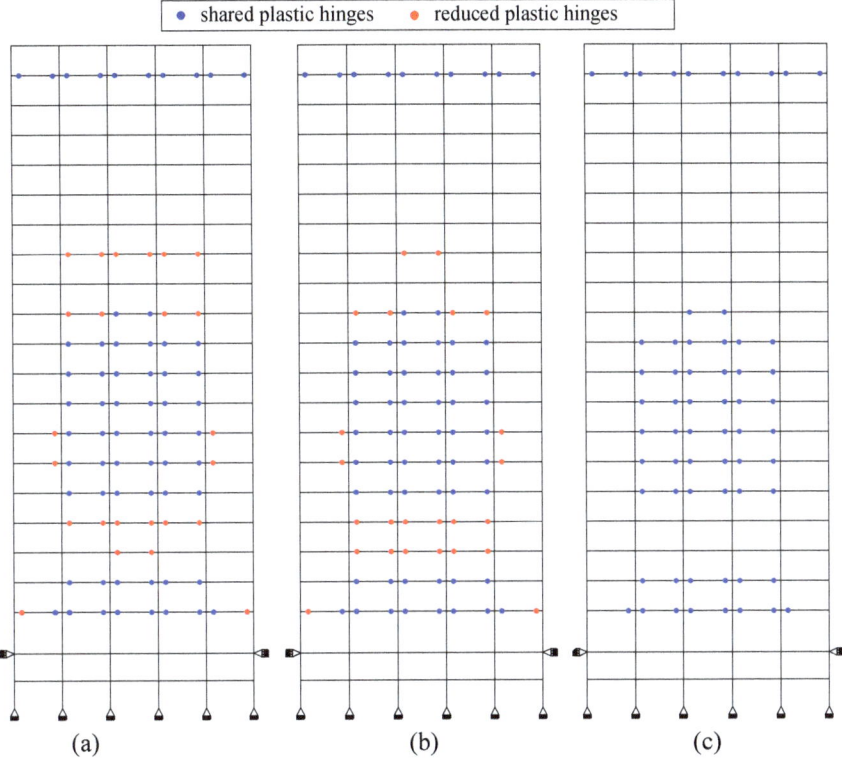

Fig. 8.29 Comparison of the distribution of plastic hinges between uncontrolled structure, conventional design and optimal design under the action of El Centro with scale 1.5: **a** uncontrolled structure; **b** with conventional TID; **c** with optimal TID

deepness that joints are in nonlinear range compared with the conventional designed TID system.

8.2.3 Reliability of Control Performance

When the actual engineering structure is completed and in the serviceability period, the structure would suffer from unpredictable earthquake excitations. In this section, the robustness of optimal designed tuned impact damper (TID) system based on finite element model is evaluated in order to validate the superiority of the TID system attached to a complex building in the real world. In the process of evaluating the reliability of control performance, a total of 40 earthquake records are selected, including 25 far-field earthquake records and 15 near-field earthquake records, which is shown in Table 8.3. The comparison of vibration control effects at the top of the main structure subjected to 25 far-field waves and 15 near-field waves, including peak

Fig. 8.30 Maximum ratio of joint curvature and yield curvature at each floor for three cases under the action of El Centro with scale 1.5

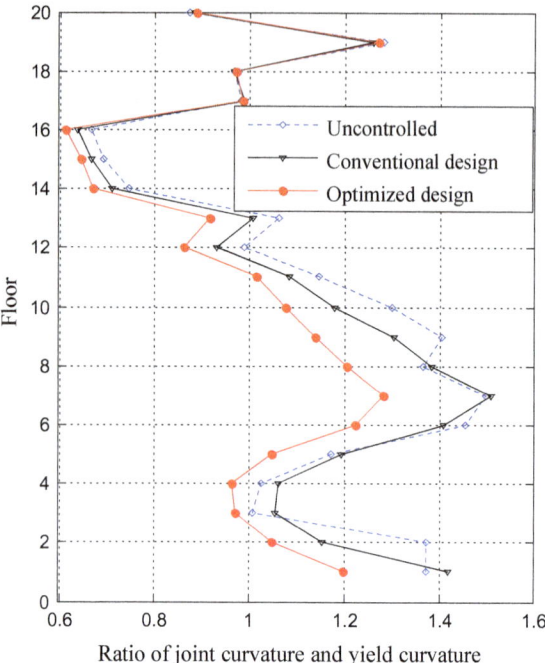

Table 8.3 The earthquake records for the reliability evaluation of control performance

Number	Station name	PGA (g)	Number	Station name	PGA (g)
1	Hector	0.150246	21	Delta	0.351119
2	Hector	0.26558	22	Tolmezzo	0.314809
3	Bolu	0.822429	23	Tolmezzo	0.351329
4	CHY101	0.440103	24	LA-Hollywood_Stor_FF	0.174179
5	CHY101	0.352887	25	LA-Hollywood_Stor_FF	0.209876
6	Duzce	0.312128	1	Sturno	0.250649
7	Arcelik	0.1499	2	Parachute_Test_Site	0.377181
8	Shin-Osaka	0.211916	3	El_Centro_Array_#6	0.438985
9	Shin-Osaka	0.243233	4	El_Centro_Array_#7	0.337514
10	Nishi-Akashi	0.502749	5	Erzincan	0.495516
11	Nishi-Akashi	0.509338	6	Erzincan	0.515256
12	Canyon_Country-W_Lost_Cany	0.482002	7	Lucerne	0.726839
13	Beverly_Hills-14145_Mulhol	0.516457	8	Lucerne	0.789157
14	Beverly_Hills-14145_Mulhol	0.415783	9	Rinaldi_Receiving_Sta	0.825195
15	Rio_Dell_Overpass-FF	0.548927	10	Sylmar-Olive_View_Med_FF	0.843306
16	Gilroy_Array_# 3	0.367362	11	Izmit	0.152082
17	Gilroy_Array_# 3	0.555024	12	TCU065	0.813816
18	Poe_Road_(temp)	0.300292	13	TCU102	0.297872
19	El_Centro_Array_# 11	0.379579	14	TCU102	0.168592
20	El_Centro_Array_# 11	0.363978	15	Duzce	0.535333

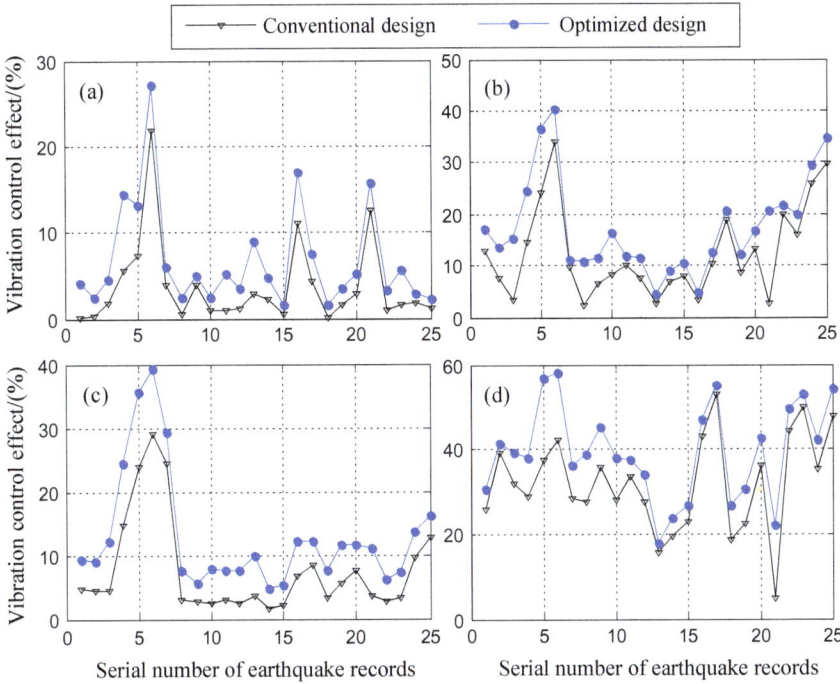

Fig. 8.31 The comparison of vibration control effects at the top of the building subjected to 25 far-field waves: **a** peak acceleration; **b** peak displacement; **c** R.M.S. acceleration; **d** R.M.S. displacement

acceleration, peak displacement, R.M.S. acceleration and R.M.S. displacement, are shown in Figs. 8.31 and 8.32, respectively. In order to further illustrate the superiority of the optimal design, six earthquake excitations are selected to compare the vibration

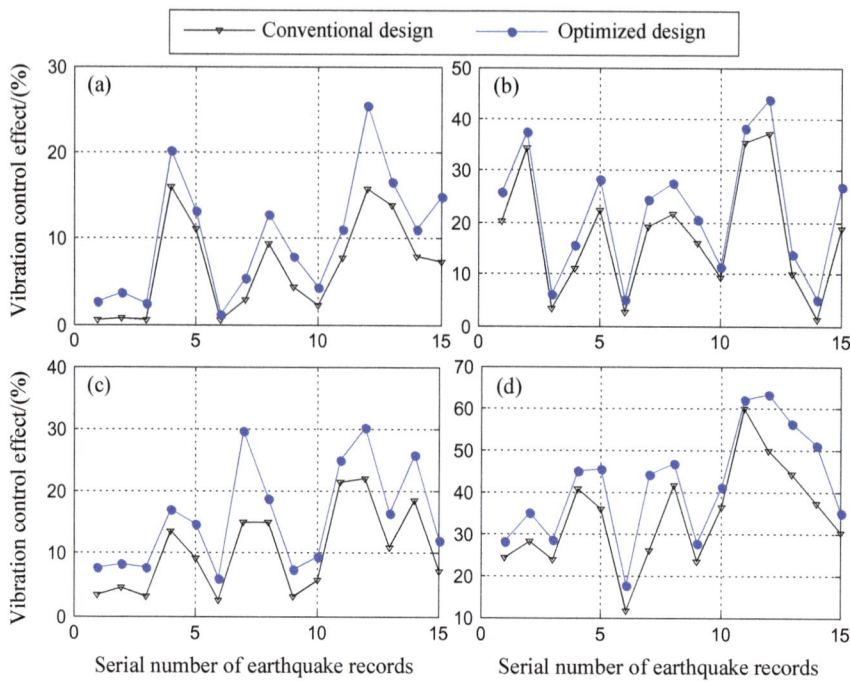

Fig. 8.32 The comparison of vibration control effects at the top of the building subjected to 15 near-field waves: **a** peak acceleration; **b** peak displacement; **c** R.M.S. acceleration; **d** R.M.S. displacement

control effects through displacement time history response. Figure 8.33 shows the comparison of displacement time history response at the top of the main structure among uncontrolled structure, conventional design and optimal design, subjected to three far-field waves, Bolu, CHY101 and El_Centro_Array_#11. Figure 8.34 shows the corresponding comparison of displacement time history response among uncontrolled structure, conventional design and optimal design, subjected to three near-field waves, El_Centro_Array_#6, El_Centro_Array_#7 and Sturno.

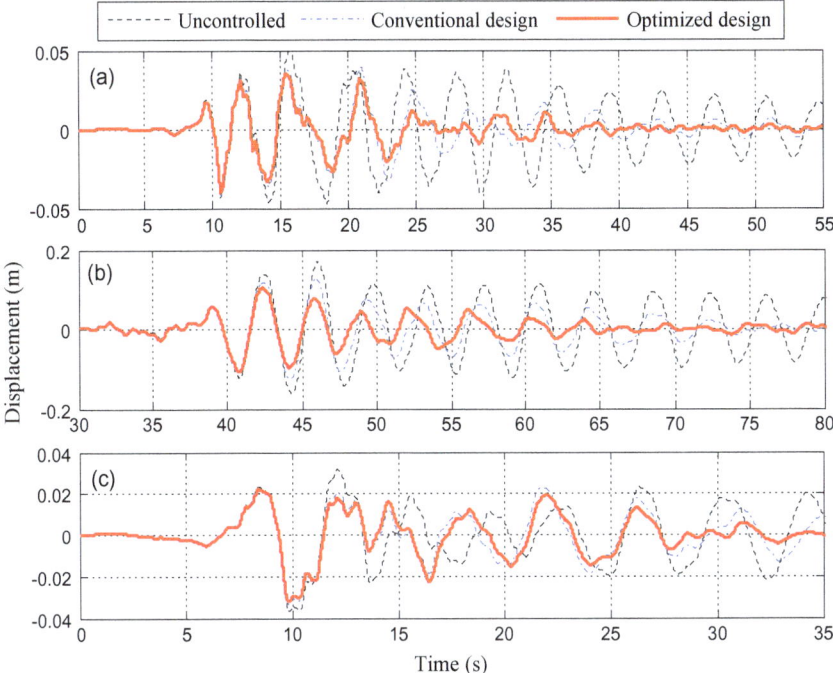

Fig. 8.33 The comparison of displacement time history at the top of the building subjected to three far-field waves: **a** Bolu; **b** CHY101; **c** El_Centro_Array_#11

It can be seen from Figs. 8.31, 8.32, 8.33, 8.34 that the TID system based on optimal design can significantly mitigate the displacement response compared with that based on the conventional method when the main structure is suffered from unpredictable earthquake excitations, especially for the near-field earthquake excitations, the maximum vibration control effect of optimal design can improve by 20% for R.M.S. displacement response and 15% for peak displacement response compared with conventional design. In general, the performance-based optimal design can lead to better robustness of the TID system, which is not very sensitive to unpredictable earthquake excitations, thus such a performance-based optimal design method of the TID system has the promising realistic engineering applications.

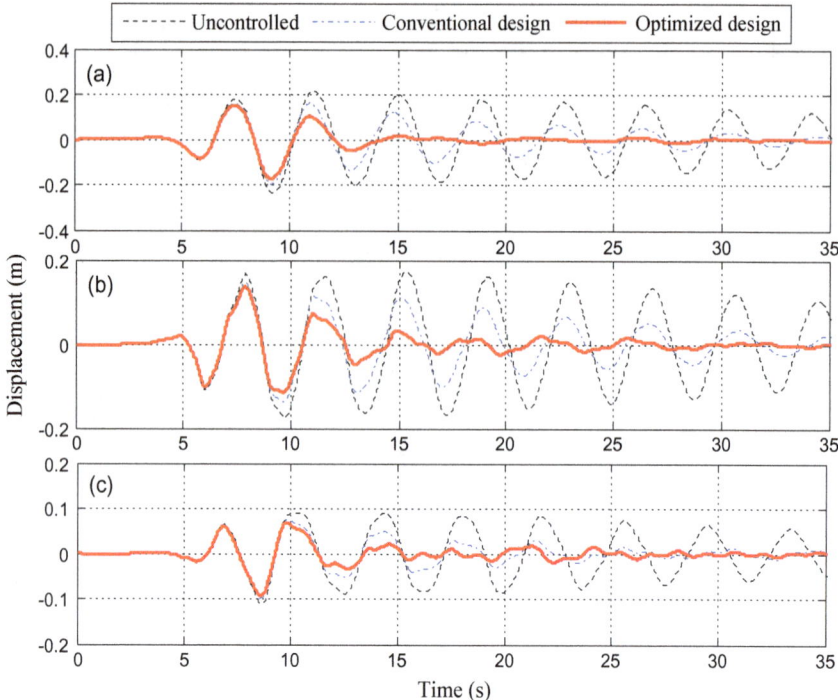

Fig. 8.34 The comparison of displacement time history at the top of the building subjected to three near-field waves: **a** El_Centro_Array_#6; **b** El_Centro_Array_#7; **c** Sturno

References

1. Shan, J., et al. 2016. Seismic data driven identification of linear models for building structures using performance and stabilizing objectives. *Computer-Aided Civil and Infrastructure Engineering* 31 (11): 846–870.
2. Storn, R., and K. Price. 1997. Differential evolution – a simple and efficient heuristic for global optimization over continuous spaces. *Journal of Global Optimization* 11 (4): 341–359.
3. Ohtori, Y., et al. 2004. Benchmark control problems for seismically excited nonlinear buildings. *Journal of Engineering Mechanics* 130 (4): 366–385.
4. Saeki, M. 2005. Analytical study of multi-particle damping. *Journal of Sound and Vibration* 281 (3–5): 1133–1144.
5. Mao, K.M., et al. 2004. DEM simulation of particle damping. *Powder Technology* 142 (2–3): 154–165.
6. Lu, Z., et al. 2014. Discrete element method simulation and experimental validation of particle damper system. *Engineering Computations* 31 (4): 810–823.
7. Wong, C., A. Spencer, and J. Rongong. 2009. Effects of enclosure geometry on particle damping performance. In *50th AIAA/ASME/ASCE/AHS/ASC Structures, Structural Dynamics, and Materials Conference*. Palm Springs: American Institute of Aeronautics and Astronautics.
8. Saeki, M. 2002. Impact damping with granular materials in a horizontally vibrating system. *Journal of Sound and Vibration* 251 (1): 153–161.
9. Ni, Y.Q., et al. 2012. SHM benchmark for high-rise structures: A reduced-order finite element model and field measurement data. *Smart Structures & Systems* 10 (4_5): 411–426.

10. Lu, Z., et al. 2017. Experimental and analytical study on the performance of particle tuned mass dampers under seismic excitation. *Earthquake Engineering and Structural Dynamics* 46 (5): 697–714.
11. Masri, S.F., and A.M. Ibrahim. 1973. Response of the impact damper to stationary random excitation. *The Journal of the Acoustical Society of America* 53 (1): 200–211.

Chapter 9
Semi-active Control Particle Damping Technology

Since semi-active control can provide adaptability comparable to active control without installing large energy sources, semi-active control devices have attracted considerable attention in recent years. In fact, many devices can operate on battery power, which is important in extreme situations such as typhoons, tornadoes, earthquakes, where structural main power supply failures occur. According to the currently accepted definition, the semi-active control device does not increase the mechanical energy (including structure and device) of the control system, but has the property of being dynamically changeable to minimize the response of the structural system. Therefore, semi-active control systems do not have the potential to reduce the stability of structural systems (in limited input/limited output frames) compared to active control systems.

9.1 Preliminary Concepts of Semi-active Control Particle Damping Technology

The semi-active control particle damping technology is formed by introducing the advanced semi-active control technology based on the previous particle damping technology. The basic concept of particle damping technology has been already introduced in detail. In this part, active control technology is briefly discussed, and then the semi-active control of particle damping technology will be introduced.   ˙

Semi-active control is a kind of vibration control technology that achieves the purpose of vibration reduction by changing the stiffness or damping of the structure and adaptively adjusting the dynamic characteristics of the structure based on the structural reaction. It has the advantage that the control effect is similar to the active control but requires less energy input. Besides, since it is a limited input/limited output system, there is no control instability problem like active control. When the energy supply breaks off, it can immediately transform to a passive control system and have control effect, thus having great application prospects.

© China Machine Press and Springer Nature Singapore Pte Ltd. 2020
Z. Lu et al., *Particle Damping Technology Based Structural Control*,
Springer Tracts in Civil Engineering,
https://doi.org/10.1007/978-981-15-3499-7_9

The semi-active control particle damping technology adopts the principle of semi-active control to adjust the parameters of the structural particle damping system in real time, which can widen the vibration band of the previous particle damper, improve the vibration reduction efficiency and durability, and provide a new way to apply particle dampers to structure engineering. At the same time energy consumption is largely reduced compared to active control, conforming to the idea of resource-saving sustainable development. The Chen Qian research group from Nanjing University of Aeronautics and Astronautics [1] has conducted theoretical analysis and experimental research on the damping effect of electromagnetic particle damper under DC electromagnetic field. The results show that under certain vibration intensity, it can be applied by applying DC electromagnetic field. This way promotes the momentum exchange between the particle and the vibration system to improve the vibration reduction effect. At the same time it increases the contact pressure between the magnetic particles, thereby increasing the frictional force and increasing the frictional energy consumption of the damper. The method provides a basis for expanding the application range of particle dampers and suppressing vibrations of different intensity, so that the particle damping can satisfy the requirements of different vibration environments. The results show that the particle damper can be developed from the passive vibration control to the semi-active vibration control, which provides a useful exploration for this potential filed. Masri and Miller [2] and others have conducted a preliminary experiment on a nonlinear semi-active control particle damping device, which optimizes the linkage parameters by using an adjustable gap impact damper. A simple but effective method is proposed for the problem of nonlinear vibration control of multi-degree-of-freedom systems caused by stochastic dynamic environments. It uses a nonlinear additional mass damper with adjustable constraints and arranges the damper at specific locations based on the entire nonlinear system. Numerically simulation is conducted. The efficiency of this semi-active control particle damping technique can be obtained from simulation results. The results are as follows:

It can be seen from Fig. 9.1 that the passive particle damper weakens the response of the structure to a certain extent, but its working performance is limited because it cannot adapt to the various basic structure response. Therefore, by arranging the

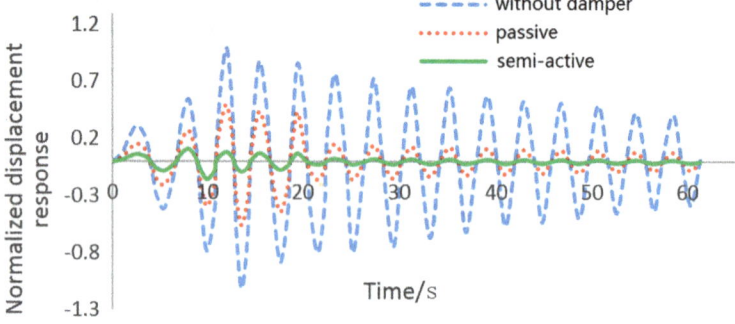

Fig. 9.1 Comparison of system response under different working conditions

semi-active control particle damper to the same basic structure, it can be seen that it performs significantly better than the passive particle damper. As the gap is adjusted in real time, the forces caused by the collision can act on the structure at the most favorable time. This linkage control algorithm does not need an accurate data of the structural characteristic parameters, but only needs to measure the structural position where the damper is arranged, so that the self-adjusting method is used to determine the gap size of each damper to achieve the purpose of optimizing the vibration reducing performance of each device. Masri [2] has studied the control effectiveness of the relative displacement on the top position with different damper arrangement for a three-degree-of-freedom system. The results show that if the other conditions are consistent, the best layout will be at the top of the structure, which is a beneficial guidance to future research. At the same time, Masri carried out stability analysis, numerical simulation and mechanical model test on this control method, and the analysis results show that the principle of this semi-active control particle damping technology is feasible, reliable and promising.

It can be seen that theoretical research and experimental researches conducted by Masri et al. have shown that the effect of the semi-active control system is significantly better than that of the passive device, and the structural vibration response can be effectively reduced in the conditions of a wide range of dynamic load.

9.2 Semi-active Control Algorithm and Numerical Simulation Studies

9.2.1 Semi-active Impact Dampers (SAID) Attached to Linear Structures

In this section, a simple, yet efficient method is presented for the on-line vibration control of nonlinear, multi-degree-of-freedom systems responding to arbitrary dynamic environments. The procedure uses nonlinear auxiliary mass dampers with adjustable motion-limiting stops located at selected positions throughout a given nonlinear system. A mathematical model of the system to be controlled is not needed for implementing the control algorithm. The degree of the primary structure oscillation near each vibration damper determines the damper's actively-controlled gap size and activation time. By using control energy to adjust the damper parameters instead of directly attenuating the motion of the primary system, a significant improvement is achieved in the total amount of energy expended to accomplish a given level of vibration control. In a related paper, the direct method of Lyapunov is used to establish that the response of the controlled nonlinear primary structure is Lagrange stable. Numerical simulation studies of several example systems, as well as an experimental study with a mechanical model, demonstrate the feasibility, reliability, and robustness of the proposed semi-active control method.

Fig. 9.2 Model of arbitrary, nonlinear MDOF system under directly applied dynamic loads $p(t)$ and/or interface motions $^0x(t)$ that is provided with a number of active vibration controllers. Vectors $^1x(t)$ and $^2x(t)$ define the absolute displacements of the primary and secondary system, respectively. The mass of actuator number I is designed by ^2m_i

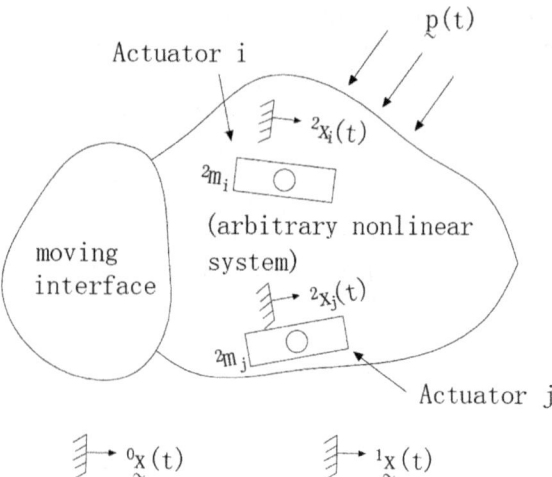

1. Formulation

Consider the arbitrary, nonlinear multidegree-of-freedom system shown in Fig. 9.2 under directly applied dynamic loads $p(t) = (p_1, p_2, \ldots, p_{n_1})^T$ and/or support motions $^0x(t) = (^0x_1, ^0x_2, \ldots, ^0x_{n_0})^T$. The equation of motion for such a system can be expressed as

$$M_{11}^1\ddot{x} + C_{11}^1\dot{x} + K_{11}^1x + M_0^0\ddot{x} + C_{10}^0\dot{x} + K_{10}^0x + f_N(x, \dot{x}, t) = p(t) \qquad (9.1)$$

where

$^1x(t) = (^1x_1, ^1x_2, \ldots, ^1x_{n_1})^T$ is the system displacement vector,
$^0x(t) = (^0x_1, ^0x_2, \ldots, ^0x_{n_1})^T$ is the support displacement vector,
M_{11}, C_{11}, K_{11} = matrices, possibly function of time, each of order $(n_1 \times n_1)$, that characterize the inertia, damping and stiffness forces associated with the n_1 system degrees-of-freedom,
M_{10}, C_{10}, K_{10} = matrices, possibly function of time, each of order $(n_1 \times n_0)$ that characterize the inertia, damping, and stiffness forces associated with the interface motions,
$f_N(x, \dot{x}, t)$ = an n_1 column vector of nonlinear nonconservative forces involving $^1x(t)$ as well as $^0x(t)$, and
$p(t)$ = an n_1 column vector of directly applied forces.

Assuming, without any loss of generality, that the system mass matrix is diagonal allows Eq. (9.1) to be expressed in the form

$$^1m_i^1\ddot{x}_i + ^1f_i(^1x, ^0x, ^1\dot{x}, ^0\dot{x}) = p_i(t); i = 1, 2 \ldots, n_1 \qquad (9.2)$$

where

^1m_i is mass associated with the system degree-of freedom i, and
^1f_i is the "restoring force" associated with system DOF i arising from passive interactions.

If the nonlinear system under consideration is now provided with a number (n_2) of active vibration dampers distributed throughout the vibrating structure, then the n_2 equations of motion of the involved DOFs will change from the form of Eq. (9.2) to

$$^1m_j^1\ddot{x}_j +^1 f_j\left(^1x,^0 x,^1 \dot{x},^0 \dot{x}\right) -^2 f_j(z_j,\dot{z}_j) = p_j(t); j = 1,2\ldots,n_2 \qquad (9.3)$$

Additionally, the passive system's n_1 equations of motion will have to be augmented by n_2 equations that govern the motion of the active vibration dampers:

$$^2m_j^2\ddot{x}_j +^2 f_j(z_j,\dot{z}_j) = p_j(t); j = 1,2\ldots,n_2 \qquad (9.4)$$

where

$^2x(t) = \left(^2x_1,^2 x_2,\ldots,^2 x_{n_1}\right)^T$ = auxiliary masses displacement vector,
$z_j(t) =^2 x_j(t) -^1 x_j(t)$ is the displacement of damper ^2m_j relative to ^1m_j,
^2m_j = mass of auxiliary damper j, and
^2f_j = interaction forces arising from the presence of damper j

It is seen from Eqs. (9.1) through (9.4) that the following convention is followed in the choice of notation: left superscript (0) pertains to the n_0 interface (support) DOFs, left superscript (1) pertains to the n_1 passive system DOFs, and left superscript (2) pertains to the n_2 auxiliary mass dampers' DOFs.

Consider now the class of nonlinear auxiliary dampers that resemble dynamic vibration neutralizers (DVN) with resilient motion-limiting stops. The performance of such devices under a variety of excitations is available in the work of Masri [3]. The influence of this class of devices on the primary system to which they are attached can be expressed as:

$$^2f_j(z_j,\dot{z}_j) =^2 m_j^2\ddot{q}_j(z_j,\dot{z}_j,^j \theta); j = 1, 2, \ldots, n_2 \qquad (9.5)$$

where ^2q_j, the normalized force associated with damper ^2m_j, is given by

$$^2q_j(z_j,\dot{z}_j,^j \theta) = g_j(z_j,d_j) + h_j(z_j,\dot{z}_j,d_j) + r_j(z_j,\dot{z}_j) \qquad (9.6)$$

and the three terms appearing on the right-hand side of Eq. (9.6) are:

$g_j(z_j,d_j)$ = nonlinear conservative force arising from the contact of damper ^2m_j with its constraining (limiting) stops of characteristic dimension d_j,
$h_j(z_j,\dot{z}_j,d_j)$ = nonlinear nonconservative force arising from the contact of damper ^2m_j with its stops, and

$r_j(z_j, \dot{z}_j) =$ nonlinear nonconservative forces arising from the coupling mechanism between damper $^2 m_j$ and its attachment location $^1 m_j$ when the motion-limiting stops are not engaged.

To help interpret the various force terms appearing in the general damper representation of Eq. (9.6), consider the following special cases.

Case (1); Dynamic Vibration Neutralizer (DVN): This widely used linear damper (also known as the "vibration absorber" or "Frahm damper") employs a linear elastic element and, quite often, a linear viscous damping element to couple the auxiliary mass to the oscillating structure. Thus, for this case

$$^2 q_j(z_j, \dot{z}_j, {}^j \theta) = {}^j \theta_1 z_j + {}^j \theta_2 \dot{z}_j \tag{9.7}$$

where

$^j \theta_1 =$ stiffness coefficient of the coupling spring, and
$^j \theta_2 =$ damping coefficient of the coupling dashpot.

Comparing Eq. (9.7) to the general form of Eq. (9.6) yields

$$g_j(\cdot) = 0 \tag{9.8}$$

$$h_j(\cdot) = 0 \tag{9.9}$$

$$r_j(\cdot) = {}^j \theta_1 z_j + {}^j \theta_2 \dot{z}_j \tag{9.10}$$

Case (2); Nonlinear Vibration Neutralizer: In this class of dampers, the coupling element between $^1 m_j$ and $^2 m_j$ has nonlinear characteristics involving the stiffness and/or damping terms. For example, when polynomial-like nonlinearities exist, if the spring has hardening stiffness and the damping forces are a quadratic function of the relative velocity, the three generic components of the damper force appearing in Eq. (9.6) become:

$$g_j(\cdot) = 0 \tag{9.11}$$

$$h_j(\cdot) = 0 \tag{9.12}$$

$$r_j(\cdot) = {}^j \theta_1 z_j + {}^j \theta_2 \dot{z}_j + {}^j \theta_3 z_j^3 + {}^j \theta_4 \dot{z}_j^2 \tag{9.13}$$

Case (3); Impact Damper: In an ideal impact damper which is moving freely in a container with a stiff, resilient stops, the components of Eq. (9.6) assume the form

$$g_j(\cdot) = {}^j \theta_1 \left[z_j - sgn(z_j) {}^j \theta_3 \right] u(|z_j| - {}^j \theta_3) \tag{9.14}$$

$$h_j(\cdot) = {}^j\theta_2 \dot{z}_j u\big(|z_j| - {}^j\theta_3\big) \tag{9.15}$$

$$r_j(\cdot) = 0 \tag{9.16}$$

where

${}^j\theta_1$ = stiffness of the slightly resilient damper stops,
${}^j\theta_2$ = equivalent viscous damping coefficient involved during impacts,
${}^j\theta_3$ = impact damper clearance, equal to one half of the total gap size in the passive damper,
$sgn(\cdot)$ = indicates the algebraic sign of its argument, and
$u(\cdot)$ = unit step function defined by:

$$u(a) = \begin{cases} 1 \ if \ a > 0 \\ 0 \ if \ a \leq 0 \end{cases}$$

Notice that, in this case, no coupling exists between the colliding masses when the relative displacement of the auxiliary mass is less than the available gap; consequently, the coupling force $r_j(\cdot)$ is zero.

Case (4); Nonlinear Vibration Neutralizer With Motion-limiting Stops: This device combines features of the conventional DVN and the impact damper. In the terminology of Eq. (9.6), it is responsible for the following forces:

$$g_j(\cdot) = {}^j\theta_1\Big[z_j - sgn(z_j){}^j\theta_3\Big]u\big(|z_j| - {}^j\theta_3\big) \tag{9.17}$$

$$h_j(\cdot) = {}^j\theta_2 \dot{z}_j u\big(|z_j| - {}^j\theta_3\big) \tag{9.18}$$

$$r_j(\cdot) = {}^j\theta_4 z_j + {}^j\theta_5 \dot{z}_j \tag{9.19}$$

where it is recognized that forces g_j and h_j are identical to the corresponding terms. in Case (3), and force r_j has the same form as in Case (1). Notice that, here, parameters ${}^j\theta_1, {}^j\theta_2$ and ${}^j\theta_3$ govern the performance of the damper in its nonlinear range of motion (i.e., when the available gap is exceeded), while parameters ${}^j\theta_4$ and ${}^j\theta_5$ determine the behavior of the damper within its linear range.

Consider again the nonlinear system whose oscillations are to be attenuated:

$$ {}^1m_i^1 \ddot{x}_i + {}^1f_i = p_i(t); i = 1, 2 \ldots, n_1 - n_2 \tag{9.20}$$

$$ {}^1m_j^1 \ddot{x}_j + {}^1\ddot{x}_j + {}^1f_j - {}^2f_j = p_j(t); j = n_1 - n_2, \ldots, n_1 \tag{9.21}$$

$$ {}^2m_k^2 \ddot{x}_k + {}^2f_k = 0, k = 1, 2, \ldots, n_2 \tag{9.22}$$

Let $y(t)$, an n_1 column vector, denote a measure of the primary system response of interest. For example, if the structural deformations with respect to a moving base are of concern, $y(t)$ can be composed of a combination of the primary system relative displacements and velocities. On the other hand, if peak deformations are of interest, the entries in y can correspond to the maximum deformations of designated locations. Hence, the response of the dynamic system with dampers whose motion is governed by the $(n_1 + n_2)$ equations given in Eqs. (9.20)–(9.22) can be expressed as:

$$y(t) = y\left({}^1\theta, {}^2\theta, \ldots, {}^{n_2}\theta\right) \tag{9.23}$$

Let the cost function to be minimized be

$$J\left({}^1\theta, {}^2\theta, \ldots, {}^{n_2}\theta\right) = \int_{t_0}^{t_0+T_{opt}} y^T(t)Wy(t)dt \tag{9.24}$$

where W is an arbitrary weighting matrix.

In principle, the optimization task is now reduced to seeking the set of damper parameters which will minimize J over the response segment T_{opt}. When this optimization is performed once "off-line" for the whole response record, the result is an optimized set of passive damper parameters. However, passive dampers, even when optimally designed for a particular situation, may have limited effectiveness when operating under wide-band excitations.

On the other hand, the continual optimization and adjustment of the damper parameters (fully active control) requires the "on-line" solution of Eq. (9.24) and the continuous feedback of the results to the control actuators. This approach, while mathematically appealing, is not feasible for a variety of reasons, the leading one of which is the demanding analytical and computational effort required to determine (let alone adjust) the optimum damper parameters in a small fraction of the structure time constant.

This study presents a compromise solution of the two control options discussed above: (1) passive dampers initially optimized off-line, and (2) fully-active optimized dampers with continuous feedback control. The alternate option of this part is to trade degraded optimization (i.e., open-loop, suboptimal control) for ease of implementation in real life engineering situations with actual hardware.

The motivation behind the proposed control algorithm is the observed behavior of passive impact dampers configured in the form of dynamic vibration neutralizers with motion-limiting stops. When one such passive damper is attached to an oscillating primary system undergoing transient excitations, the auxiliary mass will sustain repetitive (possibly chaotic) impacts on different sides of its container. The number, location, and intensity of these irregular impacts is a highly nonlinear function of

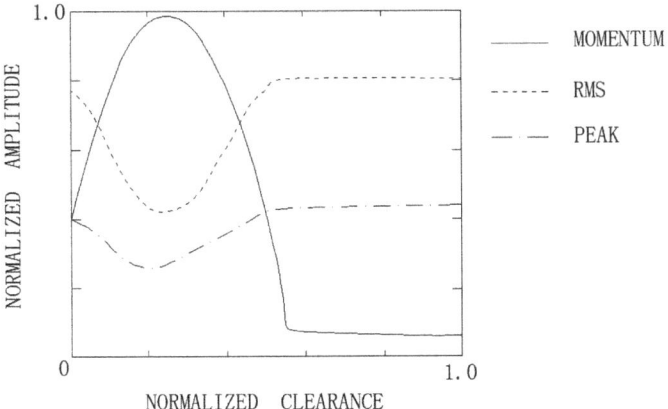

Fig. 9.3 Transient response of a linear SDOF system provided with a passive impact damper and subjected to swept-sine excitation. A representative segment of the primary system response between two consecutive impacts separated by a time period T_{opt} is considered. The plotted curves show the variation of the indicated quantity with the gap sized, all other parameters remaining the same. The primary system ratio of critical damping is 0.05. The impact damper mass ratio is 0.10 and its coefficient of restitution is e ≈ 0.75. Notice that the time increment T_{opt} varies with the gap size. **a** Momentum transfer between the colliding masses at the end of the observation time segment from t_0 to $t_0 + T_{opt}$; **b** peak value of the primary system displacement; and **c** R.M.S. value of the primary system displacement

the system characteristics and the nature of the excitation. The ensuing plastic deformations, Coulomb friction, and momentum transfer between the two masses during collisions tend to reduce the vibrations of the primary system.

If the time of occurrence of one of these impacts is used to define a reference time t_0, then the variation of the peak and R.M.S. levels of the primary system response with the gap size that governs the time of occurrence of the succeeding impact will be as indicated in Fig. 9.3.

Since the predominant mechanism that governs the interaction between ^1m_i and ^2m_i is momentum transfer, it is reasonable to expect a strong dependence of the criterion function $J(\cdot)$ on the discontinuity in the velocity of $^1\dot{x}_i$ and/or $^2\dot{x}_i$ during the impact process. This expectation is borne out by the results depicted in Fig. 9.3, where the value of the momentum transfer is superimposed on the graph of the constituents of $J(\cdot)$.

It is thus clear that, at least for the example problem shown in Fig. 9.3, optimizing $J(\cdot)$ is practically identical to seeking an extremum value of the momentum transfer involved in the impact process. For the class of problems under discussion, this condition is equivalent to having an impact occur when the primary system's velocity is at its peak value.

The preceding discussion established the guidelines for a procedure to optimize the operation of semi-active impact dampers configured as mentioned above. To maximize the efficiency of an impact damper between two consecutive impacts, the gap size d should be adjusted so that the following conditions are satisfied:

- For each damper mass ^2m_j an impact is made to occur when the velocity of the corresponding primary system mass ^1m_j has reached its peak value. This instant corresponds to the zero crossing of the corresponding primary system displacement.
- The velocities of the various set(s) of two colliding masses must be opposite to each other at the time of impact. This condition insures that the impact process(es) will stabilize the motion of the primary system.

On this basis, the following control algorithm for on-line implementation of the damping device(s) is proposed. The control strategy consists of detecting the displacement from the neutral position (absolute, or relative to a moving support) zero crossings of the oscillating structure damper locations, and generating sets of impulsive control forces by inducing a collision between each of the auxiliary masses and their corresponding structure locations. The essential features of this approach can be summarized as follows:

- Virtually no on-line information regarding the global dynamic system characteristics is needed.
- Whether the primary system is linear or nonlinear has no bearing on the algorithm.
- Monitoring of only the system relative displacements at the dampers' locations is required.
- The on-line computation of the optimum clearance distances is reduced to a simple detection process.

To illustrate the application of this approach, a representative segment of the motion of a linear SDOF oscillator being controlled by such a semi-active damper is shown in Fig. 9.4. The primary system is harmonically excited at resonance. These graphs represent the absolute and relative state variables of the system and the nonlinear conservative and nonconservative control functions, g and h. The amplitude of all time histories in this figure have been normalized to lie between -1.0 and $+1.0$. The length of time segment shown is approximately two natural periods.

In Fig. 9.4a, the solid line represents $^1x(t)$ the absolute displacement of the primary system $^1m_1$, while the dashed line represents $^2x_1$, the absolute displacement of the secondary system $^2m_1$. Similarly, in Fig. 9.4b the solid and dashed lines represent $^1\dot{x}(t)$, the absolute velocity of $^1m_1$ and $^2\dot{x}(t)$, the absolute velocity of $^2m_1$, respectively. The time histories of the relative displacement $z_1 = {}^2x_1 - {}^1x_1$ and relative velocity $\dot{z}_1 = {}^2\dot{x}_1 - {}^1\dot{x}_1$ are represented by the solid and dashed lines, respectively, in Fig. 9.4c. The time history of the nonlinear stiffness force $g_1(t)$ generated by the contact of the oscillating mass $^2m_1$ with its resilient "stops" is shown in Fig. 9.4d. Similar results for the nonlinear damping force $h_1(t)$ arising during the impact process is shown in Fig. 9.4e. As seen from these results, suitable control impacts are applied twice every fundamental period of the system. The total control energy exerted on the structure during an impact is the sum of the areas under the g and h functions.

The only significant disadvantage of this technique is the lack of consideration for possible hardware delays in the activation of the impacting mechanism. Two possible provisions may be adopted to overcome this inadequacy:

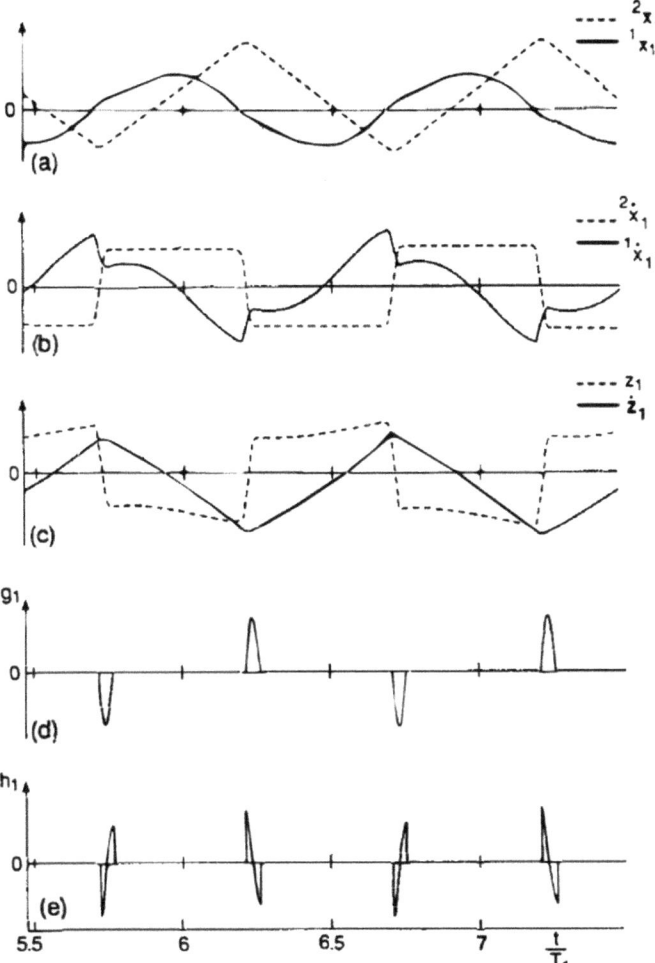

Fig. 9.4 Time history of a representative segment of the steady-state motion of a SDOF system, that is harmonically excited at resonance and provided with a semi-active impact damper having a mass ratio of 0.1. The time segment shown covers approximately two natural periods T_1 of the primary system. For clarity, the amplitude of all the plotted quantities have been normalized. **a** Absolute displacements of the primary and secondary systems; **b** absolute velocity of the primary and secondary systems; **c** relative displacement and velocity between the primary and secondary systems; **d** nonlinear stillness force; and **e** nonlinear damping force

- The first obvious choice is to design a high speed activation system with delays that are small when compared to the fundamental period of the structural system.
- The alternative solution is to anticipate the system response and thus activate the impacting mechanism when the displacement of the structure has crossed a certain prescribed threshold level.

2. *Stability Analysis*

A stability analysis of the device under discussion has been performed and is available in the work of Karyeaclis and Caughey [4]. Using Lyapunov's approach it is shown that, under fairly permissive conditions, the response of a system provided with the SAID under discussion is bounded.

3. *Numerical Simulation*

The efficiency of the proposed control strategy is demonstrated by presenting numerical simulation results for several SDOF and MDOF models with diverse characteristics, subjected to deterministic and stochastic dynamic environments.

Example (1): SDOF System Under Swept-Sine Excitation The results shown in Fig. 9.5 correspond to a linear, viscously damped SDOF system consisting of a mass $^1 m_1$ having a ratio of critical damping $\zeta = 0.01$, and initially at rest, that is subjected to swept-sine excitation $F(t)$ given by

$$F(t) = F_0 \sin[\Omega(t)t] \tag{9.25}$$

The time variation of the exciting frequency, Ω, is of the form

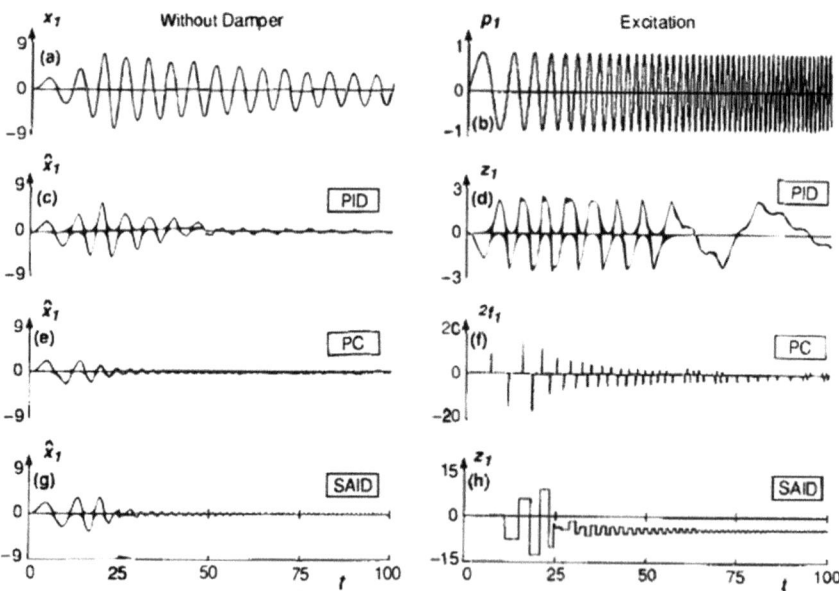

Fig. 9.5 Swept-sine excitation of a SDOF system with a variety of damping devices. **a** primary system response in the absence of any dampers; **b** excitation; **c** response with a passive impact damper; **d** relative displacement between the passive impact damper mass and the primary system; **e** response when using on-line pulse control; **f** pulse control forces; **g** response when using a semi-active impact damper; and **h** evolution of the gap size in the semi-active damper

$$\Omega(t) = at + b \tag{9.26}$$

If this linear system is subjected to a swept-sine excitation, shown in Fig. 9.5b of amplitude F_0 that varies according to Eq. (9.26) between the frequency limits $\Omega(0)/\omega_1 = 0.5$ and $\Omega(T_s)/\omega_1 = 1.5$ in sweep time $T_s/T_1 = 25$, the transient response shown in Fig. 9.5a is obtained.

Suppose now that the primary system under consideration is equipped with a conventional impact damper having an auxiliary mass ratio $\mu = 0.1$. Assume that the damper stops are relatively stiff and have impact plastic deformation characteristics equivalent to a coefficient of restitution $e \approx 0.8$ (within the range provided by hardened steel). Let the damper clearance ratio $d^* = [d/^1x_{1max}]$ be optimized in accordance with the response characteristics of such nonlinear devices thus yielding an optimum clearance of $d^*_{opt} \approx 2.0$ [5].

The normalized response of the primary system with an optimized passive impact damper (PID) will then be as shown in Fig. 9.5c. Notice that, in this case, the peak amplitude is attenuated by the factor ≈ 0.7 relative to the corresponding peak response in Fig. 9.5a. It is seen in Fig. 9.5d that, due to the nature of the passive impact damper, the relative displacement between the colliding masses is constrained to remain within the fixed gap size of $\pm d/2$.

It is clear from the results shown in Fig. 9.5c that, while the optimized PID did attenuate the peak response to some extent, its efficiency was limited because it could not adapt to the transient nature of the primary system response. This problem can be easily remedied by using an active on-line pulse control procedure previously developed by the authors. When this control method is applied to the primary system under discussion, it results in the response shown in Fig. 9.5e. The control forces that are used here are governed by the following rule ("active" viscous damping):

$$F_c(t) = \begin{cases} -^2c_1^1\dot{x}_1(t), & t_0 \le t \le (t_0 + T_d) \\ 0 & \text{otherwise} \end{cases}$$

where t_0 is the pulse initiation time and T_d is the pulse duration.

The time history of the control forces generated by the actuators that are using an external energy source are shown in Fig. 9.5f. Notice that, due to the nature of this control algorithm, the actuation time of the pulse control forces coincides with the primary system displacement zero crossings (equivalent to velocity peaks). In addition, the magnitude of the control force is changed once every one-half system period to maintain a value which is a constant factor of the primary system velocity.

The attenuation in peak amplitude with the on-line pulse control is $[^1\hat{x}_{1max}/^1x_{1min}] \approx 0.3$, which is substantially better than what was achieved with the optimum passive dampers discussed above. Obviously, the cost of this added efficiency is the need to furnish an external energy source for the expenditure of the control energy.

If a semi-active impact damper is now attached to the primary system under discussion, the response shown in Fig. 9.5g is obtained. The auxiliary mass has the same ratio ($\mu = 0.10$) used by the passive impact damper discussed previously, and the damper stops have the same coefficient of restitution as for the PID. The evolution of the adjustable stops is shown in Fig. 9.5h.

As might be expected, the efficiency of the SAID (a peak reduction factor in the ratio of $= 0.4$) is better than what was achieved by the optimized passive impact damper, but not as good as the active pulse-control procedure. A clear visual explanation for the improved damping efficiency of the SAID is furnished by Fig. 9.5h where it is seen that the envelope of the optimum gap size, was being adapted to closely match that of the primary system response. Furthermore, the time-varying gap size caused the collisions between the oscillating masses to occur at a time when the interaction force $^2f_1(\cdot)$ components had the most beneficial effect (as regarding motion attenuation) on the primary system. Thus, $^2f_1$, the combined force due to g_1 and h_1, is seen to play the same role, and to have the same qualitative features, as the active control force $F_c(t)$ shown in Fig. 9.5f.

Example (2): SDOF Under Nonstationary Random Excitation This case is similar to the one in Example 9.1 except that the disturbance is a wide-band nonstationary random excitation. The identical damper parameters of Example 9.1 are used again. The performance characteristics of the various dampers are shown in Fig. 9.6. The relative efficiency of various damping devices matches the results under swept-sine excitation shown in Fig. 9.5.

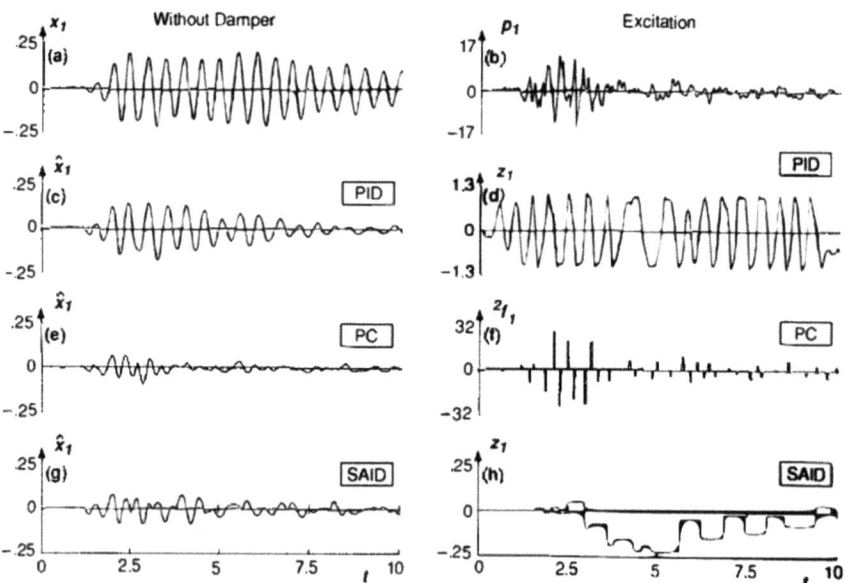

Fig. 9.6 Nonstationary random excitation of a SDOF system with the same variety of damping devices used in Fig. 9.5

The time history of the adaptive gap size is shown in Fig. 9.6h. The lack of any discernible pattern in the evolution of the optimum gap size reflects the nature of the random disturbance. The complex changes of $z_1(t)$ between impacts clearly illustrate the handicaps passive dampers have to cope with, since their initial (fixed) gap size cannot change in time to accommodate quiescent or active episodes of the random response.

Example (3): Linear MDOF System Under Nonstationary Random Excitation
Consider a MDOF linear frame structure that is subjected to wide-band random interface motion and without any directly applied loads. The chain-like nature of this example is in no way a requirement of the control algorithm under discussion; it is merely a convenient choice so as to make the system resemble, for example, a building-like structure undergoing earthquake ground motion.

The response of this structure under a simulated earthquake, operating without any auxiliary mass dampers, is shown in the LHS column of plots in Fig. 9.7. If a SAID of mass ratio $\mu_i = 0.05$, $i = 1, 2, 3$ is attached to each of the three "stories," then the controlled response would be as shown in the middle column of plots in Fig. 9.7, and the corresponding variable gap sizes are shown in the RHS column of plots.

Variable $y_i(t)$ represents the displacement of the ith level in the structure with respect to the oscillating base. $y_1(t)$ is closest to the base and $y_3(t)$ is the farthest

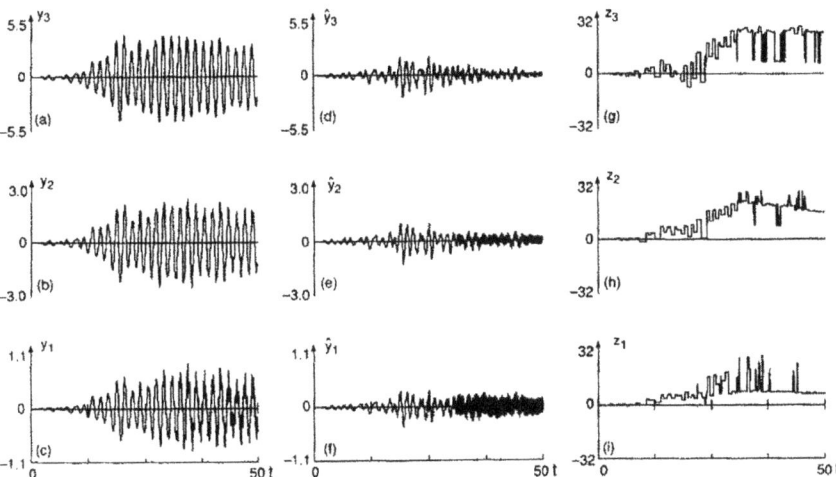

Fig. 9.7 Response of a linear frame structure, resembling a 3-story building, under nonstationary base excitation. The left-hand side, column of plots represent the transient response without augmented damping, the middle column shows the corresponding response (plotted to the same scale) when a separate SAID is attached to each level in the structure, and the right-hand column of plots shows the evolution of the dampers stops. Variable $y_i(t)$ represents the displacement of the ith level in the structure with respect to the oscillating base. The mass of each damper is 5% of the corresponding location mass

away. For clarity, different scales are used for the ordinates of the plots corresponding to y_1, y_2 and y_3 in Fig. 9.7a, b, c. However, the middle column of plots uses the same amplitude scales as the corresponding uncontrolled cases. Comparison of the controlled and uncontrolled responses of various locations indicates that nearly the same percentage vibration attenuation is achieved at each of the controlled locations.

The influence of the SAID location on the efficiency of the device is demonstrated in Fig. 9.8, where a single SAID is attached to different locations in the MDOF system under discussion. The LHS column of plots in Fig. 9.8 shows schematic diagrams of the attachment points of the SAID, the middle column of plots presents the controlled response of the top mass m_3 (not necessarily the location of the SAID), and the RHS column of plots gives the evolution if the optimum gap size for each of the three tests.

The plots in Fig. 9.8 show the effects of the SAID locations on the attenuation of the relative displacement of the top floor. Notice that evolution of the damper clearance is clearly dependent on the local oscillations in the vicinity of the damper. The same relative reduction in the response is attained for the location that are not shown in the figure. Everything else being the same, it is clear that the top floor is the best location to use if a single SAID is to be employed.

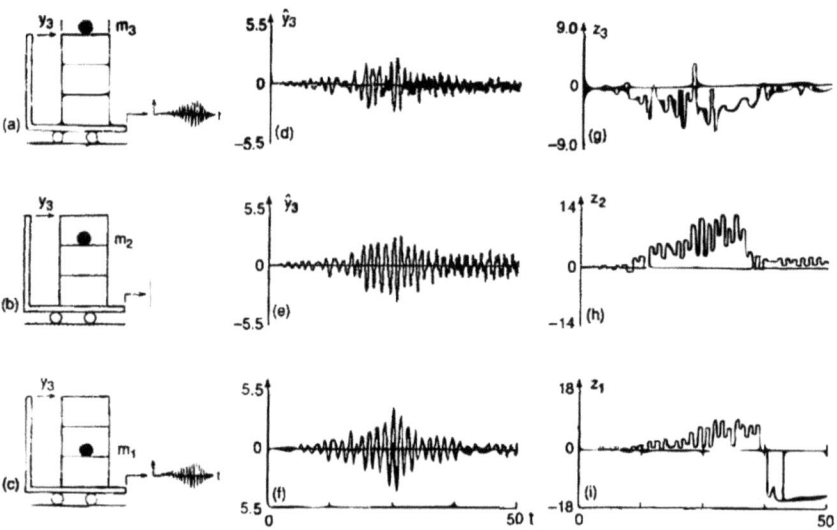

Fig. 9.8 Influence of the controller location on the response of the structure in Fig. 9.7. The top row of plots indicate in part (**a**) the location of a single SAID, whose mass equals S percent the total primary structure mass, attached to the top "floor" m_3 of the structure, the controlled response of m_3 relative to the moving base is shown in (**d**), and the evolution of the damper gap is shown in (**g**). Similar results are shown in the middle row of plots for the case where the location of the damper is moved from m_3 to m_2, and in the bottom row for the case where the damper location is moved to m_1

The applied excitation is identical to that used in conjunc-tion with the system of Fig. 9.7. The mass ratio of the single SAID used in each of the three cases illustrated in Fig. 9.8 was equivalent to that total mass ratio incorporated in the three SAID used simultaneously in Fig. 9.7.

Information about the effect of the placement of active control devices on vibrating structures is available in the works of Lindberg and Longman [6] and Chassiakos et al. [7].

9.2.2 Semi-active Impact Dampers Attached to Nonlinear Structures

The semi-active impact damper (SAID) is proposed to improve the damping effi-ciency of traditional passive impact dampers. In this part, in order to investigate its damping mechanism and vibration control effects on realistic engineering structures, a 20-story nonlinear benchmark building is used as the main structure. The studies on system parameters, including the mass ratio, damping ratio, rigid coefficient, and the intensity of excitation are carried out, and their effects both on linear and nonlinear indexes are evaluated. The damping mechanism is herein further investigated and some suggestions for the design in high-rise buildings are also proposed. To validate the superiority of SAID, an optimal passive particle impact damper (PID_{opt}) is also investigated as a control group, in which the parameters of the SAID remain the same, and the optimal parameters of the PID_{opt} are designed by differential evolution algorithm based on a reduced-order model. The numerical simulation shows that the SAID has better control effects than that of the optimized passive particle impact damper, not only for linear indexes (e.g., R.M.S. response), but also for nonlinear indexes (e.g., component energy consumption and hinge joint curvature).

1. *Configuration model and semi-active control description*

In this section, the system governing equation of a traditional particle impact damper (PID) is reviewed first, based on that, the difference between the PID and semi-active impact damper (SAID) is then discussed. Finally, the specific simulation details of SAID are set forth. Additionally, a detailed example comparison of the two control devices will be launched in Sect. 9.3.

In order to compare the vibration control effect of SAID, the 20-story nonlinear seismic benchmark building that can resist the external horizontal loads, designed for the SAC Phase III Steel Project, is used in the researches by Ohtori et al. [8]. The benchmark building is 80.77 m in height, and its plane dimensions are 30.48 m × 36.58 m. The building has five bays each in the north-south (N-S) and six bays in the east-west (E-W) direction, and the bays are 6.10 m in all directions on the center. The lateral force-resisting system consists of steel perimeter moment-resisting steel frames (MRFs). The simulation study focuses on the in-plane analysis (N-S MRFs), and then the in-plane finite element model is analyzed. The elevation of the N-S MRFs is shown in Fig. 9.9. Taking the nonlinear response of the material into consideration,

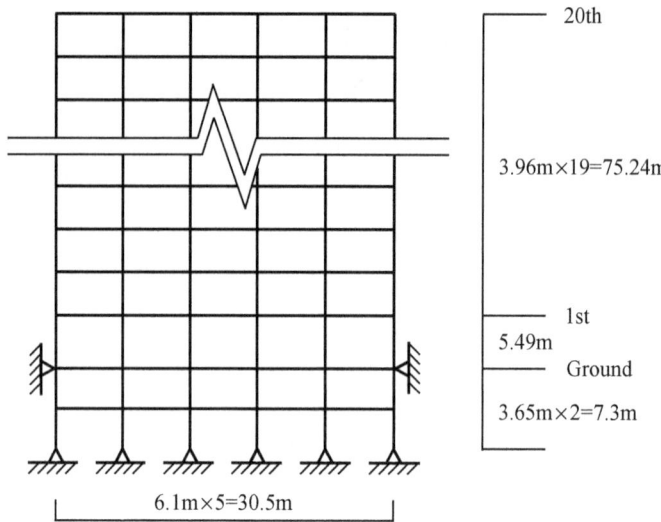

Fig. 9.9 Elevation of the N-S MRFs

involves the moment-curvature bi-linear hysteresis model for structural components, and the concentrated plasticity model is utilized, whose plastic hinges only occur at the ends of the moment resisting beam-column and column-column joints and beams and columns are assumed to be elastic. The first ten natural frequencies of the 20-story nonlinear benchmark model are: 0.261, 0.753, 1.30, 1.83, 2.40, 2.44, 2.92, 3.01, 3.63 and 3.68 Hz, respectively. The numerical simulations are carried out by MATLAB and the SIMULINK platform. The simplified diagram of the main structure with PID and SAID are shown in Fig. 9.10, respectively. The system governing equation of PID can be written as Eqs. (9.27)–(9.30):

$$\begin{cases} \mathbf{M\ddot{U}} + \mathbf{C\dot{U}} + \mathbf{KU} = \mathbf{ME}\ddot{x}_g + \mathbf{\psi} F_p \\ m_p \ddot{u}_p + F_p = \mathbf{0} \\ F_p = c_p H(y, \dot{y}) + k_p G(y) \end{cases} \tag{9.27}$$

$$\begin{cases} G(y) = \left[y - \text{sgn}(y) * \frac{d}{2} \right] * \mathrm{u}\left(|y| - \frac{d}{2} \right) \\ H(y, \dot{y}) = \dot{y} * \mathrm{u}\left(|y| - \frac{d}{2} \right) \end{cases} \tag{9.28}$$

$$\text{sgn}(y) = \begin{cases} -1 & y < 0 \\ 0 & y = 0 \\ 1 & y > 0 \end{cases} \tag{9.29}$$

$$\mathrm{u}\left(|y| - \frac{d}{2} \right) = \begin{cases} 1 & |y| - d/2 > 0 \\ 0 & |y| - d/2 \leq 0 \end{cases} \tag{9.30}$$

where $\mathbf{M}$, $\mathbf{C}$, $\mathbf{K}$ represent the mass, damping and stiffness matrices of main structure, respectively; $\mathbf{U}$, $\mathbf{\dot{U}}$, $\mathbf{\ddot{U}}$ are the displacement, velocity and acceleration vectors of main

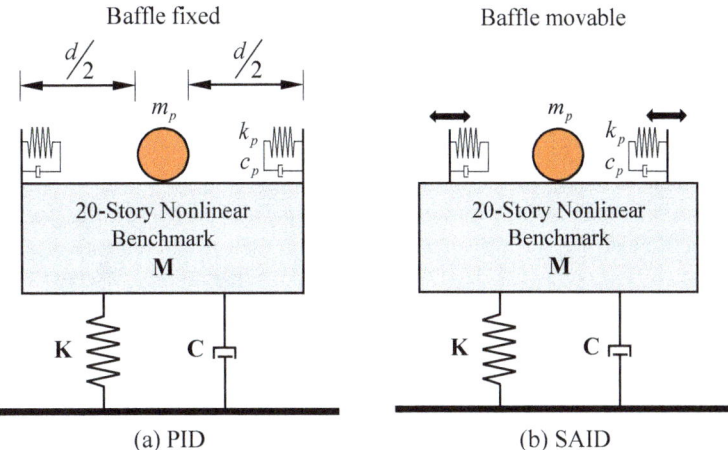

Fig. 9.10 Simplified diagram of the main structure with controller

structure, respectively, $\mathbf{E}$ is the location vector of seismic force, ψ represents the location vector of the PID (For high-rise buildings, the first order mode usually contributes the most response, both PID and SAID are installed on the top layer in this investigation thus), $y = u_p - u_{top}$ represents the relative displacement of the particle with respect to the top floor, $G(y)$ and $H(y, \dot{y})$ are nonlinear functions in Eq. (9.28), as shown in Fig. 9.11a, b, respectively. $[y - \text{sgn}(y) * d/2]$ represents the overlapping distance in the collision process and $u(|y| - d/2)$ is the collision indicator; d represents the gap clearance between the particle and the baffle, the gap clearance keeps in constant for the PID.

In addition, F_p represents the nonlinear restoring force between the main structure and the particle, including the nonlinear stiffness force and nonlinear damping force. k_p and c_p represent the stiffness and damping respectively, which can be used to

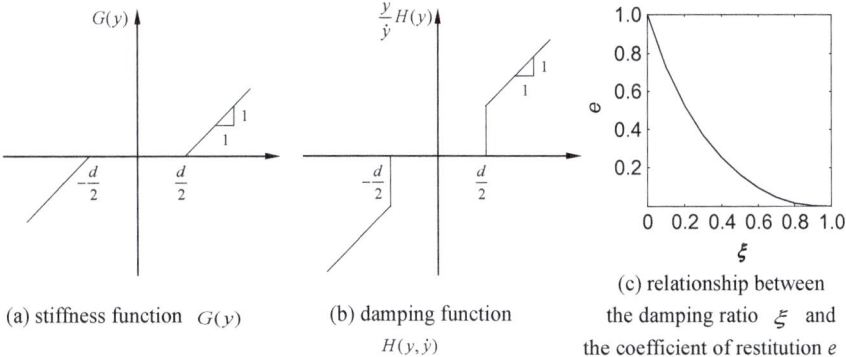

Fig. 9.11 Parameter relationship graph

simulate the interaction between the particle and the baffle: $k_p = m_p\omega_p^2$, $c_p = 2\xi m_p\omega_p$ and $\omega_p = \lambda\omega_1$, where ω_1 represents the first-order frequency of the main structure. λ represents the rigid coefficient between the particle and the baffle, which can be used to simulate the baffle by a proper value; ξ represents the damping ratio, which can be used to simulate inelastic impacts, ranging from completely plastic to elastic impacts. In fact, the Lu et al. [9] found that damping ratio is related to the well-known coefficient of restitution, as shown in Fig. 9.11c; thus, the value of any desired coefficient of restitution can be achieved by selecting the proper value for ξ.

The main difference of the SAID, compared to a conventional passive PID, is that the moment of an impact is adjustable. As mentioned above, an impact is made to occur when the velocity of the top floor of the main structure has reached its peak value, and the corresponding displacement of the top floor is equal to zero at the instant. Hence, the main governing equation of the SAID remains the same as PID (Eq. 9.27) and the definition of gap clearance changes, which will be illustrated in the following simulation details.

The impacts will occur twice in a round-trip vibration, and the stability of this control strategy can be guaranteed, while the opposite condition of the relative velocity $\dot{y}$ and the absolute velocity $\dot{u}_{top}$ can be always satisfied. Diagrams of the impact process at the different moment are shown in Fig. 9.12, in which Fig. 9.12a–d represents impact 1, Fig. 9.12e–h represents impact 2. For instance, the whole process of impact 1 can be divided into four phases, just as shown in Fig. 9.12a–d, and the detailed information of simulation is discussed as follows:

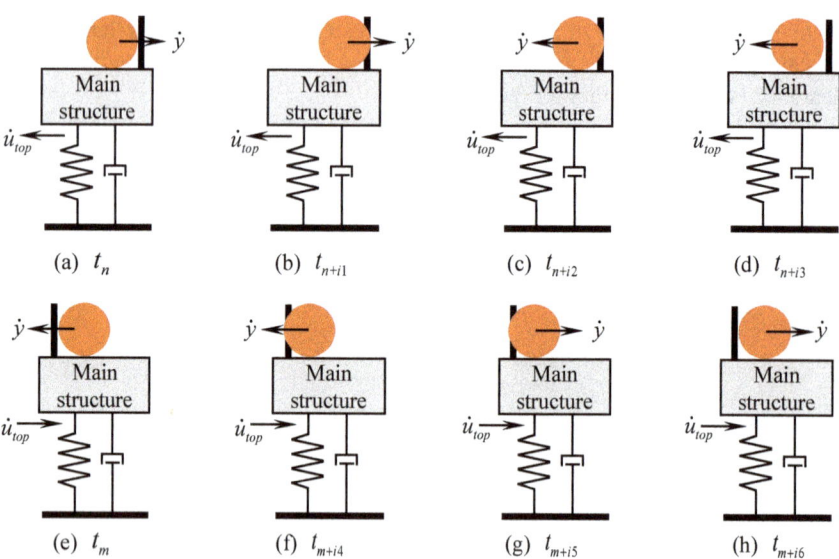

Fig. 9.12 Diagrams of the impact process at different moments: impact 1 (**a–d**); impact 2 (**e–h**)

(a) Start of the impact. The absolute displacement of the top floor is tracked, and we can determine an impact is happening when the absolute displacement cross zero, as shown in Fig. 9.12a, the collision indicator can be expressed as Eq. (9.31):

$$v\left(u_{top}(t_n), u_{top}(t_{n+1})\right) = \begin{cases} 1 & u_{top}(t_n) * u_{top}(t_{n+1}) \leq 0 \\ 0 & u_{top}(t_n) * u_{top}(t_{n+1}) > 0 \end{cases} \tag{9.31}$$

where $u_{top}(t_n)$ and $u_{top}(t_{n+1})$ represents the absolute displacements of the top floor at time t_n and t_{n+1}, respectively. The relative displacement $y(t_{n+1})$ between the particle and the top floor at time t_{n+1} is regarded as the new gap clearance in this impact.

(b) Overlapping and relative velocity $\dot{y}$ does not reverse, as shown in Fig. 9.12b. At the time $t_{n+i}(i \geq 2)$, the overlapping distance can be expressed as $[y(t_{n+i}) - y(t_{n+1})]$, which generates the nonlinear restoring force. Nonlinear functions $G(y)$ and $H(y, \dot{y})$ in the whole impact process can be rewritten as follows:

$$\begin{cases} G(y) = [y(t_{n+i}) - y(t_{n+1})] * v\left(u_{top}(t_n), u_{top}(t_{n+1})\right) \\ H(y, \dot{y}) = \dot{y} * v\left(u_{top}(t_n), u_{top}(t_{n+1})\right) \end{cases} \tag{9.32}$$

Furthermore, the nonlinear stiffness force will increase and the nonlinear damping force will decrease in this process.

(c) Overlapping and relative velocity $\dot{y}$ reverses, as shown in Fig. 9.12c. The non-linear stiffness force will decrease with the decrease of overlapping distance according to Eqs. (9.27) and (9.32), and the nonlinear damping force reverses since the relative velocity $\dot{y}$ reverses.

(d) End of the impact, as shown in Fig. 9.12d. The function $v(y(t_{n+i}), y(t_{n+1}))$ expressed in Eq. (9.33) is used to determine if the impact is over.

$$v(y(t_{n+i}), y(t_{n+1})) = \begin{cases} 1 & abs(y(t_{n+i})) \leq abs(y(t_{n+1})) \\ 0 & abs(y(t_{n+i})) > abs(y(t_{n+1})) \end{cases} \tag{9.33}$$

In conclusion, combined with governing Eq. (9.27) and Eqs. (9.31)–(9.33), the response of the main structure with the SAID can be determined.

2. *Parametric study for SAID*

In this section, parametric studies are performed on the nonlinear benchmark building under discussion, with a semi-active impact damper (SAID), under El Centro wave with intensity 1.5 to enhance the understanding of the SAID behavior. According to the governing equations, some system parameters, such as mass ratio, the damping ratio of the particle, the rigid coefficient, can influence the nonlinear restoring force, and thus influencing the seismic behavior of the nonlinear main structure. Therefore, these three parameters are investigated in this section. Additionally, some external parameters, such as excitation intensity are also investigated.

The mass ratio μ is defined as the mass of the particle to the total mass of the main structure, for practical high-rise buildings, the mass ratio is usually less than 0.01, such as the Shanghai Tower. Thus, two mass ratios, 0.005 and 0.01, are chosen in this section for discussion. The rigid coefficient λ can be used to simulate the baffle with a proper value; similar to the coefficient of restitution, the damping ratio ξ can be used to simulate both elastic and inelastic impacts. For an impact damper, a larger nonlinear restoring force would cause a larger noise; hence the smaller rigid coefficient, ranges from 6 to 16 with the interval 1, is chosen, and the value of the damping ratio of the particle ranges from 0 to 0.3 with the interval 0.03, with totally 121 data points for each mass ratio.

To evaluate the vibration control effects of the SAID with different parameters, four linear indexes, including the peak acceleration (a_{max}), peak displacement (x_{max}), root-mean-square acceleration (R.M.S. a) and root-mean-square displacement (R.M.S. x), and three nonlinear indexes, including the number of plastic hinges (N_p), the total component energy consumption (N_e) and the maximum ratio of joint curvature (N_c), are investigated in this section. Note that the response of the top floor is the largest, which will affect the four linear indexes directly. Furthermore, all values (a_{max}, x_{max}, R.M.S. a, R.M.S. x, N_p, N_e, N_c) have been normalized by dividing by uncontrolled condition (a_{max0}, x_{max0}, R.M.S. a_0, R.M.S. x_0, N_{p0}, N_{e0}, N_{c0}).

Figure 9.13 shows the peak acceleration at the top floor. It can be seen that the control effect of SAID on peak acceleration is not good due to the sudden impact. To illustrate this phenomenon, time-history curves of acceleration at the top floor without control and with SAID (parameters: $\mu = 0.005$, $\lambda = 16$, $\xi = 0.15$) are shown in Fig. 9.14, and two factors have been summarized as follows:

(1) The peak acceleration of the main structure occurs at a very early time (around the 3rd second), and sufficient collisions between the particle and the baffle would take a few seconds. As shown in Fig. 9.14, only the inefficient impact 1 occurs before it, which has little influence on the time-history curve.

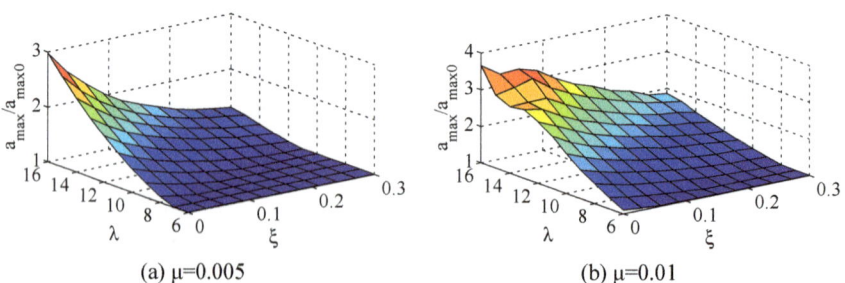

(a) μ=0.005	(b) μ=0.01

Fig. 9.13 Peak acceleration at the top floor with different mass ratio

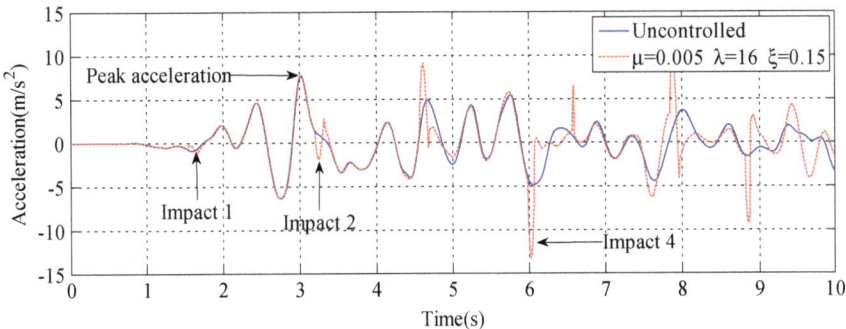

Fig. 9.14 Time-history response of acceleration at the top of the building

(2) When the nonlinear restoring force generated by the SAID with unsuitable parameters is too large, the peak acceleration can be amplified instead, such as the impact 4 in Fig. 9.14.

As the dark blue area is shown in Fig. 9.13, when the normalized value is close to 1, it indicates the factor (1) is the main factor; when the normalized value is larger than 1, it indicates the factor (2) is the main factor. In addition, the following rules can be summarized when the factor (2) is dominant:

(1) With a certain mass ratio μ, as the rigid coefficient λ increases, the peak acceleration increases; as the damping ratio ξ increases, the peak acceleration decreases; the influence of λ is more significant than ξ.
(2) With the increase of μ, the area originally dominated by the factor (1) has changed into factor (2), and the peak acceleration increases. In order to select parameters conveniently, a smaller μ is recommended.

The peak acceleration is proportional to the nonlinear restoring force at the top floor, which includes two parts: (a) nonlinear stiffness force F_{Ns}, and (b) nonlinear damping force F_{Nd}, which can be expressed as follows, respectively:

$$F_{Ns} = k_p G(y) = m_p \omega_p^2 G(y) = m_p \omega_1^2 \lambda^2 G(y) \tag{9.34}$$

$$F_{Nd} = c_p H(y, \dot{y}) = 2\xi m_p \omega_p H(y) = 2m_p \omega_1 \lambda \xi H(y, \dot{y}) \tag{9.35}$$

The influence of parameter variation on the nonlinear restoring force is mainly reflected in two aspects: (1) Direct influence, such as λ and ξ; (2) Indirect influence, such as the nonlinear function $G(y)$ and $H(y, \dot{y})$. In order to unify these two aspects into the effect of parameter variation, three sets of parameters are investigated: (a) $\mu = 0.005, \xi = 0.3, \lambda = 16$; (b) $\mu = 0.005, \xi = 0.3, \lambda = 8$; (c) $\mu = 0.005, \xi = 0.15, \lambda = 16$. Set (a) and set (b) can be utilized to investigate the influence of λ, while set (a) and set (c) can be utilized to investigate the influence of ξ. Figure 9.15 shows

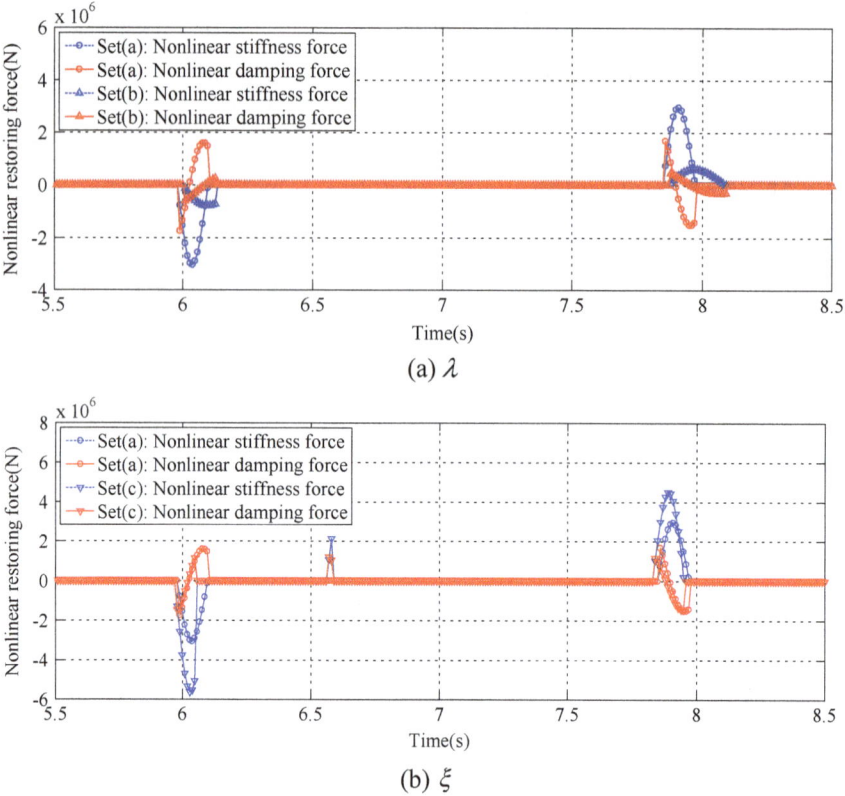

Fig. 9.15 Nonlinear stiffness force and nonlinear damping force with different parameters

nonlinear stiffness force and nonlinear damping force with different parameters. It can be summarized that:

(1) As the increase of λ, both nonlinear forces show quadratic growth;
(2) As the decrease of ξ, nonlinear stiffness force shows proportional growth, while the nonlinear damping force declines slowly.

Figure 9.16 shows the peak displacement at the top floor. It can be seen that no adverse effect on peak displacement is observed. This is because the relative velocity between the particle and the baffle $\dot{y}$ is always opposites to the absolute velocity $\dot{u}_{top}$ at the top floor, and the nonlinear restoring force attenuates the vibration with no doubt. Similarly, some rules can be summarized as follows:

(1) The SAID has a certain control effect on peak displacement.
(2) With a certain μ, as λ increases, the peak displacement decreases; as ξ increases, the peak displacement increases; the influence of λ is more significant than ξ.
(3) From the comparison of the ordinates of Fig. 9.16a, b: as the increase of mass ratio, the peak displacement decreases.

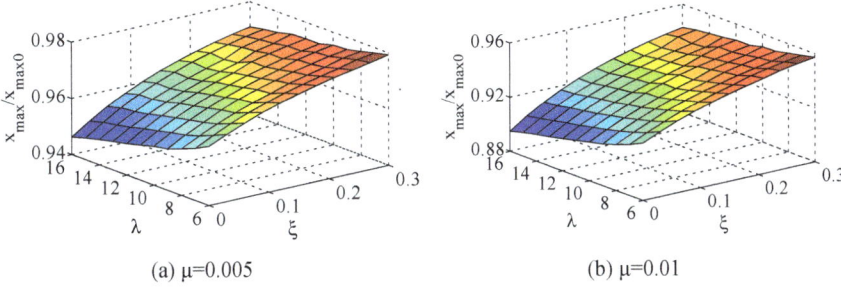

Fig. 9.16 Peak displacement at the top floor with different mass ratio

In combination with the previous discussion of peak acceleration, these rules actually indicate the fact that larger nonlinear restoring force will have a better control effect on the reduction of peak displacement.

Due to the damping mechanism and physical nature of the device performance, the basic rules of R.M.S. response remain generally the same as peak response. The R.M.S. responses are shown in Figs. 9.17 and 9.18. However, some interesting phenomena in comparison to the peak response are pointed out as follows:

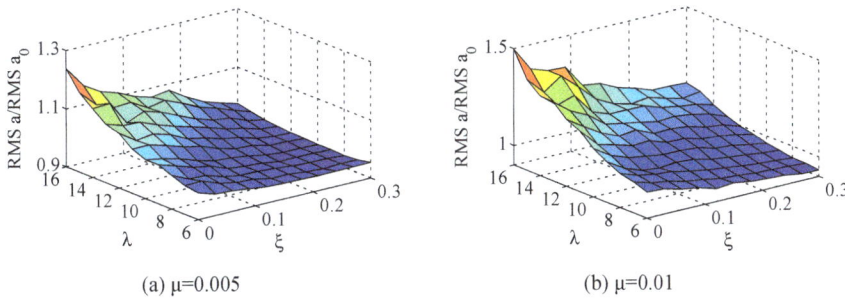

Fig. 9.17 R.M.S. acceleration at the top floor with different mass ratio

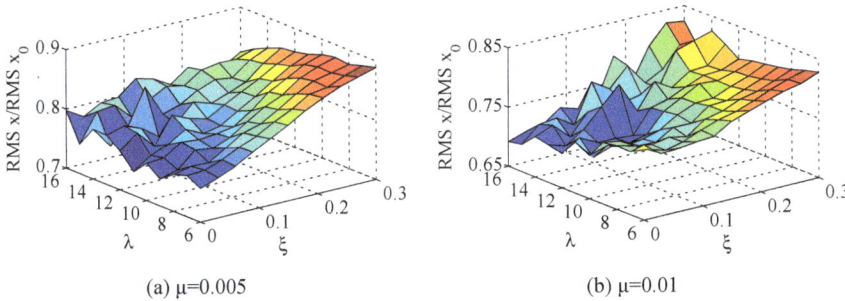

Fig. 9.18 R.M.S. displacement at the top floor with different mass ratio

(1) As shown in Fig. 9.18, the associated surfaces are not as smooth as those for the peak response; this is because the peak response is the only response at a moment, while the R.M.S. response has the relationship with all time-history response. Some impacts would have a little adverse effect on the R.M.S. response, such as fluctuating at zero, but it does not affect the general trend.

(2) Compared with the peak acceleration, it has a certain control effect on the R.M.S. acceleration, but the control effect is still not good, while the control effect on the R.M.S. displacement is satisfactory. For instance, the normalized value in Fig. 9.18b is close to 0.7 with mass ratio 0.01, which means the R.M.S. response can be reduced by 30%.

(3) Moreover, the Fourier spectrums of the time-history curve on the top floor are shown in Fig. 9.19; it indicates that the acceleration is controlled by higher-order modes, while the displacement is controlled by the first-order mode, as shown in Wongprasert and Symans's study [10–14]. Since the SAID is installed on the top floor, the control effect on acceleration is not good thus. If the damper is installed at the place corresponding to the fourth-order mode (10 and 16th floor), the acceleration control effect would be much better.

Three nonlinear indexes, including the number of plastic hinges, total component energy consumption, and the maximum ratio of joint curvature, are investigated here. The simulation results are shown in Figs. 9.20, 9.21, 9.22. Some conclusions can be summarized as follows:

(1) As the mass ratio μ increases, the control effects improve, including less number of plastic hinges, lower total component energy consumption and a smaller maximum ratio of joint curvature. However, the rate of improvement of the control effect is significantly lower than the increase rate of μ.

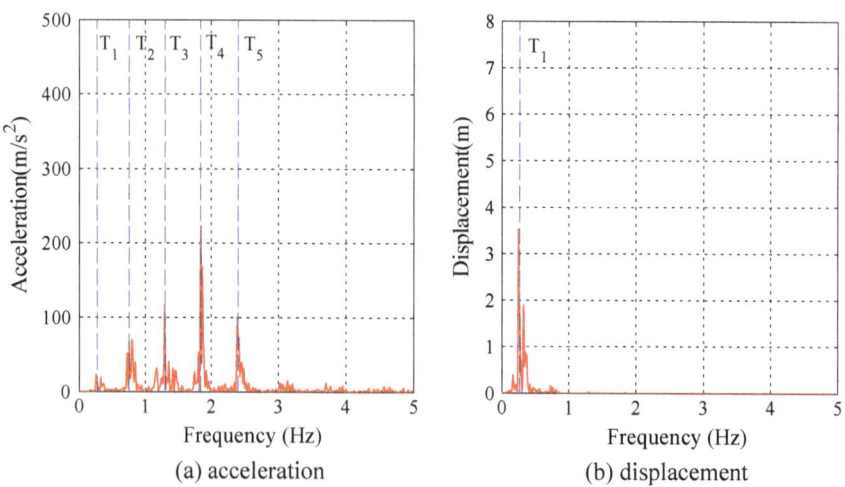

(a) acceleration   (b) displacement

Fig. 9.19 Fourier spectrums of the time-history curve at the top floor ($set(c): \mu = 0.005, \xi = 0.15, \lambda = 16$)

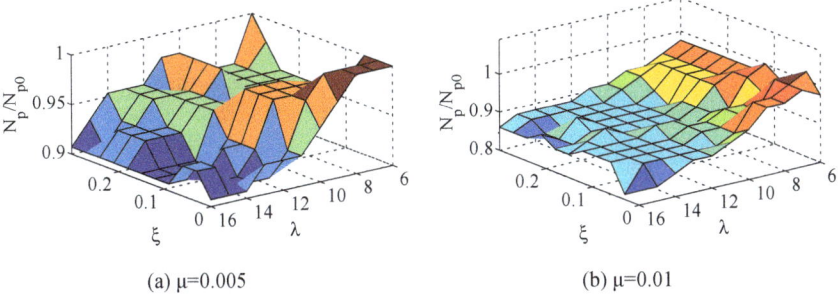

(a) μ=0.005   (b) μ=0.01

Fig. 9.20 Number of plastic hinges of the main structure with different mass ratio

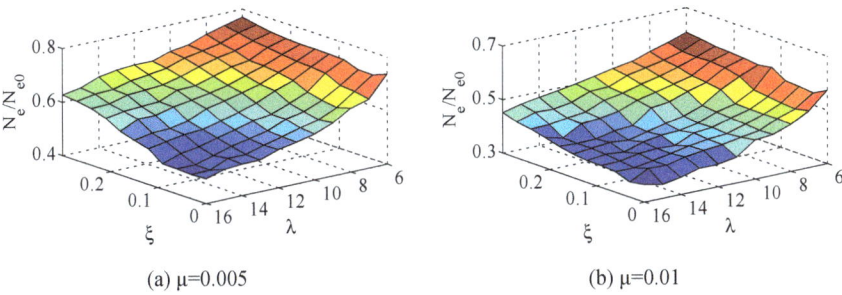

(a) μ=0.005   (b) μ=0.01

Fig. 9.21 Total component energy consumption of the main structure with different mass ratio

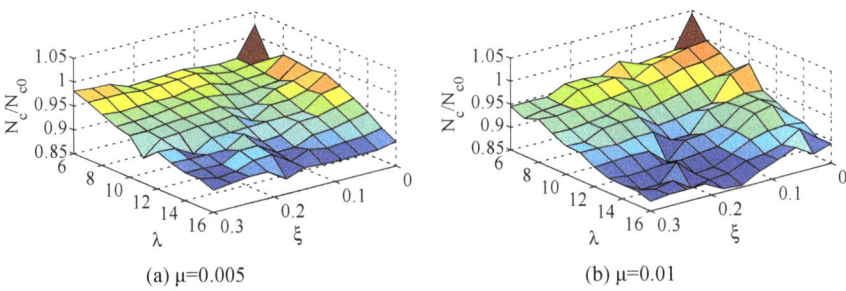

(a) μ=0.005   (b) μ=0.01

Fig. 9.22 Maximum ratio of joint curvature of the main structure with different mass ratio

(2) The number of plastic hinges and the maximum ratio of joint curvature are not sensitive to the damping ratio ξ, and mainly influenced by the rigid coefficient λ. As the λ increases, the number of plastic hinges and the maximum ratio of joint curvature decrease.

(3) Total component energy consumption is one of the most important nonlinear indexes, and it shows the same rules as linear indexes: as the λ increases and ξ decreases, this index decreases.

The contour lines of both the momentum exchange between the particle and the main structure and the total component energy consumption (shown in Fig. 9.21) of the main structure are shown in Figs. 9.23 and 9.24, respectively. The general trend of the total component energy consumption is opposite to the momentum exchange, which means effective momentum exchange can reduce the damage to the main structure, leading to a better vibration control effect. Moreover, the shape of the contour line in Fig. 9.23 is close to a straight line, while the shape of the contour line in Fig. 9.24 ($\xi < 0.2$) has a bit difference in lower damping ratio area. For instance, the points on the dashed line have similar momentum exchanges, as shown in Fig. 9.24, but the control effect on the total component energy consumption decreases with the decreasing of damping ratio, which indicates the damping ratio also plays an important role in the energy dissipation.

The analysis of the system parameters is based on the El Centro wave with intensity 1.5; thus, it is necessary to study the influence of the excitation intensity on the vibration control effect. Therefore, intensity levels of 0.5, 1.0 and 1.5 are considered in this section, and three basic methods can be used to investigate the effect of intensity:

(1) Keeping ξ and λ as constants 0.1 and 10, respectively, and changing μ;
(2) Keeping μ and λ as constants 0.01 and 10, respectively, and changing ξ;
(3) Keeping μ and ξ as constants 0.01 and 0.1, respectively, and changing λ;

All values have been normalized by dividing by the corresponding uncontrolled condition. For instance, the results of the R.M.S. displacement are shown in Fig. 9.25. The curves of intensity 0.5 and intensity 1.0 are completely overlapping, which indicates the control effects of SAID is very stable in the elastic stage. While the main

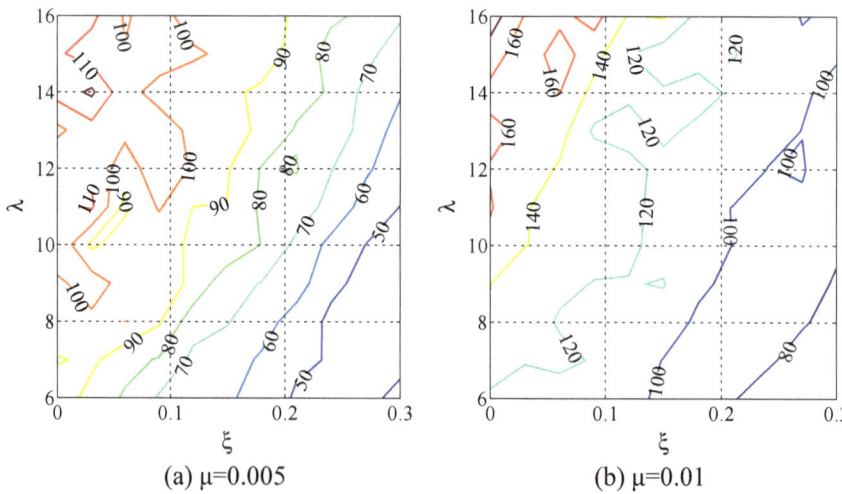

(a) μ=0.005   (b) μ=0.01

Fig. 9.23 Contour line of momentum exchange ($\times 10^5$ kg m/s) between the particle and the main structure with different mass ratio

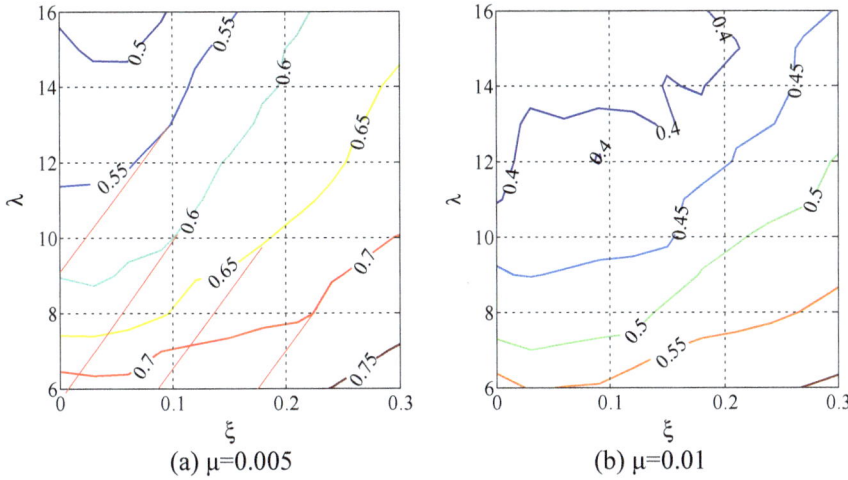

Fig. 9.24 Contour line of total component energy consumption of the main structure with different mass ratio

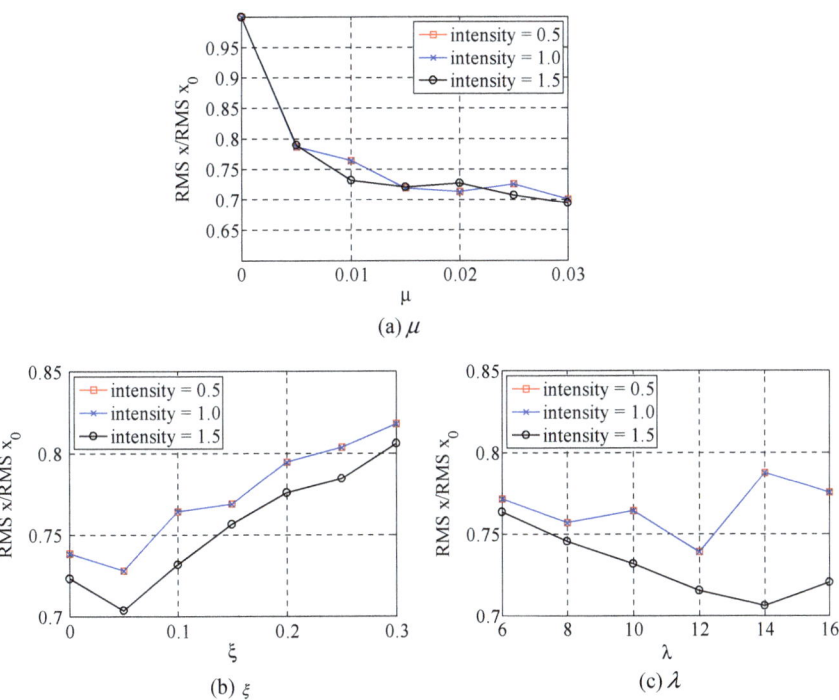

Fig. 9.25 Intensity influence on R.M.S. displacement with variable parameters

structure undergoes the nonlinear stage, the trend is consistent with the elastic stage. To summarize, the SAID is not sensitive to the increase of the excitation intensity, and certain control index even improves, such as the R.M.S. x. Additionally, Fig. 9.25a is also a supplement proof for the mass ratio. When the mass ratio is greater than 0.01, the growth rate of the control effect on the R.M.S. displacement is obviously lower than the growth rate of the mass ratio.

Based on the extensive parametric study on SAID, some damping mechanism rules indicating its physical performance can be summarized as follows:

(1) As the mass ratio increases, the control effects on displacement response and all nonlinear indexes improve, however, the efficiency of improvement is obviously lower than the increase rate of mass ratio when the mass ratio is larger than 0.01.

(2) As the damping ratio increases, the control effect on the acceleration response is better, while the control effects on the displacement and total component energy consumption are worse, and the control effects on the number of plastic hinges and the maximum ratio of joint curvature change little. Moreover, increasing the damping ratio is beneficial to energy consumption when the damping ratio is lower than 0.2.

(3) As the rigid coefficient increases, the control effect on displacement response increases as well as the nonlinear behavior of the main structure improves. This parameter is more sensitive than the damping ratio.

(4) The influence of these parameters is actually the influence on the nonlinear restoring force, which reflecting the control mechanism for the damper.

(5) The intensity of excitation has generally no influence on the control effect under the elastic phase.

Taking this nonlinear structure as an example, some suggestions for practical design steps are proposed:

(1) Choose an appropriate mass ratio. It not only determines the range of parameters selection, but also controls the cost. Therefore, regarding no adverse effects on the acceleration control as a goal, the mass ratio can be chosen in advance, and also leaving enough range for other parameters. For illustration, the mass ratio is ideally less than 0.01 in this part.

(2) For a better dissipation effect on the energy derived from momentum exchange, a suitable damping ratio is selected. For instance, the damping ratio should be larger than 0.2 in this investigation.

(3) Calculate the maximum inertia force at the top floor with a certain intensity, then the value of the rigid coefficient can be determined by meeting the condition that the nonlinear restoring force equals to the maximum inertia force.

3. *Comparison between the SAID and PID$_{opt}$*

To further understand the damping performance and also validate the superiority of the SAID, the optimal parameters of the passive impact damper (PID) are determined, while the parameters of the SAID remain the same. To evaluate the vibration control effects of the SAID, two linear indexes, including the root-mean-square (R.M.S.)

acceleration and R.M.S. displacement along the height, and three nonlinear indexes, including the number of plastic hinges, component energy consumption of the main structure and the maximum ratio of joint curvature, are investigated. In addition, the El Centro wave with intensity 1.5 is utilized in this section.

Since the original finite element (FE) model of this 20-story nonlinear building has a large number of degrees of freedom, the optimization of the particle impact damper (PID) will cost a lot of time; thus, a reduced-order model is used here, based on differential evolution (DE) algorithm. More details about the parameter identification for the reduced-order and DE algorithm can be found in the reference. The optimal parameters of PID are also determined by DE algorithm, and the optimal parameters of PID include the rigid coefficient, gap clearance, mass ratio and damping ratio of the particle. Considering that the R.M.S. displacement response at the top floor is very important for the structural design, the detailed optimization process of PID is as follows:

$$\min J(z) = abs[J_1(z) - J_{obj}] \tag{9.36}$$

$$J_1(z) = \frac{abs[RMS(u_{top}(t)) - RMS(u_{top}(z, t))]}{RMS(u_{top}(t))} \tag{9.37}$$

$$z = (\mu, \lambda, \xi, d) \tag{9.38}$$

where $J_{obj} = 0.6$ represents objective vibration control effect of the PID; $J_1(z)$ represents vibration control effect of the PID; $u_{top}(t)$ represents displacement response at the top of the main structure without control; $u_{top}(z, t)$ represents displacement response at the top of the main structure with the PID; z represents parameters vector of the PID. According to Sect. 9.2, the mass ratio is preset to 0.01. Table 9.1 shows the optimal parameters of PID while the parameters of SAID is determined the same as PID's. These parameters are also rational for SAID according to the discussion in Sect. 9.2.

To compare the performance of optimal PID (PID$_{opt}$) and SAID, the vibration control effect and the improvement rate are first defined as Eqs. (9.39) and (9.40):

Vibration control effect (VCE) =
$$\frac{\text{the RMS of structure without control} - \text{the RMS of structure with PID/SAID}}{\text{the RMS of structure without control}}$$
$$\tag{9.39}$$

Table 9.1 Parameters of PID and SAID

Controllers	λ	μ	ξ	d(m)
PID	6	0.01	0.2316	0.8192
SAID	6	0.01	0.2316	–

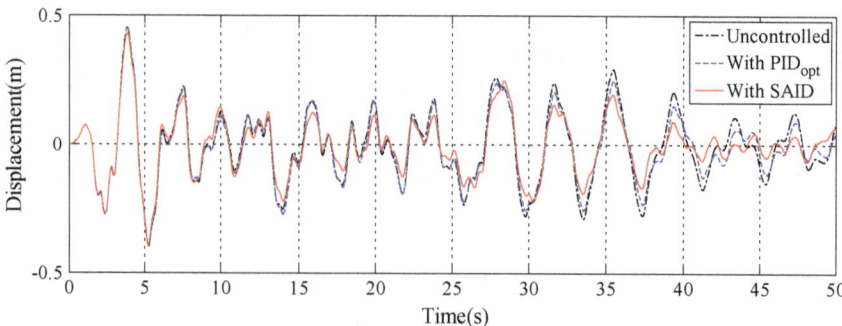

Fig. 9.26 Comparison of the displacement response at the top of the building under El Centro wave

Improvement rate =

$$\frac{\text{the VCE of the structure with PID} - \text{the VCE of the structure with SAID}}{\text{the VCE of the structure with PID}} \qquad (9.40)$$

The time-history curve of the displacement under El Centro wave at the top of the building is shown in Fig. 9.26, and three operating conditions (uncontrolled case, with PID_{opt} case and with SAID case) are compared. At the beginning of the excitation, the control effects of both PID_{opt} and SAID can be ignored, owing to the fact that sufficient impacts between the particle and the baffle would take a few seconds. As time progresses, the control effect is getting better.

The R.M.S. response of each floor of the main structure subjected to El Centro wave are shown in Fig. 9.27. It can be seen that both the PID_{opt} and SAID can reduce the response along the structural height, in which the SAID performs better. From Fig. 9.27a, it can be seen that the maximum vibration control effect on R.M.S. acceleration for PID and SAID is 3.5 and 12.8% occurred at floor 19 and 7, respectively, and the improvement rate of the SAID compared with the PID_{opt} can reach 366%. The maximum vibration control effect on R.M.S. displacement response at floor 20 for PID_{opt} and SAID can reach 7.5% and 19.2%, respectively, and the improvement rate of the SAID compared with the PID_{opt} can reach 256%, which validates the superiority of the SAID.

Figure 9.28 shows the inter-story drift ratio of each floor of the main structure subjected to El Centro wave. It can be seen that compared to PID_{opt} case and uncontrolled case, the SAID case can largely reduce the inter-story drift ratio for many stories.

Considering material nonlinear state and the damage mechanism of beam hinge, some nonlinear performance indexes are investigated, including the number of plastic hinges, component energy consumption, and the maximum ratio of joint curvature. The number of plastic hinges and component energy consumption of each floor of the main structure under El Centro wave is shown in Fig. 9.29. It can be seen that, although the number of plastic hinge for SAID case is equal to PID_{opt} case, the

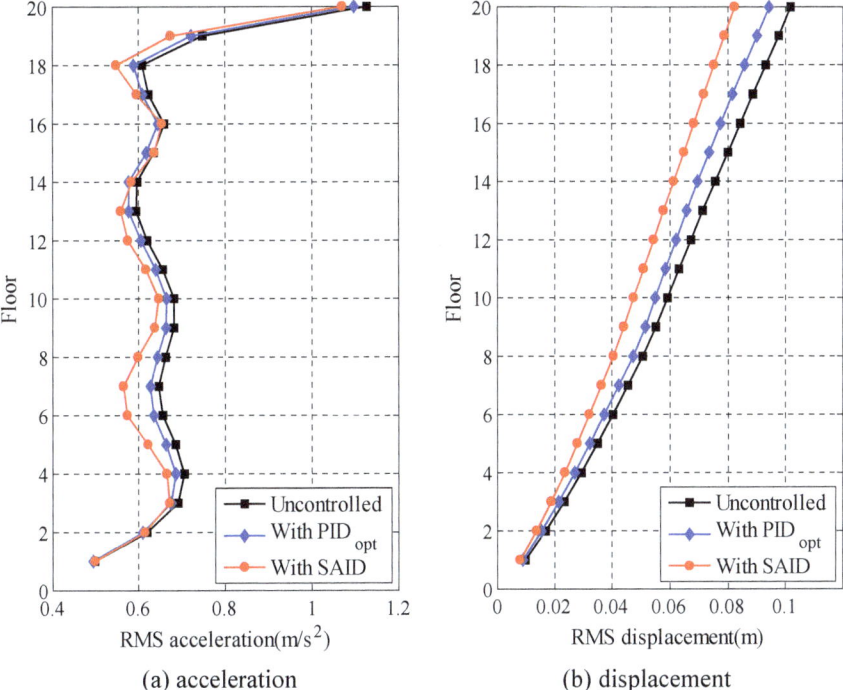

Fig. 9.27 R.M.S. response of the main structure under El Centro wave

component energy consumption is largely reduced, indicating that the input energy flows to SAID greatly, hence protecting the main structure.

Figure 9.30 shows the maximum ratio of joint curvature under El Centro wave. The results are consistent with that of component energy consumption, the structural damage under SAID case is much smaller than the PID_{opt} case, hence is much easier to be repaired after the earthquake.

9.3 Implementation Scheme of Semi-active Control Particle Dampers

Studies by Masri, Nishitani and Paulet-Crainiceanu have shown that a suitable semi-active impact damper can achieve better control effectiveness compared with the same level of passive control dampers, while the active control system using direct control force of the pulse is even better than semi-active control systems, but the effective damping capacity under wide dynamic load conditions needs to be considered.

The semi-active control particle damper employs a non-linear additional mass damper with adjustable motion restraint device and can be placed anywhere in the

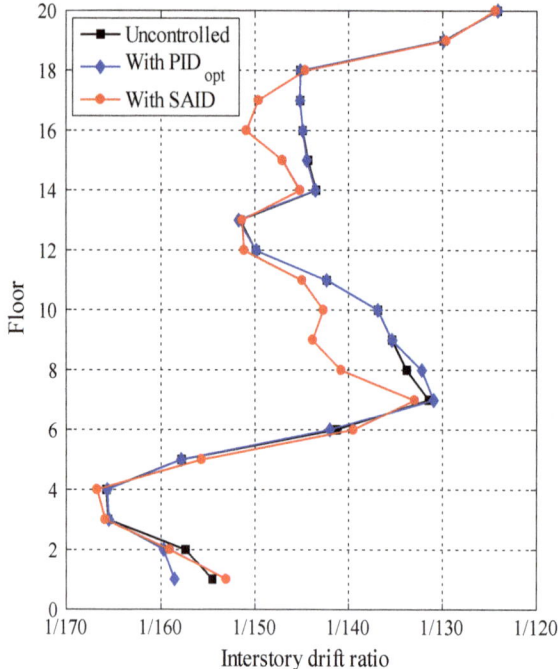

Fig. 9.28 Comparison of the inter-story drift ratio under El Centro wave

structure. Its control algorithm does not require a systematic mathematical model. The extent of vibration of the main structure in the vicinity of each semi-active control devices determines the active control gap size and collision time of the damper. The semi-active control particle damping device adjusts the key parameters through a controllable motion constraint device instead of directly applying the active control force to reduce the vibration of the main structure. At the same time, the Lyapunov direct method also proves that the semi-active control algorithm to attenuate the response of the nonlinear main structure is Lagrangian stable.

Figure 9.31 shows the experimental setup used to evaluate the effectiveness of the semi-active control algorithm. Figure 9.32 shows the normalized displacement and velocity response of the device main structure.

Figure 9.33 shows the displacement response of the above semi-active control device under free vibration. It can be seen from the attenuation curve that the semi-active impact damper is better than the passive impact damper.

Besides, in order to investigate the performance of the semi-active impact damper under actual laboratory conditions, Masri's group [2] have designed and made a single-degree-of-freedom framework model, as shown in Fig. 9.34. The entire experimental setup consists of the following components:

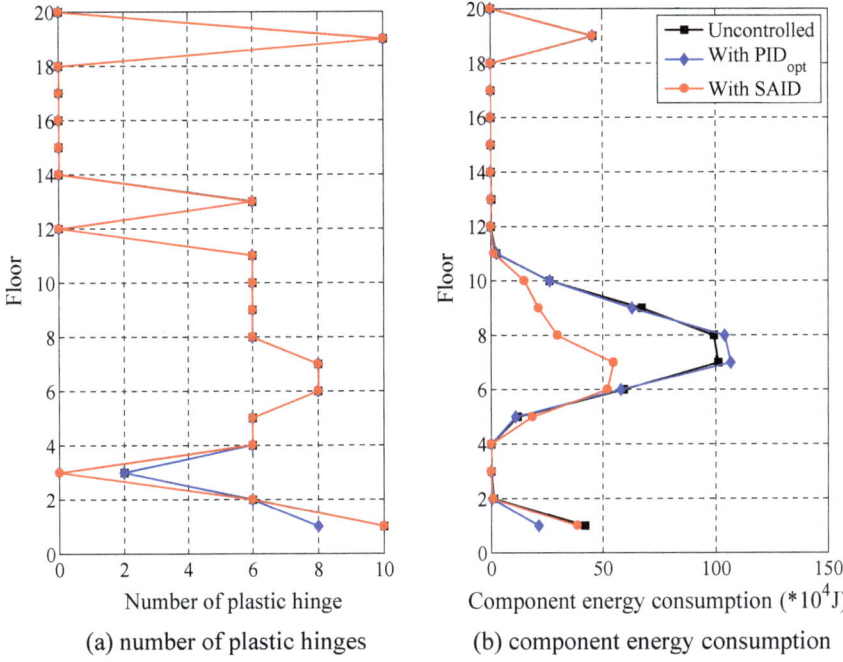

(a) number of plastic hinges   (b) component energy consumption

Fig. 9.29   Comparison of nonlinear performance indexes under El Centro wave

Fig. 9.30   Comparison of the maximum ratio of joint curvature and yield curvature at each floor under El Centro wave

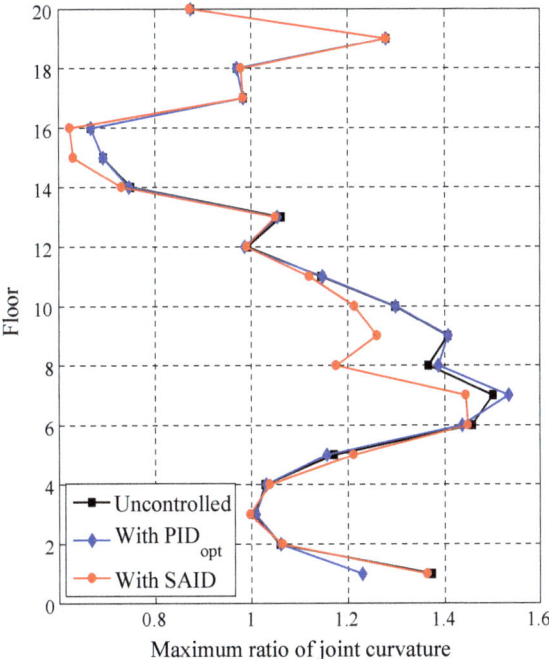

Fig. 9.31 Experimental device diagram of semi-active impact damper

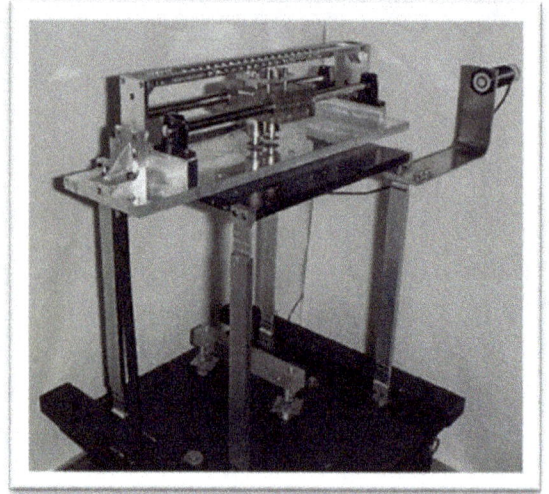

Fig. 9.32 Normalized displacement and velocity response of semi-active control device

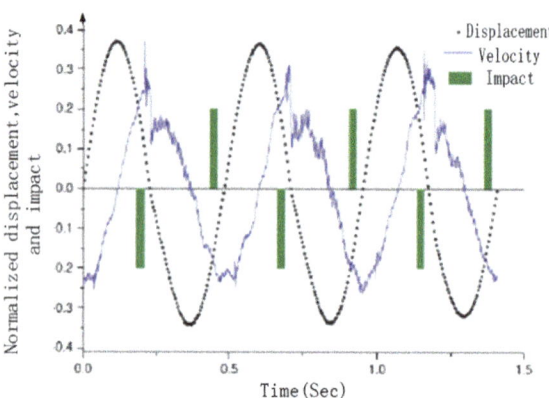

Fig. 9.33 Free vibration displacement response of a single-degree-of-freedom mechanical model with a semi-active impact damper

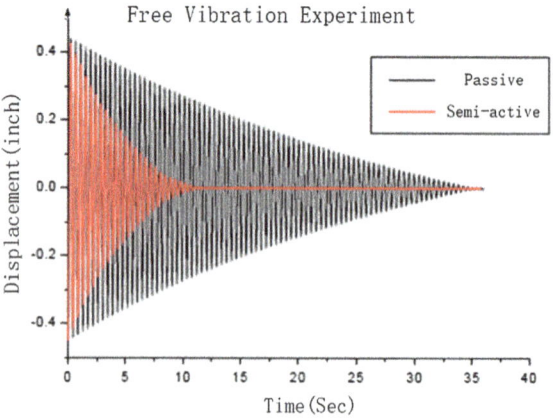

Fig. 9.34 Top view (**a**) and front view (**b**) of the model experimental setup

(a)

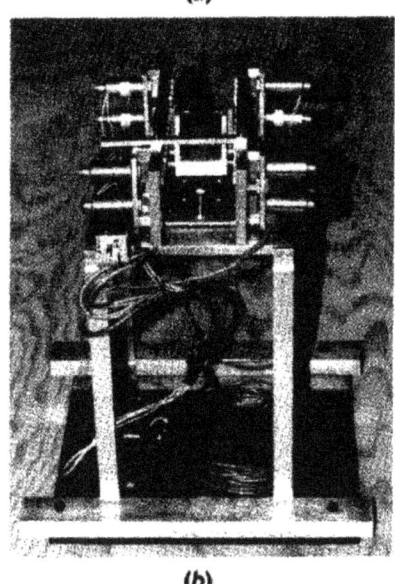

(b)

(1) A rectangular container (about 35 cm × 35 cm in size) for limiting the movement of the additional mass, fixed on basic structure.

(2) A bearing-type additional mass that can perform low-resistance movements in the groove of the container panel.

(3) Four movable panels, used to fix the collision barrier so that the mass collides with the basic structure. The four panels can be moved forward or backward along the centerline.

(4) Four sets of 16 spring-loaded collision stoppers. The stoppers are mounted on the panel by hinge wedges, separated by two centimeters. These gears are nailed to the panel so that the mass can only move freely in one direction. The upper two sets of panels allow the mass to move in one direction along the long axis

of the container, while the lower panel allows the mass to move in the opposite direction. The ratcheting action of panels are controlled by electromagnetic force.

The advantage of this device is that the sensor for detecting the motion state of the structure is eliminated. The damper startup and closing algorithm is simplified, and it is no longer necessary to use a computer to calculate the optimal gap size. The latter can greatly reduce the cost of the decision process, thus compensating for the delay caused by the mechanical startup.

The logic operation of the above system is performed by a Z-80 microcomputer using FORTH. When the computer detects that the system displacement reaches zero, it will execute a "collision command". This command will connect the relay circuit and electric current will be generated by the electromagnetic solenoid. The electromagnetic force acts to open the stopper on the panel, causing the required mass impact. The solenoid can adjust the gear to the "collision position" in 4 microseconds (approximately 1/20 of the system's natural vibration period).

The mass damper mass ratio μ in Fig. 9.34 is 0.10. When the structure is excited by a sinusoidal sweep, the peak structural response with damper is about 45% of that when it is not placed, and about 42% when random excitation is performed. It can be found that under actual laboratory conditions, the performance of the damper is lower than theoretically expected. That is mainly because:

(1) The control fixture material in the model is aluminum, and its recovery factor is smaller than that of the hard steel in theoretical analysis. When e is higher, the energy dissipation of collision can be effectively reduced, so that the damper can get better performance.

(2) The distance between the stops is not small enough, so that the mass cannot always collide at the optimal moment.

(3) The manufacturing error in the assembly of the control unit generates a large amount of bounce (non-linear dead zone), which aggravates the mechanical energy dissipation when the mass is in collision with the structure (resulting in the decrease of momentum exchange efficiency).

These problems can be solved by using more suitable materials and high-precision manufacturing processes.

In general, the semi-active control particle damping technology has its own characteristics and advantages, and will become a hot topic in research and application of structural vibration control based on particle damping in the future.

References

1. HongWei, Z., and C. Qian. 2008. Research on vibration reduction mechanism and experiment of electromagnetic particle damper. *Journal of vibration engineering* 21 (2): 162–166.
2. Masri, S.F., et al. 1989. Active parameter control of nonlinear vibrating structures. *Journal of Applied Mechanics* 56 (3): 658–666.

3. Masri, S.F. 1972. Theory of the dynamic vibration neutralizer with motion-limiting stops. *Journal of Applied Mechanics, ASME* 39 (2): 563–568.

4. Karyeaclis, M.P., and T.K. Caughey. 1987. Stability of a semi-active impact damper. *Journal of Applied Mechanics* 56(4): 926–929.

5. Masri, S.F., and T.K. Caughey. 1966. On the stability of the impact damper. *Journal of Applied Mechanics* 33 (3): 586–592.

6. Lindberg, R.E., and R.W. Longman. 1984. On the number and placement of actuators for independent modal space control. *Journal of Guidance, Control and Dynamics* 7 (7): 215–221.

7. Chassiakos, A.G., Masri, S.F., Bekey, G.A., and Miller, R.K. 1988. Optimum controller location for mitigating earthquake induced response of structures provided with point actuators. In *Proceedings ninth world conference on earthquake engineering*. Tokyo-Kyoto, Japan.

8. Ohtori, Y., et al. 2004. Benchmark Control Problems for Seismically Excited Nonlinear Buildings. *Journal of Engineering Mechanics* 130 (4): 366–385.

9. Lu, Z., et al. 2017. Experimental and analytical study on the performance of particle tuned mass dampers under seismic excitation. *Earthquake Engineering and Structural Dynamics* 46 (5): 697–714.

10. Saeki, M. 2005. Analytical study of multi-particle damping. *Journal of Sound and Vibration* 281 (3–5): 1133–1144.

11. Mao, K.M., et al. 2004. DEM simulation of particle damping. *Powder Technology* 142 (2–3): 154–165.

12. Lu, Z., et al. 2014. Discrete element method simulation and experimental validation of particle damper system. *Engineering Computations* 31 (4): 810–823.

13. Wong, C., A. Spencer, and J. Rongong. 2009. Effects of enclosure geometry on particle damping performance. In *50th AIAA/ASME/ASCE/AHS/ASC structures, structural dynamics, and materials conference*. Palm Springs, California: American Institute of Aeronautics and Astronautics.

14. Saeki, M. 2002. Impact damping with granular materials in a horizontally vibrating system. *Journal of Sound and Vibration* 251 (1): 153–161.

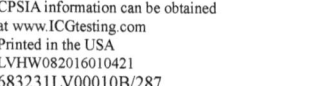

CPSIA information can be obtained
at www.ICGtesting.com
Printed in the USA
LVHW082016010421
683231LV00010B/287